Evolution for the People

From Darwin's *The Origin of Species* to the twenty-first century, Peter Bowler reinterprets the long Darwinian Revolution by refocusing our attention on the British and American public. By applying recent historical interest in popular science to evolutionary ideas, he investigates how writers and broadcasters have presented both Darwinism and its discontents. Casting new light on how the theory's more radical aspects gradually grew in the public imagination, *Evolution for the People* extends existing studies of the popularization of evolutionism to give a more comprehensive picture of how attitudes have changed through time. In tracing changes in public perception, Bowler explores both the cultural impact and the cultural exploitation of these ideas in science, religion, social thought and literature.

Peter J. Bowler is Professor Emeritus of the History of Science at Queen's University Belfast, a fellow of the British Academy, a member of the Royal Irish Academy and a past president of the British Society for the History of Science.

Evolution for the People

Shaping Popular Ideas from Darwin to the Present

Peter J. Bowler
Queen's University Belfast

Shaftesbury Road, Cambridge CB2 8EA, United Kingdom

One Liberty Plaza, 20th Floor, New York, NY 10006, USA

477 Williamstown Road, Port Melbourne, VIC 3207, Australia

314–321, 3rd Floor, Plot 3, Splendor Forum, Jasola District Centre, New Delhi – 110025, India

103 Penang Road, #05-06/07, Visioncrest Commercial, Singapore 238467

Cambridge University Press is part of Cambridge University Press & Assessment, a department of the University of Cambridge.

We share the University's mission to contribute to society through the pursuit of education, learning and research at the highest international levels of excellence.

www.cambridge.org
Information on this title: www.cambridge.org/9781009448970

DOI: 10.1017/9781009449007

When citing this work, please include a reference to the DOI 10.1017/9781009449007

First published 2024

A catalogue record for this publication is available from the British Library

Library of Congress Cataloging-in-Publication Data
Names: Bowler, Peter J., author.
Title: Evolution for the people : shaping popular ideas from Darwin to the present / Peter J. Bowler, Queen's University Belfast.
Description: Cambridge, United Kingdom ; New York, NY : Cambridge University Press, 2024. | Includes bibliographical references. | Summary: "In a compelling new study of popular evolutionism over two centuries, Peter Bowler uses the growing interest in popular science to reinterpret how evolutionary ideas have shaped modern culture. He demonstrates how Darwinism and its rivals sought public attention via mass media, and their resulting impact on popular consciousness"– Provided by publisher.
Identifiers: LCCN 2024008582 (print) | LCCN 2024008583 (ebook) | ISBN 9781009448970 (hardback) | ISBN 9781009448994 (paperback) | ISBN 9781009449007 (epub)
Subjects: LCSH: Evolution (Biology)–Social aspects–History. | Evolution (Biology)–Public opinion–History. | Evolution (Biology)–Philosophy. | Evolution (Biology) and the social sciences. | Communication in science–History. | Darwin, Charles, 1809-1882–Influence.
Classification: LCC QH366.2 .B676 2024 (print) | LCC QH366.2 (ebook) | DDC 576.8/2–dc23/eng/20240318
LC record available at https://lccn.loc.gov/2024008582
LC ebook record available at https://lccn.loc.gov/2024008583

ISBN 978-1-009-44897-0 Hardback
ISBN 978-1-009-44899-4 Paperback

Contents

List of Figures — *page* vi
Preface — ix

1 Bridging the Gap — 1
2 Before Darwin — 25
3 Reacting to the *Origin* — 56
4 Human Ancestry — 84
5 Evolutionary Epics — 104
6 Challenging Darwinism — 138
7 Reconfiguring the Ascent of Life — 167
8 Social Evolutionism — 186
9 The Evolutionary Synthesis — 217
10 Toward the Modern World — 233

Bibliography — 258
Index — 285

Figures

2.1 Mastodon skeleton on display at Rembrandt Peale's museum in Philadelphia, 1821 *page* 33
2.2 Waterhouse Hawkins' reconstruction of the dinosaur *Megalosaurus* on display at Crystal Palace 37
2.3 Diagram representing the development of the vertebrates through time, from Robert Chambers, *Vestiges of the Natural History of Creation* (1846) 50
2.4 Beaks of Galapagos finches adapted to different modes of feeding, from Charles Darwin, *Journal of Researches* 55
3.1 Varieties of domestic pigeons used to illustrate Darwin's views on the power of artificial selection 60
4.1 Profiles of racial types with comparison to ape, from Robert Knox, *The Races of Men* (1850) 89
4.2 Frontispiece to R. M. Ballantine, *The Gorilla Hunters* (1897) 95
4.3 Frontispiece to T. H. Huxley's *Man's Place in Nature* (1863) 98
5.1 The development of different embryos from a common starting point 113
5.2 'Tree of life' with a central trunk running up to humanity at the top, from Ernst Haeckel, *The History of Creation* (1879) 116
5.3 O. C. Marsh's sequence of horse fossils showing progressive specialization 122
6.1 'Tree of life' with a central trunk dividing at the top to indicate the divergence of birds from mammals, from J. Arthur Thomson, *The Outline of Science* (1922) 153
6.2 Diagrams illustrating the range of variation within a population, from A. R. Wallace's *Darwinism* (1889) 162
7.1 Cover of *Scientific American*, 8 April 1911, showing a reconstruction of the dinosaur *Iguanodon* 173
7.2 *Diplodocus* skeleton on display at London's Natural History Museum 174
7.3 Horse evolution represented as a branching tree 181

8.1 Restoration of 'Neanderthal Man' with brutal features, from H. H. Wells, *The Outline of History* (1920) 194
8.2 'An Early Pliocene Entertainment', cartoon depicting early hominids learning to walk upright, from *Punch*, 26 June 1929 197
8.3 Cartoon depicting a native Irish figure with ape-like features, from *Punch*, 29 October 1881 201
10.1 Flyer advertising a meeting organized by the Creation Resources Trust in Belfast on 1 June 1996 254

Preface

The manuscript for this book was begun as a 'lockdown project' at the start of the COVID-19 pandemic, but as I continued writing I became increasingly sure that it would have real value for scholars in fields other than the history of science. I also realized that I had some points to make that would be of interest to those who do specialize in the history of evolutionism. The basic idea – explained in more detail in Chapter 1 – is to tell the story of the long Darwinian Revolution as much as possible in terms of what was available to those outside the scientific community and the intellectual elite. Anyone interested in how the idea of evolution was being presented to the wider middle class and the general public at any point in time can gain a quick overview from this survey. Scholars in areas such as the history of ideas and of religion, social history and literary studies can tell at a glance what was 'in the air' on the broad topic of evolutionism. Historians of science are now well aware that the specialists themselves sometimes had to take into account how their findings were being presented to a wider public, so a survey focusing on the popular material also has the potential to change our ideas on key steps in the development of the evolutionary paradigm.

An important element in this survey is the inclusion of general ideas about the history of life on earth in addition to the usual focus on Darwin's theory of natural selection and its critics. There has always been substantial popular interest in the discovery of fossils – including human fossils – and by taking this into account we see how changes in the perception of evolution could be brought about by the way this material is presented. We also see that what was actually discovered and displayed could play a role in shaping those perceptions, as with the huge expansion in knowledge of the dinosaurs. The same can also be true for discoveries in natural history – I suggest, for instance, that the popular interest aroused by the discovery of the gorilla in the 1860s may have significantly influenced how ordinary people reacted to the *Origin of Species*. It wasn't just that the gorilla made a frightening candidate for a human ancestor; by focusing attention onto the ascent from ape to human it encouraged a linear mode of thinking more akin to the ancient chain of being than to Darwin's 'tree of life'.

This brings us to a hobby-horse I have been riding for some time now, as explained in a previous book, *Progress Unchained*. I think that to understand the rise of Darwinism we need to see it in terms more general than those framed by the theory of natural selection. Of course, the arguments over the struggle for existence and the survival of the fittest were important, but there was also the emergence of a new way of visualizing the progress of life on earth. The assumption that evolution must have a built-in drive toward the production of humanity was gradually undermined by a growing recognition that it has been an open-ended and less goal-directed process. Darwinism in the wider sense played a major role in this transition, and that role is as important as the new way of thinking about the actual mechanism of change. Historians of science have long recognized that Darwinism was initially challenged by alternative ideas about how evolution worked. In some respects the debate over mechanisms can be seen as a vehicle through which a less teleological vision of the world was articulated. When we see that this changing vision was gradually incorporated into how evolutionism was presented to the wider public, we gain a sharper view of just how central it has been to the emergence of the modern world-view.

This is a relatively short book for so large a topic, so necessarily I paint with a very broad bush. Detailed examples of general and popular accounts of evolution are included to give a flavour of what was being discussed and debated at each point in the story; however, these are but small samples of what could be used. As I explain in Chapter 1, we do have many individual studies of particular episodes in the reception of evolutionism and related topics in natural history and palaeontology. Footnotes guide the reader to this literature for further information. Some of the secondary literature is my own, especially in the area of early twentieth-century Britain. For other areas I am indebted to the work of numerous scholars whose studies I have mined without shame, including Constance Areson Clark, Adrian Desmond, Piers Hale, Marcel LaFollette, Bernard Lightman, Ron Numbers, Martin Rudwick, Michael Ruse and James Secord, to name but a few. Angela Schwartz' massive compilation in German, *Streitfall Evolution*, parallels this study in some areas and is especially valuable for its many images, especially for the modern period (for which reason the images I have used focus on earlier material and are mainly derived from my own collection of popular literature on the topic). Without the contributions of these many scholars this project would never have got off the ground – being in my mid-seventies when I began, I was in no position to undertake primary source research across the whole field covered in the book. I hope those whose work I have pillaged think I have added something by pulling it all together. My thanks to two anonymous reviewers for suggestions that have improved several sections of the book.

The working title for this project was 'Evolution for Everyone', but I then found that David Sloan Wilson had used that title for his survey of modern ideas. *Evolution for the People* has an old-fashioned ring, but perhaps that is more appropriate for a historical survey. The name 'Darwin' in the subtitle may be taken to include Erasmus Darwin as well as his now better-known grandson.

1 Bridging the Gap

There are many books about the 'Darwinian Revolution' and its impact, so some justification is needed for adding to the list. *Evolution for the People* extends our understanding of the issues by bridging a divide within the history of science created by a significant new approach. Where once the focus was on great theoretical innovations, historians have become more interested in the interaction between science and everyday life. Studies of how key figures came to their conclusions are no longer fashionable, nor do we look solely at the impact of scientific ideas on highbrow culture. The study of popular science – that is, material aimed at a much wider audience – has become an active area of research, forcing us to recognize that what counts as 'science' for the ordinary citizen is often shaped by factors beyond the control of the scientific community. In extending our remit, though, we have to some extent lost sight of the conceptual innovations that have forced everyone, highbrow and lowbrow alike, to rethink our position in the world. The great innovators, including Darwin, remain central to public awareness of what has been achieved.

This book applies the new techniques to provide a different window through which to view a theoretical revolution whose impact still divides opinion within the general public: the emergence of evolutionism. It still makes reference to the 'big issues' identified by historians and by modern commentators, but where most histories have focused on the debate within the scientific and cultural elites, this survey concentrates on how the issues were presented to ordinary people in an ever-expanding variety of media that were never under the control of the scientific community. Here I link my own long-standing involvement with the 'Darwinian Revolution' to my new fascination with popular science to yield insights into how perceptions of evolutionism have changed over time. The variety of formats and participants central to the study of popular science may change our appreciation of how evolutionism has impacted on the public consciousness.

The new approach focuses as much on the public's ideas about how evolution is supposed to work as on the better-known debates about its implications. Historians are now aware that the story of evolutionism is much wider than that of Darwinism, and Darwinism itself is a complex and ever-changing entity composed of subsidiary concepts that have an intellectual life of their own. We may have to distinguish between the direct impact of evolutionism and the wider cultural changes within which the idea is embedded. The claim that nature is a scene of constant struggle, for instance, can be accepted without seeing the struggle as a driving force of change. Some of the cultural developments attributed to Darwinism represent more broadly based transformations of attitudes and beliefs, and the emergence of Darwinism can sometimes be seen as a particular manifestation of wider trends.

Darwinism may be in part a product of wider cultural developments, but it has its own ways of shaping how the issues are perceived. To sharpen the focus, this study concentrates on media discussions in which some form of evolutionism is the explicit topic. While not ignoring the subsidiary components, it distinguishes between those elements and the debates over whether biological evolution occurs, how it might work and how it applies to humans. Visions of what the idea of evolution means for our wider beliefs and values are shaped by what sort of process people think evolution actually is, which includes broader components beyond the mechanism of change. A focus on the ever-changing popular representations of evolutionism adds to our understanding of wider cultural developments that may or may not have required commitment to Darwinian natural selection.

Revealing changes in the public perception of how evolution works yields a better understanding of its cultural significance when viewed against the backdrop of wider trends. The new perspective offered by studying popular reactions confirms that the immediate impact of the *Origin of Species* was less dramatic than older historical accounts imply. The initial debates often reflected pre-Darwinian developments that people were belatedly starting to absorb, so that Darwin became a symbol for ideas his own approach was trying to supersede. Historians are aware of the difficulties that the Darwinian theory faced during the period of its so-called eclipse around 1900, and it will hardly be a surprise that the alternatives found their way into the mass media of the time. Related areas such as fossil discoveries also played a role in shaping the popular image of how life has evolved. Revealing what was available to non-specialists at each phase of the ongoing debates allows a better appreciation of how the science was perceived and how it impacted other areas such as religion and politics.

Changing Priorities

The relationship between science and society is mediated by a variety of actors, not all of whom are scientists or even share the interests of the scientific community. Expanding channels of communication have introduced a host of commentators who presented science news in ways that the scientists themselves sometimes did not like. Historians of science have now begun to pay more attention to the way science was presented to the public in past eras, challenging the traditional approach focused on the great conceptual revolutions that have defined the modern world-view. The original technique, soon to be dismissed as 'internalist', concentrated on the role of observation and experiment in shaping the thinking of the key innovators. Influences from outside the realm of direct scientific investigation were largely ignored. External factors could at best only slow down the acceptance of a new theory. Any suggestion that the emergence of a new perspective might be influenced by external forces was repudiated. Claims that evolutionism had been used to justify the harsh policies known as 'social Darwinism' were countered with the assertion that this was a misuse of scientific knowledge.[1]

Accounts of the Darwinian Revolution highlighted the observations that inspired Darwin's discoveries, acknowledged the opposition of religious thinkers, and then pushed on to show how the emergence of genetics resolved technical issues that Darwin himself had been unable to deal with. Loren Eiseley's *Darwin's Century* of 1958, for instance, treated the rise of modern Darwinism as a triumphant, step-by-step approach toward the modern position. There were, however, already some more nuanced studies that suggested that the interaction with religion had been a two-way process. John Green's *The Death of Adam* in 1959 highlighted the theory's challenge to the biblical world-view but noted that religious beliefs shaped the views of some naturalists who had made significant contributions. The barrier between internal and external factors was already beginning to crumble.

In the academic year 1965–6 I was lucky enough to study under a scholar who threw a metaphorical bomb into the cosy world of Darwin studies. This was Robert M. Young, who insisted that science could not be understood as something isolated from the culture and society within which it functioned. Social Darwinism was not a misuse of science because Darwinism itself reflected values characteristic of the community within which Darwin operated. Seizing on Darwin's appeal to Thomas Malthus' principle of population expansion, Young argued that ideological factors lay at the heart of the theory. Natural selection was used to justify free-enterprise individualism because that political philosophy was built into its conceptual foundations. There were howls of protest from internalist

[1] For more details on how the history of science has developed, see the introductory chapter of Bowler and Morus, *Making Modern Science*.

historians, but younger scholars saw that Young had to be taken seriously. Soon the ideological dimensions of Darwinism were being expanded to include its impact on race relations and the eugenics movement.[2]

Darwin's *Origin of Species* offered a radical explanation of how species become adapted to their environment, intended to replace the view that they were designed by a benevolent God. It aroused some interest among naturalists, and its appeal to the 'struggle for existence' certainly caught the popular imagination. By offering a new initiative consistent with the views of a younger generation of scientists, it triggered a fairly rapid conversion to evolutionism that soon spread into the wider world. But many – including 'Darwin's bulldog' Thomas Henry Huxley – found it hard to believe that the emergence of major animal types could be due to nothing more than an endless sequence of minor adaptive transmutations. Many wanted to preserve the belief that the world was designed to achieve a morally significant goal, in which case the outcome of any struggle for existence would have to be the emergence of 'higher' types, a view that even Darwin allowed for in his conclusion.

A wider Darwinism thus challenged the traditional view of 'man's place in nature' to create an ideology of progress, usually assumed to include moral and spiritual progress. Herbert Spencer's hugely influential philosophy of evolutionism, only partly Darwinian in nature, was the most characteristic expression of this movement. Recognition of the ideological dimension chimed with new initiatives in other areas. The work of James Moore and others showed that far from simply resisting Darwinism, religion had to be seen as an array of positions, some of which welcomed evolutionism in the name of progress, while others hardened their opposition. My own work on scientific evolutionism showed that Darwin's theory of natural selection was challenged by a number of rival explanations, some of which became popular at least in part because they were more congruent with liberal Christian values. In the conventional picture of the Darwinian Revolution the decades around 1900 were an embarrassment – a period when the scientific community had been unable to appreciate the true value of the selection theory. Yet this was the period when the general idea of evolution took hold of the public imagination, suggesting that the enthusiasm could not have been inspired solely by the theory of natural selection, however powerful Darwin's impact was on the culture of his time.[3]

[2] Young's articles are collected in his *Darwin's Metaphor*.

[3] On the religious debates, see Moore, *The Post-Darwinian Controversies*, and my own *Monkey Trials and Gorilla Sermons*. For non-Darwinian evolutionism, see my *Eclipse of Darwinism* and *The Non-Darwinian Revolution*.

Parallel developments were taking place in the study of the pre-Darwinian period. Seeking for the 'forerunners of Darwin' was no longer fashionable – the basic idea of evolution was being promoted before the *Origin of Species* was published, but not in forms that anticipated Darwin's theory. Adrian Desmond uncovered a network of activists within the medical profession who were looking for ideas that undermined the claim that the social order was established by God. James Secord's study of the pre-Darwinian evolutionary tract *Vestiges of the Natural History of Creation* confirmed that it had alerted many to the idea's subversive implications. Material aimed at a broad, non-specialist readership had played an important preliminary role in the Darwinian Revolution, shaping the way in which Darwin's theory would be perceived.[4]

Historians now began to focus on those who published material about science aimed at the ordinary citizen. Alvar Ellegård's *Darwin and the General Reader* had addressed the topic as early as 1958 but was focused mainly on the cultural elite. Bernard Lightman and others have since revealed how the expansion of mass-market publications in the nineteenth century created an audience for material about science that the scientists themselves were unable or unwilling to satisfy. Science writers with their own agendas influenced what people believed and could adapt their presentations to what would sell. It turned out that much of the literature about evolutionism did not always reflect the ideas and values of the scientists. The same techniques are now being extended into the twentieth century, taking in the new media of film, radio and television and paving the way for the world of the internet.[5]

Studies of popular science now provide significant insights into the way evolutionism was presented to the general public, but the information is scattered among sources confined to particular time periods or formats. There have been few surveys of material on a single topic over a longer period. This may be a consequence of historians' desire to demote the study of big theories from the – admittedly overstated – position it once held. Now may be the time to bring the two areas together. Our understanding of the rise of evolutionism is now spread

[4] Desmond, 'Artisan Resistance of Evolution in Britain, 1819–1848' and *The Politics of Evolution*. On *Vestiges*, see Secord, *Victorian Sensation*.

[5] Lightman, *Victorian Popularizers of Science*; see also Fyfe and Lightman, eds., *Science in the Marketplace*, and Geoffrey Cantor et al., *Science in the Nineteenth-Century Periodical*. For general surveys, see Jane Gregory and Steve Miller, *Science in Public*, and Peter Broks, *Understanding Popular Science*. On twentieth-century developments, see, for instance, Bowler, *Science for All*; Timothy Boon, *Films of Fact*; Marcel Chotkowski LaFollette, *Science on the Air*; and Constance Areson Clark, *God – Or Gorilla*. For those who read German, Angela Schwartz, ed., *Streitfall Evolution*, provides a survey strong on visual imagery that extends through to the age of the internet.

over a longer time span than the original Darwinian Revolution, and it would surely benefit from the insights that can be provided by the popular science perspective.

Arenas of Discourse

Much of the public interest in evolutionism has been driven by the theory's wider implications, but the arguments were never purely about science. Chambers' *Vestiges of Creation* had already made the link between evolution and social progress before Darwin published. The *Origin of Species* could be read by anyone with a reasonable education. The now-famous debate between Thomas Henry Huxley and Bishop Samuel Wilberforce took placc before a packed crowd at a meeting of the British Association for the Advancement of Science and was commented on widely in the press. Herbert Spencer turned the link between evolution and progress into one of the most active ideological movements of the late nineteenth century. Liberal Christians promoted a very similar message in the hope of modernizing the faith. All were looking to influence public opinion in an age when the reach of the communications media was expanding at a rapid pace. In the age of the 'press barons', their newspapers had enormous influence, while new media such as radio and later television would soon expand the scope of outreach immensely.

The pattern of expansion was the same on both sides of the Atlantic, although the debates in America played out differently as they were influenced by its unique social and religious culture. Asa Gray's clash with Louis Agassiz mirrored that between Huxley and Wilberforce, while the popularity of Spencer's philosophy identified it with a particularly intense capitalist ideology. Books and articles were frequently reprinted on both sides of the Atlantic, not always by permission. Huxley lectured to huge audiences on an American tour, while Spencer too was lionized on a visit. By the end of the century, American science was gaining more influence, typified by the publicity associated with the discoveries of fossil dinosaurs in the American West. But outright opposition to the idea remained active among evangelical Christians, culminating in the campaign that led to the prosecution of John Thomas Scopes for teaching evolution theory in 1925. The efforts of the press to shape how the 'Monkey trial' was viewed in the big cities are a central feature in historians' accounts of these events.[6] The flow of ideas and information

[6] There are many studies of evolutionism in America, including Cynthia Eagle Russett, *Darwin in America*, and Ronald Numbers, *Darwinism Comes to America*. On the era of the Scopes trial, see Clark, *God – Or Gorilla*, and Edward J. Larson, *Summer for the Gods*.

across the Atlantic was now increasingly from west to east, and commentaries on the re-emergence of Darwinism by American scientists such as George Gaylord Simpson were as widely circulated as those by Julian Huxley and other Europeans.

Conventional histories of the Darwinian Revolution focus on commentaries in the expensive books and magazines read by the cultural elite. The key issues to be debated were identified by those who had or sought cultural authority. For the vast majority of the population other means of communication were more relevant and may reveal other issues that were of greater interest. Cheaply printed books, magazines and newspapers led the way in the nineteenth century, supplemented by public lectures often given by itinerant promoters of topics that were capturing public attention. The means of communication expanded in the twentieth century to include film, radio, television and the internet. For the topic of evolution, displays in museums, zoos and exhibitions also became important. Direct and indirect references to evolutionism also appeared in popular fiction.

Ellegård's classic *Darwin and the General Reader* surveyed responses to Darwin's *Origin of Species* and *Descent of Man* in the British periodical press, identifying a range of topics and attitudes. But Ellegård focused on publications aimed at a well-educated readership and suggested that it took time for the debates to reach a wider audience. An appendix lists his sources and classifies their religious or political leanings and the level of sophistication they assumed on the part of their readers. Titles are identified as conservative, liberal or radical, while the readership is categorized as highbrow, middlebrow or lowbrow. The latter area illustrates the concentration on the elite and well-to-do: of the 113 titles listed, 44 (39%) are highbrow, 53 (47%) are middlebrow and only 16 (14.2%) lowbrow. Of those 16, four are 'not included in statistics' and none have more than a handful of references in the text. Politically, only 6 titles (5.3%) are identified as radical and 5 (4.4%) as liberal/radical.[7]

Since Ellegård's pioneering work, research has drawn attention to the emerging radical press and to popular material on evolution aimed at the rising middle class. We also know that there was strong public interest in spectacular discoveries of fossils such as the dinosaurs, which inevitably attracted comment on how these extinct forms could have appeared and disappeared. There was a huge expansion in the publication of popular natural history texts, some of which began to include brief references to how life on earth might have changed. This material must be taken into

[7] Ellegård, *Darwin and the General Reader*, pp. 40–1 and 26–7; the figures are given in his appendix 2, pp. 368–84.

account to appreciate how ordinary people were brought face to face with the topic of how life evolved. We need to reverse Ellegård's ratios, concentrating on the low- and middlebrow publications while including specialist and highbrow literature only when it attracted wider attention.

The categories of high-, middle- and lowbrow are an impressionistic way of recognizing that publications demand different levels of education and cultural sophistication on the part of their readers. The study of popular science avoids the highbrow literature, including technical scientific journals and periodicals aimed at the well-educated. It pays more attention to material aimed at readers with only a casual interest in the implications of scientific theorizing. Much of the historians' attention now focuses on the practicalities of how this material is produced and consumed. Lowbrow texts must be accessible by someone with limited education, and to be really popular they must also be affordable to those on a small income. Literacy increased dramatically in the nineteenth century, but the level of education available to the poorest classes remained limited. The greatest impact was made by technical innovations that allowed printed material to be produced in greater quantities and much more cheaply. By the end of the century, books, magazines and newspapers were available at prices even the ordinary worker could afford.

We also need to be aware of non-print modes of dissemination, including public lectures, museums and exhibitions. In early nineteenth-century Britain, lecturers toured the country promoting popular topics such as phrenology, often accompanied by demonstrations. The fashionable world heard lectures by experts at venues such as the Royal Institution. In the later nineteenth century, many well-known figures engaged in lecture tours on a wider scale – T. H. Huxley was one of many British scientists who ventured across the Atlantic to promote their work this way and earn some extra cash.[8] This activity continued into the early twentieth century, but the means of reaching the public now extended to include film and radio, a trend taken further by television and the internet. Natural history museums had also been created in many cities – in the United States they were often funded by industrial tycoons anxious to display their public spirit. Research in areas such as fossil-hunting was funded, but the emphasis was now on increasing the engagement with the public.

In the area of print alone we can see how new technologies such as the steam press expanded the production of books and magazines and reduced their price. Books by elite scientists and intellectuals were still published at

[8] See Diarmid Finnegan, *The Voice of Science*; Huxley's tour is also discussed in Chapter 5 below.

prices that put them out of reach for all but the well-to-do. Mid-nineteenth-century books on fossils were often priced at several pounds and would take up a month or more of the income of a lawyer's clerk, by no means the lowest paid worker.[9] Even a mid-priced book would take a week's wages, so to reach a mass market publishers needed to print really cheap books at a shilling (one twentieth of a pound) or less. Admittedly, there were libraries where expensive books were available to those who could not afford to buy them, and a thriving trade in used books.

Magazines and newspapers were gradually coming down in price and becoming more widely available. Elite scientific societies published their transactions, but in the early nineteenth century there were efforts by enthusiasts and commercial publishers to found magazines that would appeal to all levels of expertise. Even *Nature* (founded in 1869) began with this aim, only later to morph into an in-house journal for the professionalized scientific community. Natural history was an obvious target for this kind of initiative because it attracted enthusiasts with a wide level of expertise, although it proved increasingly difficult to balance the needs of the amateur enthusiast with those of the serious expert, let alone the professional biologist. Many ordinary readers would have got their information about new discoveries and new ideas from articles in more general magazines and newspapers, increasingly written by authors who were not themselves engaged in research.[10]

These changes can be seen in the print runs of books and the circulation figures for magazines and newspapers. Some of the most important scientific books of the period sold only a few hundred copies, while the circulation of highbrow magazines was only a few thousand. Only when cheap reprints of classic texts – including the *Origin of Species* – became available toward the end of the century did sales reach the tens of thousands. But long before this, books on science were being written explicitly for the less sophisticated reader and sold at prices that were affordable. By the middle of the century, there were magazines and newspapers reaching hundreds of thousands of readers, and by 1900, newspapers with circulations in the millions gave enormous power to the press barons who controlled them.

The cheap material on popular science was often written to supplement the limited scope of formal education available to the working

[9] See Ralph O'Connor, *The Earth on Show*, pp. 222–3. For details of how the old British system of pricing worked, see 'A Note about Money' in Bowler, *Science for All*, p. xi.

[10] See Gowan Dawson et al., eds., *Science Periodicals in Nineteenth-Century Britain*, and Bernard Lightman, *Victorian Popularizers of Science*. These developments are discussed in more detail in the relevant chapters below.

classes. Among the texts were books and articles that directly addressed the topic of evolution, while others might introduce it indirectly. Natural history was a favourite topic in the popular and semi-popular magazines, and while much of this literature focused on observation, references to evolution would come in when dealing with areas such as fossils and the distribution of species. Darwin himself contributed articles on his botanical work to the *Gardener's Chronicle*, which reciprocated by reviewing his books. There was a steady expansion of natural history museums and zoological gardens, which provided information on the significance of their displays. Some of this material was aimed at juveniles – dinosaurs acquired an aura of excitement for children as well as adults, opening up questions about origins even for the younger reader.

Political and ideological literature often referred to evolution: Spencer's work and the later eugenics movement both drew inspiration from evolutionism. Books and magazines on religious topics also provided commentaries. Although initially critical of Darwinism, the liberal Christian media increasingly favoured a progressionist form of evolutionism. Ideas about evolution were also influenced by their inclusion in fictional material. H. G. Wells' *The Time Machine* addressed the future evolution of humanity, while the alien Martians of *The War of the Worlds* suggested that humanity is not the intended goal of evolution. Science fiction has become a powerful way of extending discussions of complex issues raised by theories such as evolution to readers who might never turn to a popular-science text.

Media developments transformed the way information and opinion is presented. The ability to supplement text with images improved dramatically, especially when photographs could be reproduced directly. In the early twentieth century, colour illustrations gradually became the norm. Parallel developments allowed lecturers to make their presentations more attractive through increasingly sophisticated visual material. Similar transformations took place in the presentation of displays in museums and exhibitions. As natural history museums began to attract people from the lower social classes, the need to make their collections more attractive as well as informative led to much more colourful displays, eventually including dioramas that could give the visitor a sense of actually witnessing a scene from the past.

Radio allowed larger audiences to be reached but initially retained the limitations of print. The normal presentation was a 'talk' by an expert, but even here the possibilities of dramatization slowly became apparent. *The War of the Worlds* was famously presented on radio by Orson Welles and has become the basis of several movies. Film and television allowed real-life actions to be presented in a way that enhanced any message that the director wanted to emphasize. Visual presentations had the power to shape how a complex topic such as evolution would be perceived, not always in the most obvious ways. We shall see how new representations of the dinosaurs, for instance, changed ideas about the shape of the tree of life.

Engagement and Control

As the availability of material aimed at a mass market expanded, the question of who controlled the flow of information and commentary became ever more crucial. In Britain, the press was often subject to state control, and radio broadcasting was limited to what the centralized authority of the BBC would allow. The American media were always more commercial and more populist, and references to science often involved far more than a desire to educate a passive audience. Not that the theme of education is irrelevant, since for some writers and publishers there was a genuine intention to provide news about the latest developments. But who decided what counted as valid information, especially when the scientists themselves could not agree? What happened if the scientists themselves did not want to get involved in addressing the public? Who would take on the job, and what would their attitude toward the scientific community be?

Some scientists have always thought it part of their duty to communicate with the public, but when the science itself is in dispute, rival positions could be presented, each backed up by expert authority. Expertise could be gained by various routes, especially in a field such as natural history where dedicated amateurs could rival professional biologists in their knowledge of a particular area. Outside the realm of educational literature, many scientists and naturalists felt uncomfortable engaging with the sensationalism encouraged by the media. A growing body of science journalists and correspondents began to take over the job, and their interests did not necessarily coincide with those of the scientific community. They had various agendas, in some cases openly critical but always including the desire to reach as wide an audience as possible. This meant that although they had the potential to shape attitudes on controversial issues, they also had to take account of what could gain traction with public opinion. We now have a body of literature charting the emergence of the popular science communicator; rather than summarizing them in detail here, they will be explored at the start of each of the following chapters.

Many of those involved – scientists and non-scientists alike – had agendas stretching beyond the desire to communicate the latest discoveries. Historians now recognize that the original debate over Darwinism was driven in part by the newly professionalized scientific community seeking to displace the old ecclesiastical and cultural elites. This was what led T. H. Huxley to promote Darwinism even though his ideas differed from those of Darwin himself. But there was no unanimity among the biologists, and even in the early twentieth century there were eminent scientists who argued that evolutionism was compatible with the belief that the world has a moral purpose. This was the position taken by one of Britain's most prolific popular science writers, the Aberdeen professor of natural history J. Arthur Thomson, whose works also sold well across the Atlantic. The fact that evolutionism raised issues in fields as diverse as theology, morals and politics opened the way for commentators in all of these fields to get involved, although their engagement with the science was not always sympathetic. Popular writers including Samuel Butler and G. K. Chesterton attacked Darwinian theory by claiming that any intelligent person was competent to assess its validity, a view echoed by the religious opponents in the United States who led the campaign to exclude it from the schools. H. G. Wells is an example of a popular writer more sympathetic to the scientists and to Darwinism.[11]

Evolution and Its Cultural Matrix

The sheer breadth of media coverage makes it difficult to assess evolutionism's engagement with and impact on popular opinion. The problem is particularly acute for areas outside the realm of popular science communications, such as fiction and various forms of visual presentation. A substantial body of scholarship analyses the impact of evolutionary themes on forms of expression beyond the realm of non-fiction texts. Literary scholars led by Gillian Beer and George Levine have found Darwinian themes embedded in Victorian fiction and poetry, a project recently extended by Michael Ruse. Other scholars find similar links in the visual arts, early photography and even music. Darwinian elements are occasionally detected in material created before the *Origin of Species* was published.

The anomaly of finding Darwinian impacts in the period before Darwin published highlights a problem of interpretation noted in Bernard Lightman's analysis of the popular science literature of the time. He draws attention to an article by the writer Grant Allen in 1888 that claimed that evolutionary themes had become pervasive in all areas of

[11] See Bowler, *Science for All* and *Monkey Trials and Gorilla Sermons*, and, on America, Ronald L. Numbers, *The Creationists*.

discourse, although often presented in ways that showed little understanding of the science. Curiously, Allen suggests that the upsurge of interest had arisen only recently – although this is over a quarter of a century after the *Origin* is supposed to have sparked debate on the topic.[12]

How do we reconcile this last opinion with the modern scholarship that detects Darwinian themes throughout mid-nineteenth-century culture? One possibility is the sheer complexity of the idea of evolution and of the various explanations of the process, including Darwinism. The rival theories raise a series of underlying issues: Are the changes gradual or discontinuous? Is the overall process driven by some kind of progressive trend or law, or by pressures arising from the local environment? If there are law-bound trends, do they have a predetermined goal? All of these questions (explored in more detail below) imply wider issues that are not necessarily confined to debates over the process of biological evolution.

The Darwinian theory makes use of concepts and themes that can have independent roles within a wide spectrum of debates about nature and human nature. The 'struggle for existence' may be the driving force of natural selection, but it can be applied in several different ways depending on whether the competition is supposed to be between individuals or groups. It can also be invoked, as in Herbert Spencer's thinking, as the spur to self-improvement serving as the basis for cumulative change through the Lamarckian process of the inheritance of acquired characteristics. The basic notion of struggle can also be highlighted by the relationship between predators and prey, as in Tennyson's famous line about 'Nature, red in tooth and claw'. The phrase 'struggle for existence' occurs in Thomas Malthus' work on population pressure, although it appears in a discussion of warring tribes, not in the context of individual competition. Neither Malthus nor those who claimed that want and misery are inevitable saw the struggle for resources as the driving force of progress. Darwin drew on the individualism that lay at the heart of Malthus' thinking, but he extended it in ways that few of his contemporaries anticipated.

Another concept with a life of its own is Darwin's image of the 'tangled bank', used to illustrate the complexity of what we now call ecological relationships.[13] But again the belief that nature is complex had gained traction in Victorian thinking without necessarily implying that there will be long-term changes in the species involved. Ecology and evolutionism

[12] See Lightman, 'The Popularization of Evolution and Victorian Culture', pp. 286–7.

[13] *The Tangled Bank* is the title of Stanley E. Hyman's classic study of parallels in the thinking of Darwin, Marx, Fraser and Freud.

are sciences that interact but are not co-extensive, and the same is true of their cultural applications.

Even the relationship between humans and animals, so central to the Victorian debates on 'man's place in nature' can be explored in ways that do not imply an evolutionary connection. It had always been obvious that our bodily and some of our mental functions parallel those of animals, a point driven home by the pre-Darwinian interest in phrenology, the claim that the brain is the organ of the mind. It was by no means inevitable that those who were impressed by Darwin's illustrations appreciated the evolutionary implications of the texts in which they were embedded. The much-discussed 'missing link' between humans and apes was something that might have been predicted from the ancient concept of the chain of being, in which all intermediate forms were supposed to exist. When members of non-white races or individuals with excessive body hair were exhibited as living examples of the link, they challenged the Darwinians' prediction that the common ancestor from which apes and humans diverged could be found only in the fossil record. Cartoons implying that we evolved directly from gorillas presented evolution as a more-or-less linear process with a predictable goal, not at all what Darwin's theory implied. The theory of sexual selection was an important extension of Darwin's thesis, but it incorporated existing stereotypes of male and female behaviour into the evolutionary mechanism. Darwin did not create those preconceptions, although he may have encouraged his contemporaries to be a bit more self-conscious about them.

The reception of Darwin's theory was embedded in a cultural matrix that was in flux, although the changes were driven by multiple factors and took place more slowly than we used to think. Darwin's works enhanced the credibility of the component elements such as the struggle for existence, allowing the term 'Darwinism' to become a convenient label to denote ideas and images that were part of the theory – although they could also function independently. His writings may have highlighted and even transformed perceptions of components that pre-existed in the culture of the time, but his core theory was not the only driving force. Evolutionism was a product of its cultural matrix as well as a significant addition to it. The theory of natural selection had radical implications if taken as the sole or even the main explanation of how evolution works. Many found those implications hard to accept, and even Darwin conceded that his theory could be seen as the basis for a belief in the inevitability of progress. It would take time for the world to come to terms with a view of nature in which humans were not the goal of creation.

This gives us a way to make sense of Allen's 1888 claim that evolution was only just beginning to impact the public consciousness. I shall argue that the – admittedly intense – debates over the *Origin of Species* were still to some extent centred on pre-Darwinian notions that had already challenged conventional religious views but which people were only just beginning to come to terms with. They were already grappling with these issues when Darwin published, and his intervention was seen as a radical application of positions already emerging, rather than a significant new perspective that adopted some of the old ideas but challenged others. Darwin became a symbol for ideas that his viewpoint had already transcended, his vision of the 'tree of life' modified to imply a central line of progress toward humanity.

On this model, Allen's comments may reflect a recognition that newer images of evolution including those proposed by both Darwin and Herbert Spencer were at last starting to take hold toward the end of the century. The possibility that evolution was not a process with a preconceived end-point represented by humanity was now starting to sink in. The non-Darwinian theories proposed in the decades around 1900 show that even in science, an older developmental version of evolutionary progress still persisted, and it is no surprise that it found its way into the popular science of the time. The tensions are visible in related areas whose subject matter was able to attract media attention, such as spectacular fossil discoveries of dinosaurs and early humans. Fossil hominids were routinely fitted into linear patterns of development even in the early twentieth century. The mass-media coverage of evolution into the twentieth century represented an ongoing battle between an open-ended Darwinian model and more teleological interpretations of the history of life on earth.

Alternative Theories

A starting point for analysis of the issues raised by evolutionism is provided by the identification of the main alternatives that emerged in the debates. Listing the rival theories, however, provides only a bare framework and will require further unpacking of a range of subsidiary elements and assumptions. Some of these components are not obvious from a bald statement of the theories' basic explanatory mechanisms, and some are relevant for more than one theory. This situation generated much room for confusion when the ideas were simplified for presentation to an audience not aware of the technical issues involved. We can identify six basic ways of trying to explain how new species appear, whose popularity fluctuated in the course of time. Darwin's hypothesis of natural selection is the best-known today since it is the basis of most modern thinking on the topic, but it became dominant only in the mid-twentieth century, and before then several alternatives had been proposed. Some of these challenged even the most basic principles of the Darwinian approach. In roughly chronological order they are listed as follows.

Ascending the Chain of Being. Even before the concept of evolution emerged, naturalists had ranked the species into a linear hierarchy stretching upward from the simplest forms to humanity. Some early ideas about the development of life on earth assumed that the links in the chain had appeared one after the other, giving rise to a progressive ascent toward humans as the goal of creation. In principle, the model provided by a linear scale was abandoned in the nineteenth century and replaced by a branching 'tree of life', but the expectation that there must be a central trend leading to ourselves remained popular especially in accounts aimed at a non-specialist audience.

Theistic Evolution. Early accounts often presumed that the ascent of life unfolded in a teleological, or goal-directed, manner following laws imposed directly by the Creator. The production of new characteristics was thought to be somehow predisposed to advance in a meaningful direction. The goals might be adaptation to changing environments, progress to higher levels of organization, or even the production of beautiful colours and structures. This was a useful compromise for those with the appropriate religious beliefs, but was rejected as unscientific by materialists who required an explanation in terms of laws that did not anticipate future goals.

Lamarckism. This is a term applied to a mechanism proposed by the French naturalist J. B. Lamarck in the pre-Darwinian period (although he also invoked an inherently progressive force). It became popular as an alternative to natural selection in the late nineteenth century. The mechanism involves the 'inheritance of acquired characteristics' and, like natural selection, is based on the adaptation of species to changing environments (this is sometimes called the utilitarian approach, since it assumes that the new characters are useful to the organisms). It is known that if living things are exposed to a new environment, they can modify themselves in order to accommodate the change. In the case of animals, they will change their habits and modify their physical characteristics accordingly. Lamarckism assumes that those modifications can be transmitted to the offspring (a possibility subsequently denied by genetics). This allows them to accumulate over successive generations to change the species. Since all members of the population can learn the new habit, there is no need to invoke a struggle for existence, although Herbert Spencer thought that competition stimulated the animals' efforts to improve themselves.

Natural Selection. Darwin's mechanism of evolution is also a process of adaptation but operates in a very different way. He noted that the individuals making up a population vary slightly among themselves in ways that seem to have no purpose. The normal form of the species is adapted to the prevailing environment, but if conditions change, any individual variation that is by chance better adapted to the new situation will survive and reproduce better than the old norm, while any disadvantaged variants will be exterminated. Darwin knew that the variant characters could be transmitted to the offspring (as demonstrated by animal breeders using artificial selection) and will accumulate to change the species. He argued that the process of natural selection was driven by the 'struggle for existence' – the competition for resources necessitated by the fact that reproduction tends to produce more individuals than the environment can support. Since the original variations impose no direction, there can be no predetermined course of evolution – change occurs only because some variants cope better with the local environment. If a population is divided and exposed to different conditions, the subpopulations will evolve in different directions and the species will divide (hence his vision of the tree of life). There is no implication that the new characters must be 'higher' in any meaningful way, and some Darwinians have rejected the whole concept of evolutionary progress.

Orthogenesis. Many naturalists could not accept that all evolutionary changes are adaptive; indeed, they thought that many of the characters that distinguish related species had no adaptive purpose. To explain the existence of non-adaptive characters, they postulated a process of 'definitely directed variation', sometimes known as orthogenesis. This presumes that variation within the species is neither random nor guided by the organism's efforts but is pushed in a single direction by forces arising from within the organism. The character that emerges need not provide any advantage in dealing with the environment and may even be harmful to some extent, driving the species to extinction when fully developed. Since closely related species might share the same predetermined trend, they would advance in the same direction to give parallel evolution.

Saltationism. The belief that species change by abrupt jumps, or saltations, to a new form was also invoked to explain the existence of what were supposed to be non-adaptive characters. The original meaning of the term 'mutation' referred to the sudden appearance of totally new characters within a population, and it was supposed that if several individuals with the same mutation appeared, they would form a new breeding population and thus found a new species. Unlike orthogenesis, saltationism did not imply that the new changes added up in a consistent directions, although the two models were sometimes combined since both supposed that the transformations were directed by forces internal to the organism.

Issues, Concepts and Confusions

While the scientists debated these various alternatives, evolutionism could be presented to the public in ways that did not necessarily focus on the mechanism of change. We need to identify the alternative positions that can be taken up on a series of key issues that define how the history of life on earth has unfolded but do not map directly onto the rival theories. The alternatives are seldom clear-cut because the extreme positions were often blurred in presentations for the non-specialist. But recognizing the extremes can be useful, if only to position the intermediates between them. We can then appreciate what was driving the complex arguments that underpinned even the most sensationalized or over-simplified accounts of the theory and its implications. Knowing why people were being encouraged to imagine or visualize evolution or its results in a particular way will help us chart a course through a maze of popular images even more convoluted than the formal debates of the scientists.

The most basic distinction for the theory's religious implications is whether the evolutionary process is natural or designed. For religious thinkers used to believing that the world was created by God, the most obvious fallback position was theistic evolutionism, in which the laws governing the process ensure that the Creator's overall purpose will eventually be achieved. The naturalistic alternative – best illustrated by Darwinism – accepts that the laws of nature (even if instituted by a Creator) do not recognize or work towards a future goal. For Darwin, the laws governing the production of variations within a population cannot foresee what the species might need at some time in the future if the environment changes – that is why they seem so trivial and pointless. Natural selection is needed to make use of the few that by chance turn out to be useful, and although the result looks like design, it is achieved by a combination of purely mechanistic processes. Lamarckism too is in principle naturalistic, since before the emergence of genetics the inheritance of acquired characters seemed a plausible mechanism.

Generations of liberal religious thinkers have insisted that there must be a way in which the Creator's ultimate purpose is built in to the system. Many Lamarckians sought to do this on the scale of local adaptations by arguing that organisms have the power to modify their behaviour and structure in response to environmental challenges. In effect, God has delegated his creative powers to life itself. But could evolution build a wider purpose into the overall development of life on earth? If the laws of mechanics and gravitation are enough to turn a cloud of dust into a solar system, surely the laws governing evolution could have a similar purpose embodied within them. If intelligence and cooperative behaviour are important factors shaping how organisms respond to challenge, there might be an overall tendency for those factors to be enhanced in every engagement. This would drive progress toward higher mental and moral powers, culminating in humanity. Suggesting that the mental powers of organisms played a positive role in shaping their species' evolution was favoured by the opponents of materialism. They accused Darwinism of encouraging the belief that animals were just mechanisms operating at the mercy of variations they had no power to control.

Any theory implying open-ended change must, however, suggest that there are many ways in which life could advance to higher levels. The laws of evolution would not be aimed at some particular goal, but would allow for different ways of achieving higher levels of sophistication. Darwin himself supposed that evolution would be progressive, at least in the long run, but his theory implied that many branches of the tree of life have advanced to some extent. Perhaps the human form is not the only way in which higher mental and moral powers could become manifest (a point we shall return to later).

This way of thinking about the tree of life points to a distinction that can be drawn within the range of naturalistic theories: Is evolution governed by rigid developmental trends, or is it an open-ended process that can at any point be driven in many possible directions depending on the local circumstances? This is linked to the question of whether variations are produced by forces within the organism or in response to environmental impact – or perhaps some combination of the two. Utilitarian theories, those based on adaptation as the driving force, necessarily assume that whatever the source of variations, only those conferring an advantage in dealing with the environment will survive and reproduce. In principle any mechanism that depends on adaptation must be open-ended because populations can become divided by migration and thus become exposed to different conditions.

Undirected macromutations could also generate a diverse range of forms, in this case depending on the internal forces that produce the new characters. Many of the naturalists who appealed to this model openly rejected the utilitarian basis of Darwinism. They assumed that new characters could establish themselves if they conferred no advantage – or even if they were positively maladaptive. Orthogenetic theories also invoked internal factors to explain supposedly non-adaptive trends seen in the fossil record. Some Lamarckians imagined that variation could become constrained by the lifestyle habits adopted by the first members of a lineage, imposing a trend eventually producing a species so specialized that it would be unable to cope with any major change in its environment.

Palaeontologists who thought they could see rigid trends in evolution claimed that they were modelled on or even controlled by the process of individual development. In the period before genetics transformed our understanding of heredity and variation, it was widely assumed that stages in the past evolution of a species could be inferred from the sequence observed in the embryological development of the modern organism. This was the recapitulation theory, particularly favoured by Lamarckians. The embryological, or developmental, analogy was extensively promoted in popular accounts of evolution, helping to create the impression that evolution has a built-in goal. Evolution was seen as a ladder to be climbed toward a predetermined end state, not as the constantly diverging tree depicted by Darwin.

The assumption that the fossil record reveals an ascent from the simplest forms of life up to humanity had emerged in the early nineteenth century, originally in a discontinuous form that saw each new class introduced suddenly after a geological catastrophe. The geologists who promoted this vision used to be dismissed as conservatives anxious to defend the reality of Noah's flood. We can now see that their legacy lingered on, with the catastrophes gradually downgraded into occasional episodes of relatively rapid evolution. In the modern world the theory of catastrophic mass extinctions has re-emerged – to say nothing of the Young Earth creationists' continued reliance on the biblical flood.

The palaeontologists had to come to terms with the growing evidence that each new class soon divided itself into a range of divergent forms. Popular representations of the overall tree of life often depicted it with a central trunk running through the sequence of vertebrate classes (fish, reptile, mammal) and having the human form at its top, like the angel on the top of a Christmas tree. All the side branches were of lesser significance since they had dropped off the main line of development. This representation of the tree of life still gave the impression that there was a ladder to be climbed toward a final goal, as well as a series of less important branches multiplying the diverging forms within each stage. There were also efforts to argue that following an initial episode of divergence, each family followed its own rigid developmental trend.

The vision of a ladder of stages leading toward humanity is the most extreme version of the more general belief that evolution is an essentially progressive force. The alternative more open-ended tree of life makes it much less easy to define an overall progressive trend and has even been used to argue that the whole idea of progress is misleading. Each branch of the tree has done its own thing, and is obviously successful if representatives of it are still alive today. To preserve some aspect of evolutionary progress within an open-ended model, it has to be defined not by a single goal to be reached but by stages of relative complexity or sophistication that can be reached in different ways. If life began from very simple forms, then many branches have become 'higher' in at least some respects. If humans are higher than any other form of life, that is because our branch has managed to reach a stage above that gained by any other, not because we are the inevitable goal of the process. It was also recognized that many adaptive transformations are not improvements in any absolute sense – only rarely do they create some genuinely new structure or function. In some cases a lineage may actually degenerate, as when parasites adapt to a lifestyle that no longer requires them to use some of their original structures. Progress is at best only a by-product of adaptive evolution, but convincing the public of this has always been something of a problem for Darwinians.

A topic widely debated in the late nineteenth century centred on the role played by the 'struggle for existence' in shaping both the adaptation of species to their environment and the general progress to higher states. This was only an issue for those who supported utilitarian theories in which the environment determines what can survive. Struggle could be invoked by both Darwinians (as the driving force of natural selection) and Lamarckians (as the spur to self-improvement). But there was also much confusion over the level at which competition acted. The primary mechanism of natural selection works on individual variants within a population, but there were many who preferred to see the struggle operating between groups, so that one species, for instance, might drive a rival to extinction. The debate about the validity of group selection goes on still today, but in the nineteenth century it had vital relevance given the widespread assumption that all levels of evolution, biological and social, must operate by the same process. The various forms of what has become known as 'social Darwinism' depended on different levels of competition. If between individuals, then free-enterprise capitalism seemed the natural social equivalent, but if between varieties or species, an imperialist model of racial or national dominance might be endorsed.

In the later twentieth century it became gradually less popular to assume that biological evolution should provide a 'natural' model for social change. The emergence of genetics suggested that a rigid distinction should be drawn between characteristics that can be inherited and those that are acquired through experience or education. Genetics provided an explanation of how random variations can be inherited and was eventually synthesized with Darwinism. Biological evolution working by natural selection became distinct from social changes operating in a manner analogous to the Lamarckian process of the transmission of acquired characters. This moved the debate over the social implications of evolutionism into the area known as the 'nature versus nurture' distinction. To what extent are organisms, and especially humans, constrained in their abilities and actions by their genetic inheritance and hence by their evolutionary ancestry? Huge controversies have erupted around this question, including those centred on the eugenics movement and its claim that ability is predetermined by heredity and, more recently, on the social applications of sociobiology. All of these debates involve issues related directly or indirectly to the theme of evolution.

Changing Perspectives

The chapters that follow offer a roughly chronological survey of what was available to the wider public on the topic of evolution from the pre-Darwinian period almost to the present. The communicators include figures such as T. H. Huxley and H. G. Wells whose names are still known today and many more who have been forgotten except by specialist historians. There are scientists such as J. Arthur Thomson who were famous at the time, included here because they wrote for the wider public, along with popular science writers such as Grant Allen and Joseph McCabe whose names have also faded to obscurity. There is also a mass of journalists, radio broadcasters and the like who produced the more ephemeral material that sought to bring evolutionism into every home.

The original purpose behind the construction of this survey was to show scholars in fields outside the history of science what was under discussion in the wider public domain at any point in the extended revolution by which Darwinism came to its current position of influence. But the work has also revealed points at which the material clashes with the conventional image of the process. One challenge already recognized by historians is the demonstration that evolutionism was already under discussion before Darwin published, but not in forms that anticipated his particular explanation. The earlier concepts involved different mechanisms of change, but also presented very different models of what later became known as the 'tree of life', usually imagining it with a central trunk or spine leading toward humanity. What follows in Chapter 2 is further confirmation of this insight.

The same point also emerges when we turn to the debate immediately following the publication of the *Origin of Species*, where it has often been assumed that Darwinism was quite rapidly established on the cultural map. The problematic nature of this assumption has already been raised in the context of Grant Allen's claim in 1888 that the theory had only just begun to gain a dominant influence. I am not the only historian who has claimed that the original impact of Darwin's book has been misinterpreted. It certainly initiated a debate that fairly rapidly led to general acceptance of evolutionism, and some elements of his theory attracted wide attention. But the wider we spread the net by which we capture examples of the discussion, the more it becomes plain that the most radical implications of the selection theory were sidelined in favour of older, more teleological models of development. It would take time for both scientists and the wider public to come to terms with the possibility that life has not developed according to predetermined trends. Theologians and moralists did see the disturbing implications of Darwinism, but their warnings were overtaken by a consensus that turned evolution into a process that ensured the triumph of humanity and Western civilization.

Literary critics have identified reactions to the more disturbing elements of the Darwinian perspective in contemporary poems and novels, but their insights do not map easily onto the topic of how the theory was disseminated. Thomas Hardy's pessimistic poem 'Hap', for instance, was written under the influence of Darwin in 1866, but it offers no description of natural selection and was in any case not published until later in the century. Nor did Hardy himself always reflect the same level of pessimism.[14] Joseph Conrad's *Heart of Darkness* reflects a similarly pessimistic view of the world in the age of imperialism, but again a reader unfamiliar with the science of the time would hardly recognize the Darwinian elements. To see the link such a reader would have to be guided to a work on popular science. This survey works in the opposite direction to the search for influences behind the scenes – it looks at texts and other sources that were intended to inform or influence people's views on the process of biological change and how it has shaped the creatures we see around us, ourselves included. This will certainly include fiction, but only when it directly addresses the theme of evolution.

Other cases where the focus on popular material dealing with science changes our perception of the events will also emerge. The full implications of Darwin's theory were missed in part because the debate over the *Origin of Species* coincided with the wave of popular excitement aroused by Paul Du Chaillu's description of the gorilla. This generated a plethora of cartoons and bad verse implying that this allegedly ferocious beast was the ancestral form from which we ourselves evolved. Darwin's close supporters struggled to get across the message that if the apes were only our cousins, they could not provide direct evidence of the common ancestor. The earliest hominid fossils, to say nothing of the 'lower' races,

[14] On Hardy, see, for instance, John Holmes, 'The Challenge of Evolution in Victorian Poetry', pp. 53–4.

were also routinely depicted as the 'missing links' in a linear ascent from ape to human more reminiscent of the old chain of being.

A similar point emerges from a survey of the popular literature and displays of fossils such as those of the dinosaurs. When first discovered, these were often depicted as gigantic lizards or crocodiles, merely distorted versions of familiar stages in the ascent toward the mammals and humanity. It was the discovery of bizarre forms in the American West during the final decades of the nineteenth century that revealed a lost world of forms unlike anything alive today. The dinosaurs became a major branch of the tree of life that was lopped off to make room for the flowering of the mammals. The implication that evolution was an open-ended process subject to unpredictable external impacts became obvious to anyone who visited a natural history museum or looked at a popular book on the topic – even one aimed at children. When the conventional story of Darwinism's synthesis with genetics is supplemented with this and other less obvious sources of information, we gain a more nuanced and perhaps more convincing impression of what was involved in the theory's eventual triumph.

2 Before Darwin

Charles Darwin was not the first naturalist to suggest the basic idea of evolution. From the mid-eighteenth century onwards there were speculations about natural processes that might create new forms of life by naturalists later dubbed the 'forerunners of Darwin'.[1] Historians now avoid this term because it assumes that the suggestions pointed the way toward Darwin's own approach to the problem. In fact, these speculations envisioned processes that had little resemblance to natural selection and often pictured the history of life on earth in ways that do not resemble our modern viewpoint. J. B. Lamarck's explanation of adaptive change, later taken up as an alternative to natural selection, was embedded in a system that assumed the inevitability of progress and denied the possibility of extinction. The idea of natural changes in the earth's population was emerging, but not in ways that resembled the Darwinian perspective. These early ideas are crucial for understanding how the wider public began to abandon belief in the Genesis account of creation. But far from preparing the way for Darwinism, they established preconceptions that distorted the reception of Darwin's initiative.

The most radical thinkers of the eighteenth-century 'Age of Enlightenment' openly challenged the Genesis story with their materialist speculations. More enduring developments came from new ways of seeking to understand both the relationships between living species and the fossil remains of ancient forms. The ancient concept of a 'chain of being' was undermined by naturalists trying to classify the diversity of species revealed by exploration. Toward the end of the century geologists began to realize the immense age of the earth and chart the succession of the rock formations through time. The first reconstructions of the fossil record provided an outline of the history of life on earth. It was widely accepted that there had been a progression from the simplest forms through to the final appearance of humanity,

[1] *Forerunners of Darwin* is the title used for the 1959 collection edited by Bentley Glass et al.

although the inhabitants of each period had to be adapted to the conditions of the time. Some interpreted the successive populations as a series of divine creations, preserving the belief that the adaptation of species to their environment was an indication of the Creator's benevolence.

Renewed efforts to imagine naturalistic alternatives to supernatural design were made by political radicals, some of whom took up the ideas of Lamarck. In 1844 a less threatening interpretation was promoted to the rising middle classes as the basis for an ideology of social progress by Robert Chambers in his *Vestiges of the Natural History of Creation*. Like some of the earlier proposals, his theory involved a non-Darwinian vision of the ascent of life driven by predetermined trends. His book may have absorbed some of the opposition that would otherwise have been directed against the *Origin of Species*, but it also shaped public opinion in ways that made it harder for people to comprehend Darwin's more radical proposals.

Publishing Innovations

Chambers' writing on science was undertaken in the course of a successful career as a publisher and journalist. This was a period of massive change in the way ideas and information could be circulated, creating a complex relationship between an increasingly coherent scientific community, a growing body of amateur naturalists, and a wider public whose interest was only captured by the most exciting discoveries. New printing technologies transformed the publishing industry and vastly expanded its reach into a population that was also becoming more literate. Books were cheaper, making them available to the growing middle classes and even to better-off workers. Newspapers and magazines proliferated, some aimed at the cultural elite but many at the ordinary reading public. Publishers such as William and Robert Chambers specialized in reaching the middle-class audience to promote calls for social reform and scientific innovation. Some magazines were aimed at particular readerships, including amateur naturalists and the workers in the new industries who met in the Mechanics Institutes founded in many towns. There were news-sheets spreading radical political ideologies and willing to exploit any scientific idea that challenged orthodoxy.

Those who wrote for these new formats were seldom experts, but some members of the increasingly professionalized scientific community began to realize the importance of promoting their views more widely. The new means of publication were used both to disseminate information about popular topics such as fossil discoveries and to debate their implications for religion and social values. Here was a thriving marketplace for the discussion of new political, social and scientific views, always alert to the possibility of utilizing a new idea for wider purposes. This brought in the theories of evolutionary change, and a wide range of naturalists, jobbing writers and political activists were willing to get involved.[2]

Across the Atlantic the new American republic was also surging ahead, seeking to develop its own publications and local institutions to promote learning in the larger cities. There were now magazines such as the *New York Review* and *North American Review* to alert readers to the latest developments in areas of science and culture, including natural history. To begin with, much of this material was based on secondary reports and reviews written by figures who Nathan Reingold calls 'cultivators of science'.[3] *Scientific American* was founded in 1845, but was at first devoted almost exclusively to reports on new inventions. By the 1860s, though, there were many periodicals willing to include responses to controversial ideas including Darwinism.

Enlightenment Legacies

During the eighteenth-century there were many *savants* willing to challenge Christian beliefs, some becoming outright atheists. Materialist views on the nature of life encouraged the rejection of any notion that species were immutable divine creations, leading to speculations about how they might originate and change through time. Some of these speculations involved the transformation of species, but others did not – it was still possible, for instance, to imagine new forms being spontaneously generated from disorganized matter.[4] Most of the radical texts were written in French and would be accessible in the English-speaking world only to those who read the language (not unusual in the upper classes). A wider readership would have been enjoyed by two works translated into English, the comte de Buffon's *Natural History* and the baron d'Holbach's *System of Nature*. Both included speculations about the nature and history of species raising the possibility of

[2] For information on these developments, see Lightman, *Victorian Popularizers of Science*; Fyfe and Lightman, eds., *Science in the Marketplace*; Geoffrey Cantor et al., *Science in the Nineteenth-Century Periodical*; Secord, *Victorian Sensation*; Fyfe, *Steam-Powered Knowledge*; and Susan Sheets-Pyenson, 'Popular Science Periodicals in London and Paris'.

[3] See Nathan Reingold, 'Definitions and Speculationsa', esp. p. 40. This chapter appears in Alexandra Oleson and Sanborn C. Brown, eds., *The Pursuit of Knowledge in the Early American Republic*, which explores the development of institutions to promote science in this period.

[4] For details of these theories, see Roger, *The Life Sciences in Eighteenth-Century French Thought*; Bowler, 'Evolutionism in the Enlightenment' and *Evolution*, chap. 3.

transmutation, but neither attempted a comprehensive theory of evolution. A bolder speculation along these lines came from a homegrown author at the very end of the century, Erasmus Darwin.

Buffon's lavish twenty-four volume survey of the animal kingdom was published from 1749, ending with supplementary volumes including a survey of earth history in 1778. A translation by James Smith Barr was published in ten volumes in 1792, reissued in1797 with five more volumes on the birds added in the following year. A more authoritative translation was by the naturalist (and compiler of the original *Encyclopaedia Britannica*) William Smellie, issued in eighteen volumes in 1795, with a new edition in 1812. Both began with the theory of earth history published in Buffon's first volume, at the time a radical challenge to the Genesis account of creation (his later volume on the topic was not translated at the time). Scattered through the subsequent descriptions of animal species were a number of more speculative articles, included in both translations, which led some later commentators to hail Buffon as a founder of evolutionism.

Modern historians are less willing to see Buffon as a genuine advocate of evolution.[5] His second 'View of Nature' makes it clear that there can be no large-scale transmutations because 'Species are the only existences in Nature: for they are equally ancient and permanent with herself.'[6] The first 'originally created' forms serve as models for all their descendants. Another article, 'Of the Degeneration of Animals', seems at first to contradict this position by arguing that closely related modern species have diverged from a common ancestor under the influence of external circumstances. The zebra and the ass have descended from a horse-like ancestor, and the lion, tiger and leopard from an original big-cat form. But Buffon goes on to challenge his great rival Linnaeus' definition of species: the lion, tiger and leopard are not good species incapable of interbreeding – they are merely local varieties of the big-cat type. The real 'species' is what Linnaeus called the genus (in this case *Panthera*), which is permanently defined in nature so that no change can proceed beyond fixed – if rather generous – limits.[7]

How many readers in Britain would have been aware of Buffon's views? His speculations are buried in multi-volume surveys of the animal kingdom that would have found their way only into the libraries of the

[5] For details of Buffon's thought and influence, see Roger's survey and his *Buffon: A Life in Natural History*; also Bowler, 'Bonnet and Buffon'; Paul Farber, 'Buffon and the Problem of Species'; and Philip R. Sloan, 'The Buffon-Linnaeus Controversy'.

[6] Buffon, *Natural History* (trans. Smellie), 7: 89–108, quotation from p. 89; see also *Barr's Buffon*, 10: 343–66, p. 343, which uses very similar words.

[7] Buffon, *Natural History*, 7: 392–452, and *Barr's Buffon*, 9: 315–55.

wealthy. A one-volume abridgement published in 1791 did not include the more general articles, and there is little evidence of early nineteenth-century radicals citing Buffon as an inspiration – his reputation as a forerunner of Darwin was created only after the *Origin of Species* was published.

Buffon's survey was one of the sources used by Oliver Goldsmith for his *Animated Nature* of 1774, a literary rather than a scientific description of the animal kingdom that offered no speculations on the origin of species. Goldsmith arranged the species in a descending hierarchy with humanity at the top, leading some later commentators to see him as a champion of the 'chain of being', the ancient belief in a linear hierarchy of creation leading from the simplest forms up to humanity. Goldsmith realized, however, that there were many forms that could not fit in to such a neat arrangement. This point was central to the most influential development in the eighteenth century: Carl Linnaeus' system for classifying species by their degree of physical similarity. Buffon was scornful, but the system transformed the way species were named and their relationships determined, paving the way for a model of nature based on diversity rather than a geometric pattern such as the chain. Relationships were arranged as groups within groups and best represented in two dimensions rather than one – like the countries on a map rather than a linear hierarchy.[8]

Nevertheless, the image of an ascent toward humanity retained its power and remained central to most efforts to understand the development of life on earth. The theory of evolution propounded by Erasmus Darwin, grandfather of Charles, suggests how the issues of progress and diversity would continue to interact. He was a leading medical practitioner and a prominent member of the Lunar Society, a group of scientists and entrepreneurs who were at the forefront of the Industrial Revolution. The claim that all living things had arisen over a vast period of time from primitive original forms was advanced in the discussion of reproduction in his biomedical treatise *Zoonomia* of 1794. Darwin was also one of the country's leading poets, his *Botanic Garden* attracting wide attention, while *The Temple of Nature* of 1803 described his evolutionary system along with references to the 'war of nature' that have encouraged claims that he anticipated the theory of natural selection.

[8] Goldsmith, *History of the Natural World*; see Winifred Lynskey, 'Goldsmith and the Chain of Being', although James Pitman's *Goldsmith's Animated Nature* makes no reference to the chain. The history of the chain concept is described in Arthur O. Lovejoy, *The Great Chain of Being*; for more details of how it was challenged by Linnaeus and others, see Bowler, *Evolution*, chap. 3.

Darwin's explanation of how species are transformed depended on the inheritance of acquired characteristics. Animals respond to new environments by modifying their behaviour and their physical structure, and if these modifications can be inherited they will accumulate to change the species. There was no implication that the advance must lead to a predetermined goal; Darwin applied Buffon's point that the underlying similarities of related species suggest that they have diverged from a common ancestor to create the whole animal kingdom. All the higher forms have advanced beyond their distant ancestry, but in many different ways, with humanity being the outcome of the most successful line of development. For Darwin the process was the outcome of powers granted to nature by its Creator. Given the vast periods of time involved, he asks, 'would it be too bold to imagine, that all warm-blooded animals have arisen from one living filament, which THE FIRST GREAT CAUSE endowed with animality, with the power of acquiring new parts, attended with new propensities, directed by irritations, sensations, volitions and associations; and thus possessing the faculty of continuing to improve by its own inherent activity, and delivering down these improvements by generation to its posterity, world without end?'[9] The *Temple of Nature* is equally explicit in its vision of progress:

> ORGANIC LIFE beneath the shoreless waves
> Was born and rais'd in Ocean's pearly caves.
> First forms minute, unseen by spheric glass,
> Move on the mud or pierce the watery mass;
> These, as successive generations bloom,
> New powers acquire and larger limbs assume;
> Whence countless groups of vegetation spring,
> And breathing realms of fin, and feet and wing.[10]

These statements may have had only limited impact at the time, however. Darwin's most popular poem, *The Botanic Garden*, contains only a brief hint of evolutionary speculation, and by the time *The Temple of Nature* was published posthumously he was no longer in fashion. He became a target of abuse because of his support for the French Revolution, and his poetry came to be seen as outdated – the Romantic movement was now in vogue.[11] Medical professionals would have read *Zoonomia* – it went

[9] Erasmus Darwin, *Zoonomia*, 1: 507; 3rd ed., p. 397. The third edition has an appendix with another claim that evolution is a process of perpetual improvement; see p. 437. On the medical origins of his thinking. see Roy Porter, 'Erasmus Darwin: Doctor of Evolution?'

[10] Darwin, *The Temple of Nature*, lines 295–302.

[11] Jenny Uglow's *The Lunar Men* explores Darwin's social world, and Patricia Fara's *Erasmus Darwin* charts the declining fortunes of his poetry.

through several editions in both Britain and the United States – but the book was too technical for the general reader and the evolutionary material was lost in an extensive chapter on generation (reproduction). Darwin was remembered in the Derby Philosophical Society, where the young Hebert Spencer may have been influenced by his evolutionary writings. But here there was a local connection – Darwin had been one of the society's founders – and historians focusing on radical evolutionism find little evidence of an influence elsewhere.[12]

The radicals of the early nineteenth century were more inclined to materialism, and their source on the natural sciences was neither Buffon nor Darwin, but the baron d'Holbach, whose *Système de la nature*, published under an assumed name in 1770, became known as the 'Bible of atheism'. An English translation as *The System of Nature* appeared in 1797, but it was a cheap edition published by the radical activist Thomas Davison in 1820 that allowed d'Holbach to be recognized as a hero by the working class. D'Holbach's reformist view of human nature was founded on a materialism that saw all living things as structures assembled by nature's active powers (analogous to chemical combination). There was no sense of a divine plan; indeed, the world was portrayed as an endlessly changing chaos. Although the *System of Nature* did not advance a coherent theory of organic change, its philosophy left no room for the idea of fixed species. New forms were supposed to be constantly appearing either by spontaneous generation or by the accidental modification of existing types.[13]

Early nineteenth-century radicals combined this materialist view of the world with a vision of progress generated by the efforts of living things to cope with their environment. Change was natural, but it also advanced toward a goal that offered the hope of a better life. This ideology was constructed by activists who became a target for state oppression as the French Revolution encouraged the elite to reject all notions of gradual change and endorse the traditional claim that species are designed by God.

Picturing the Past

The most popular exponent of this more conservative position was the Anglican clergyman William Paley, whose *Natural Theology* of 1802 argued

[12] On the Derby Society, see Paul Elliot, 'Erasmus Darwin, Herbert Spencer and the Origins of the Evolutionary Worldview in British Provincial Scientific Culture, 1770–1850'. There are few references to Darwin in Desmond's 'Artisan Resistance and Evolution in Britain' and Secord's *Victorian Sensation*.

[13] D'Holbach, *The System of Nature*, see especially vol. 1, chaps. 3 and 6.

that natural processes cannot explain the complexity and adaptations of living things – they must be the product of supernatural design. Paley used only living examples, ignoring the latest evidence that the world and its inhabitants had changed over time. Geologists were working out the sequence in which the strata of the earth's crust were laid down and insisted that the planet was far older than the few thousand years implied by Genesis. By 1830 an outline of the geological periods we still accept today had been constructed, and palaeontologists were amassing fossil evidence of the species that had inhabited each period. The record suggested a step-by-step advance from an age in which fish were the only forms of vertebrate life, through what was soon called the 'Age of Reptiles' to later periods dominated by mammals.[14]

Geologists used invertebrate fossils to identify the strata, but for most people it was the remains of large vertebrate species that excited attention. Some of these remains were spectacular in both size and character, and were so different from modern species that it was hard to believe that they were still alive in the world today. Popular accounts of the discoveries emphasized their exotic and even frightening appearance, yet this air of excitement was imposed on reconstructions that were often based on familiar animals. Fearsome they might be, but they were not the totally bizarre forms revealed by discoveries made later in the century. It was as though creation (or nature) had undergone bursts of extravagance when each major advance on the scale occurred, followed by transitions to the more sober forms of today. Each category had its equivalents (or descendants) within the assembly of modern species, allowing the history of life to be seen as a series of additions accumulating to produce a world fit for humankind.

Georges Cuvier in France first revealed the true extent of these transitions and confirmed that the earlier species must have become extinct. He studied the remains of mammal species including giant relatives of modern forms such as the elephant-like mammoth. Some of the most spectacular finds were made in North America, where they attracted much attention and became a component of the fledgling United States' efforts to assert its cultural independence. Giant bones found in Ohio in the 1760s were identified as the remains of elephants and were used by Thomas Jefferson to discredit Buffon's assertion that the creatures of the New World were inferior to those of the Old. The American Philosophical Society promoted further searches, and more bones were

[14] For the most detailed account of these developments, see Martin Rudwick, *Bursting the Limits of Time* and *Worlds before Adam*, and for an overview the same author's *Earth's Deep History*. For a brief survey, see my own *Evolution*, chap. 4.

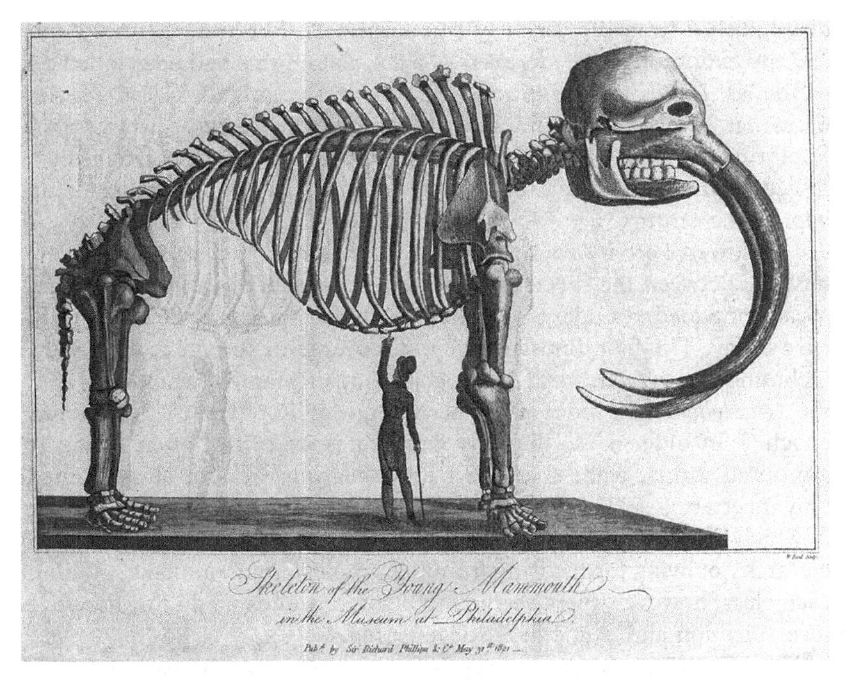

Fig. 2.1 'Mammoth' skeleton (actually a mastodon) displayed at Rembrandt Peale's museum in Philadelphia, 1821. The tusks were pointed downwards under the (mistaken) impression the creature was carnivorous. From E. Montulé, 'A Voyage to North America and the West Indies', in Sir Richard Phillips, ed., *New Voyages and Travels*, vol. 5 (London, 1821), between pp. 8 and 9. Reproduced by kind permission of the Syndics of Cambridge University Library.

unearthed in New York State, with large crowds attending the excavations. Charles Wilson Peale displayed them in Philadelphia's Philosophical Hall in 1801, where the public could see them by paying 50 cents. Peale's brother Rembrandt created a museum in Baltimore to display these and other wonders to the public (see Fig. 2.1).[15]

The American remains were only loosely related to living elephants and the woolly mammoths found in Siberia. As exploration proceeded it became clear that the creatures were no longer to be found alive, helping Cuvier to argue that the earth had once been inhabited by species now extinct. In 1806 he christened the American species the mastodon to

[15] See Paul Semonin, *American Monster*, and Stanley Hedeen, *Big Bone Lick*; also Keith Thomson, *The Legacy of the Mastodon*, chap. 6; John C. Greene, *The Death of Adam*, pp. 94–116; and Claudine Cohen, *The Fate of the Mammoth*, pp. 85–104.

distinguish it from the genus of true elephants. For him, the mastodon and the mammoth were members of a population that had been killed off by the last geological revolution – but this was merely the last in a series of earlier populations stretching back into the earth's distant past. Exploring the fossils to be found in the more ancient rocks would be a project centred largely in Europe until the American West was opened up later in the century.

The lower Tertiary rocks contained the remains of mammals intermediate between the specialized modern species. The earlier Secondary rocks contained abundant remains of reptilian species unlike those still alive today. The first dinosaurs were discovered in Britain in the 1820s, including *Megalosaurus* and *Iguanodon.* Gideon Mantell, who discovered the *Iguanodon,* later coined the term 'Age of Reptiles' to denote this epoch.[16] In older rocks the only vertebrates were fish, often strangely armoured forms, while the oldest fossil-bearing rocks of all contained only invertebrates. This seemed to be evidence of a progressive sequence in which life had ascended a scale resembling the chain of being's linear hierarchy of living forms rising from the simplest up to humanity. Within each class, however, there seemed to be other trends, including adaptive diversification and, sometimes, degeneration.

The process by which the revelations of geology and palaeontology were made available to the public forms a crucial backdrop to the emergence of evolutionism. Because they attracted popular interest, the fossils encouraged speculation about the sequence of development and its implications. Historians have shown that expectations created in the popular imagination helped to define what could be assimilated by the general reader and could even shape scientific opinion.[17]

Cuvier's survey of the fossil record was adapted for British readers by Robert Jameson to give the (false) impression that the French *savant* had associated the sequence of geological periods with the biblical days of creation. Published as *Essay on the Theory of the Earth* in 1813, it was updated in four subsequent editions through to 1827. Other prominent figures in the scientific community, including William Buckland and Richard Owen, extended their reputations by publishing surveys of the record for the wider public. Buckland's *Bridgewater Treatise* of 1836 was a contribution to a series commissioned to display 'the power, wisdom and goodness of God as manifested in the creation'. It included descriptions of many fossils and a pull-out chart displaying an idealized section of the

[16] Mantell, 'The Geological Age of Reptiles' and his *Medals of Creation*, e.g., p. 731.
[17] See Martin Rudwick, *Scenes from Deep Time*; James Secord, *Victorian Sensation*; Ralph O'Connor, *The Earth on Show*; and Gowan Dawson, *Show Me the Bone.*

earth's crust, noting the creatures associated with each formation. Owen delivered reports to the British Association for the Advancement of Science, in the second of which he coined the term 'dinosauria'. Circulation of these works would have been limited – Buckland's cost £1 15s, equivalent to over three weeks' wages for a clerk. But they were frequently cited by more accessible works, including William Broderip's articles for the weekly *Penny Cyclopaedia* (actually costing ninepence), published between 1833 and 1843 and selling up to 75,000 copies per issue.[18]

Juvenile readers were also introduced to the 'monsters' that once inhabited the earth in works such as Samuel Clark's *Peter Parley's Wonders of the Earth, Sea and Sky*, published in 1837 and frequently revised into the 1860s.[19] This literature usually promoted a theistic vision of creation through time designed to soften concerns about abandoning Genesis. The United States too had literature aimed at a younger audience, but here the British format of educated adults explaining science to juvenile readers was replaced by a technique designed to promote learning by experience. Clark's 'Peter Parley' was adapted from a successful series of works published in the United States by Samuel Griswold Goodrich, which included explorations of natural history and geology with a less 'top-down' approach. They were reprinted in Britain, leading Charles Kingsley to mock Goodrich as 'cousin Cramchild' in his *Water Babies*.[20]

Formal descriptions of fossils included illustrations of the skeletons, though rarely with representations of what the animals might have looked like. This was an appropriate caution in the many cases where only fragmentary remains were available, and it was in principle possible for the public to see the actual remains displayed in museums, although the working classes would have found access problematic. Gideon Mantell placed his collection, including the famous *Iguanodon* remains, on display at his home in Brighton and then sold them to the British Museum in 1838. The Museum moved to new buildings in the 1840s and began to attract a wider public. Mantell's *Petrifactions and their Teachings* of

[18] On the cost of Buckland's book (and many others), see O'Connor, *The Earth on Show*, pp. 222–3. On the wider use of the Bridgewater Treatises, see Jonathan Topham, 'Beyond the "Common Context"' and 'Science and Popular Education in the 1830s'. On the *Penny Cyclopaedia*'s contributions, see Dawson, *Show Me the Bone*, p. 108.

[19] *Peter Parley's Wonders* was published under the name of the 'Rev. T. Wilson', see O'Connor, *The Earth on Show*, pp. 235–40. For a facsimile of the first edition, see Samuel Clark, *Peter Parley's Wonders of Earth, Sea and Sky*.

[20] On 'Peter Parley' and other American literature, see Katherine Pandora, 'The Children's Republic of Science in the Antebellum Literature of Samuel Griswold Goodrich and Jacob Abbott'. For Kingsley's remarks, see his *Water Babies*, pp. 59, 70 and 157.

1851 was offered as a guidebook to the Museum's displays, although it was later replaced by shorter guides written by Museum staff. The specimens were usually placed in glass cases with only very basic labels. Things would gradually change as attendance at museums expanded in the course of the century, although the move to the dedicated Natural History Museum in South Kensington did not take place until the early 1880s. By this time, the staff – like those of many similar institutions around the world – were more conscious of the need to attract public attention by making their displays attractive and easier to understand.[21]

Popular works, however, made frequent use of illustrations depicting what the extinct 'monsters' might have looked like in real life. The scenes were printed in a sequence corresponding to the geological periods, giving an overview of the whole history of life. Joshua Trimmer's *Practical Geology* of 1841 had such a sequence even though the book had a mainly utilitarian focus.[22] The artists who drew the scenes had significant control over what they represented, and the scientific experts were in their hands if they wanted to reach a wider audience. John Martin's image of struggling dinosaurs, used as the frontispiece to Mantell's *Wonders of Creation* and often reprinted, deliberately invokes the mythical dragons of antiquity. Here legend overtook the expert's desire to keep within the bounds of anatomical possibility to ensure that the Age of Reptiles would be seen as a period dominated by monsters derived from the world of Gothic horror.

The most impressive reconstructions were the life-sized models made by Benjamin Waterhouse Hawkins for the park surrounding the reconstructed Crystal Palace at Sydenham, south London, opened in 1854 amid a hail of press coverage. Huge crowds came to see them, and they continue to attract visitors today. Although Richard Owen's name was associated with the reconstructions to emphasize their reliability, Hawkins was actually far more in control of the finished product (see Fig. 2.2).[23]

The same point is valid for references to prehistoric monsters in fiction, including both novels and poetry. Authors used the past for their own purposes, and popular non-fictional accounts often used picturesque

[21] On Mantell and the British Museum, see Dennis R. Dean, *Gideon Mantell and the Discovery of Dinosaurs*, pp. 164–78. On the Museum's *Megatherium*, see Dawson, *Show Me the Bone*, pp. 175–6. More generally, see Carin Berkowitz and Bernard Lightman, eds., *Science Museums in Transition*, and on guidebooks Aileen Fyfe, 'Reading Natural History at the British Museum and the *Pictorial Museum*'. The later expansion of museums will be discussed below.

[22] Rudwick, *Scenes from Deep Time*, on Trimmer's book see pp. 84–7.

[23] See Dawson, *Show Me the Bone*, pp. 179–88.

Fig. 2.2 Waterhouse Hawkins' reconstruction of the dinosaur *Megalosaurus* at Crystal Palace, south London.
Photograph by the author c. 1980.

language to make the past come alive. The structure of travel narratives was adopted to lead the reader through the sequence of geological periods.[24] When poets and novelists referred to extinct monsters, they did so for effect. The reference in Tennyson's *In Memoriam* to the extinction of the 'Dragons of the prime / that tear each other in their slime' invoked mythology to shape an image of dinosaurs that highlighted his vision of 'nature, red in tooth and claw'.[25] The opening paragraph of Dickens' *Bleak House* (published in serial form 1852–3) uses a dinosaur to create his more mundane but equally depressing image of the London weather in November: so much mud in the streets 'that it would not be wonderful to meet a *Megalosaurus*, forty feet long or so, waddling like an elephantine lizard up Holborn Hill'.

Dicken's reference to an 'elephantine lizard' suggests that despite the 'Gothick' element the extinct creatures were often presented as merely bigger and distorted versions of creatures already familiar from the real

[24] O'Connor, *The Earth on Show*, chap. 7, and on wider literary allusions chap. 8.

[25] Tennyson, *In Memoriam*, section LVI. Rudwick notes that some depictions of ancient life stress the alien nature of the creatures, but concedes that other popular accounts use analogies with modern species; see *Scenes from Deep Time*, p. 241.

world or from mythology. Cuvier's mammoth was obviously a huge elephant, while even his pterodactyl could be seen as a reptilian equivalent of a bat. The other ancient reptiles were also reconstructed and depicted as variants on modern forms. In the 1840s the 'Age of Reptiles' was based on only three dinosaurs, along with the marine plesiosaurs and ichthyosaurs. The dinosaurs were known only from isolated bones, so the experts had to extrapolate from their knowledge of living creatures to imagine what their extinct counterparts would have looked like. Even where more complete skeletons had been found (as for the marine reptiles), there was a tendency to invoke comparisons with familiar forms to help the non-specialist visualize the creature. The Crystal Palace reconstructions provide perhaps the most visible example: it was later realized that both *Iguanodon* and *Megalosaurus* were bipedal and hence unlike any living reptile, but Owen and Dawkins saw them as gigantic lizards walking on all fours. From further back in geological time the labyrinthodont amphibians were depicted as huge frogs.[26]

Such comparisons with the familiar were rife in popular written accounts, often borrowing directly from the experts. The *Penny Cyclopaedia*'s article on 'saurians' called the dinosaurs 'extinct lizards', and its article on *Megalosaurus* quoted Buckland on this 'immense extinct lizard'. Mantell exploited a link to crocodiles as well – the dinosaurs were the 'gigantic crocodile-lizards of the dry land'. The marine reptiles were also interpreted in terms of the closest modern equivalents. The *Penny Cyclopaedia* borrowed from both Buckland and Mantell to depict the ichthyosaur as a reptilian grampus (killer whale) or porpoise. Owen was quoted on the Plesiosaur as a 'lizard of the sea'. Mantell tried to help his readers visualize the long neck of this creature as 'a serpent threaded through the shell of a turtle'.[27]

The tendency to interpret the past in terms of the present constrained the first efforts to make sense of the history of life. The progress from fish to reptile and then to mammals seemed obvious to most, and the lack (at the time) of any hominid fossils confirmed that humanity was the last step in the ascent. Presenting the ancient species as variants on modern forms also encouraged the view that the ascent represented the preparation of the world for human occupation, a step-by-step creation of the components of the animal kingdom we know today. The more bizarre

[26] This aspect of the Crystal Palace reconstructions is emphasized by W. J. T. Michell, *The Last Dinosaur Book*, pp. 66–7.

[27] Mantell, *The Medals of Creation*, pp. 708–9 on the ichthyosaur and p. 715 on the Plesiosaur. See also *Penny Cyclopaedia*, on saurians, No. 20: 458; on the plesiosaur, No. 18: 25–8; on the ichthyosaur, No. 12: 430–3; on *Megalosaurus*, No. 15: 63–4. The *Penny Cyclopaedia* was a serial work issued from 1833 to 1843.

variants might have gone extinct, but neither the scientists nor the general public grasped the possibility that most life in the past had developed in ways that had little to do with the emergence of the modern world.

Genesis Updated

Many geologists sought to reconcile the fossils with Paley's claim that every species must have been designed by God. In effect, they updated the Genesis story by assuming that each 'day' of creation represents a discrete geological period. Buckland's *Bridgewater Treatise* argued that each fossil population was adapted to the environment of the time, while the succession of climates made the earth steadily more suitable for human habitation. To avoid any suggestion of a continuous progressive trend that might reduce humanity to the status of merely the highest animal, he suggested that each period ended with the mass extinction of all its inhabitants. The extinctions were followed by creative acts that produced the replacement species, so that each population represents an example of the Creator's handiwork. The abrupt transitions between the formations were the result of gigantic convulsions of the earth's crust far outstripping anything witnessed during human history. Buckland's *Reliquiae diluvianae* of 1823 presented the remains of hyenas embedded in cave deposits as evidence of a recent deluge. But he acknowledged that this would only be the last of many such convulsions, the position that came to be known as 'catastrophism'.

The element of discontinuity remained popular into the late nineteenth century, and even today we recognize that there have been interludes of relatively rapid – and occasionally catastrophic – change. Only a few critics, most notably Charles Lyell and Charles Darwin, tried to maintain the alternative 'uniformitarian' position in which all changes are slow and gradual, the apparent breaks in the sequence being due to gaps in the record caused by erosion. This position required greater periods of time to explain the huge transformations in the earth's crust – although even the catastrophists accepted a time span measured in millions of years. The problem with the uniformitarian approach was that it encouraged speculation that the episodes of divine creation might be replaced with a model of gradual change in the organic as well as the physical world.

The apparently discrete nature of the main upward steps in the ascent of life was used to checkmate any attempt to claim that later populations were merely the modified descendants of their predecessors. Owen challenged Lamarckian speculations by arguing that the dinosaurs were more mammalian than the modern reptiles, so the class had actually declined in the course of its history. The prolific stonemason-turned-geologist Hugh Miller compared the history of each class's history to the rise and fall of human empires.[28]

American geologists also promoted the vision of the ascent of life through divine creations, also with hints that the overall pattern had an underlying direction. Edward Hitchcock of Amherst College published an *Elementary Geology* in 1840, which went through several editions over the next decade. It included a tree-like diagram illustrating the lines of development leading through the plant and animal kingdoms, although the tree had parallel trunks more like ivy.[29] The claim that the patterns in life's development unfolded in a series of discrete creations was also the view of the Swiss naturalist Louis Agassiz, who emigrated to become a dominant figure in the American scientific community. He visited the country in 1846 and then took up permanent residence, hailed in the press as a significant boost to the promotion of science.

Agassiz was soon lecturing at the Lowell Institute in Boston on 'The Plan of Creation in the Animal Kingdom' to audiences of thousands. In 1848 he co-authored with Augustus A. Gould a survey intended also as a textbook, *The Principles of Zoology*, which subsequently went through numerous editions and appeared (under a different title) in Britain. The concluding survey of the development of life on earth made plain Agassiz's commitment to the idea that the pattern unfolded according to a divine plan and that the progression was aimed at the production of the human form: 'Man is the end towards which the animal creation has tended, from the first appearance of the Palaeozoic fishes.'[30] Agassiz also noted that the sequence of creation followed the same pattern as the development of the human embryo toward its goal. This was the 'law of parallelism' in which the fossil record, the hierarchy of classes and the development of the embryo all followed the same sequence. Among his followers in the post-Darwinian period it would become the basis for the claim that the development of the embryo recapitulates the evolutionary history of its species.

Agassiz suggested that the human embryo passes through phases in which it is in effect a fish, then a reptile and finally a mammal, only at the

[28] See Desmond, *The Politics of Evolution*, pp. 326–7 and 348, and on the 'rise and fall' image Bowler, *The Invention of Progress*, chap. 6.

[29] See J. David Archibald, 'Edward Hitchcock's Pre-Darwinian (1840) "Tree of Life"'; Hitchcock's diagram is reproduced in Pietsch, *Trees of Life*, p. 81.

[30] Quotation from the English edition, Agassiz and Gould, *Outlines of Comparative Physiology*, p. 418. For details of Agassiz's move to America, see Edward Lurie, *Louis Agassiz: A Life in Science*, chaps. 4–6.

end acquiring its distinctive specific characters. The lower classes were, in effect, stages in the ascent to humanity. But in the 1850s Owen and William Benjamin Carpenter noted a different parallel between the history of life and a new model of embryological development. Karl Ernst von Baer had shown that a better way of representing the process across the whole animal kingdom was a branching tree, each species gradually acquiring the specialized characters that defined it. The early human embryo may look like an embryonic fish or reptile, but not like an adult member of a previous class. Each embryo acquires progressively more specialized characters as it develops. Richard Owen noted that the same process seems to occur in the fossil record – the first members of each class tend to be generalized forms, while the later ones open up a diversity of specialized adaptations to various ways of life. Potentially, at least, this offered a model that could be incorporated into Darwin's theory of evolution based on the 'tree of life'.[31]

Agassiz and Owen occupied similar roles in the debates in their respective counties, although in the end Owen was prepared to consider a divinely planned version of evolutionism. Both saw the human form as the goal of creation, and both recognized that there were also patterns of divergence within each class (Agassiz had noted this among the fossil fish in the 1830s). The two patterns could be fitted together by imagining a main line of development from which divergent specialized groups branched out at levels corresponding to the successive classes. This would give a 'tree of life', albeit one with a central trunk defining the main line of succession.

There was, however, another possibility consistent with the tendency to compare extinct species with their closest modern equivalent. Perhaps separate lines of development within each class independently continued on into the next one, creating a series of parallel lines running through the history of life – this seems to be the implication of Hitchcock's ivy-like tree. This notion of parallelism would eventually become a major alternative to the Darwinian view.

Although most naturalists continued to insist on the discontinuity of the ascent of life, there were occasional remarks conveying the impression of a more continuous process of development. Mantell warned of the dangers of generalizations based on limited evidence, but conceded that 'by slow and almost insensible gradations we arrive at the present state of animate and inanimate nature'.[32] The example most frequently

[31] See Bowler, *Fossils and Progress*, chap. 5, and Dov Ospovat, 'The Influence of Karl Ernst von Baer's Embryology, 1828–1859'.

[32] Mantell, *The Medals of Creation*, 2: 875.

quoted by modern historians is from Owen, who edged closer to the kind of theistic evolutionism he would later adopt in opposition to Darwin. In a public lecture given at the Royal Institution in London and published as *On the Nature of Limbs* in 1849, he concluded as follows:

> To what laws or secondary causes the orderly succession and progression of such organic phenomena may have been committed we are as yet ignorant. But if, without derogation of the Divine power, we may conceive the existence of such ministers, and personify them by the term 'Nature', we learn from the past history of our globe that she has advanced with slow and stately steps, guided by the archetypical light, amidst the wreck of worlds, from the first embodiment of the Vertebrate idea, under its old Ichthyic vestment, until it became arrayed in the glorious garb of the human form.[33]

The response was a chorus of disapproval that led Owen to abandon such imagery until Darwin's initiative had broken the deadlock on the topic.

Radical Evolutionism

At one time historians believed that the theistic viewpoint outlined above went more or less unchallenged, but the study of popular science has shown that the conservatives were anxious to discredit transformist ideas being exploited by radical activists seeking reform or revolution. The idea of gradual transformation became the basis for a new understanding of progress in society. If there are forces in nature that allow living things gradually to improve themselves, the same process will guarantee the progress of humanity towards a better world. This had been Erasmus Darwin's vision, but now that his work had become unfashionable there were alternatives that could be called upon. D'Holbach's materialism was one resource, but new sciences being developed on the Continent offered an even more attractive model. Theories of organic change proposed by J.-B. Lamarck and E. Geoffroy Saint-Hilaire were taken up with enthusiasm by those calling for reform of the medical profession, by political activists demanding the destruction of the existing social hierarchy and even by reformers from the dissenting churches.

The radical ideology also exploited efforts to rework our understanding of human nature itself. The early evolutionary ideas worked in conjunction with other initiatives forcing the Victorians to reconsider 'man's place in nature'. The assumption that humanity is raised above the animals by the possession of an immortal soul was challenged both by the suggestion that we had evolved from animal ancestors but also by the

[33] Owen, *On the Nature of Limbs*, p. 89.

claim that the mind is merely the product of physical operations taking place within the brain. This was the position taken up by the new psychological doctrine known as phrenology.[34]

Phrenologists argued that the brain is the organ of the mind and that the structure of the brain determines each individual's mental faculties. The materialistic implications of this claim were obvious, but could be softened by acknowledging a distinct mental world that functioned only when physical processes in the brain permitted. Even so, the basic idea made the unique status of the human mind – and the immortality of the soul – less obvious. Creatures with smaller brains could still have a mental life, so the human mind was merely the most complex version of something that existed throughout nature. If living things had advanced in complexity in the course of geological time, their expanding brains would extend their mental powers until they reached the level we now enjoy. These ideas became even more visible when Robert Chambers' *Vestiges of the Natural History of Creation* of 1844 openly linked the progress of life on earth and phrenology's reinterpretation of the nature of the mind.

The two movements arrived in the English-speaking world in the 1820s, phrenology being inspired by J. C. Spurzheim's visits to Britain and the United States, while British zoologists visited Paris to mingle with Cuvier's radical contemporaries. Phrenology's claim that the brain is the organ of the mind implied that a person's abilities were predetermined by its structure, popularized in model heads marked to show the position of the parts of the brain thought to be responsible for each mental function. Reading someone's character from the shape of the skull was practiced by itinerant lecturers and ultimately brought the whole project into disrepute (although the basic principle of cerebral localization gained increasing currency). As the project was taken up by reformers such as George Combe, it was transformed into an ideology of self-development – people could find out which faculties they needed to focus on to improve themselves. This expectation could then be linked to the Lamarckian process in which acquired

[34] Robert Young's *Mind, Brain and Adaptation in the Nineteenth Century* recognized the significance of new ideas about the mind, and it was followed by a series of studies by historians inspired by the sociology of science who saw early nineteenth-century phrenology as a classic example of how scientific ideas reflect ideological perspectives. See Roger Cooter, *The Cultural Meaning of Popular Science*, and for a more recent survey John Van Wyhe, *Phrenology and the Origins of Victorian Scientific Naturalism*. Adrian Desmond's *Politics of Evolution* and James Secord's *Victorian Sensation* both stress the importance of phrenology as an independent source of new perspectives linked to evolutionism.

characteristics became the basis for the progress of the species. Combe developed these themes in his *Constitution of Man* of 1828, the impact of which was vastly extended after the publication of cheaper edition from 1835. Three American editions had already been published by that year, and the book went on to become one of the nineteenth-century's bestsellers.[35]

Combe lectured around Britain and toured America in the years 1838–40. On both sides of the Atlantic, societies and journals promoted phrenology and its ideology of hope for the future. The young Alfred Russel Wallace was impressed by demonstrations of phrenology linked to mesmerism in Leicester and became a firm convert after his own reading by a 'Professor of Phrenology' in the mid-1840s. He also encountered another lecturer who had examined and converted Herbert Spencer in Derby on 29 June 1842.[36] Both Wallace and Spencer would go on to link evolutionism with the hope of social progress.

The parallel developments in zoology derived from French naturalists who resisted Cuvier's efforts to see the advance of life as a series of discrete steps. Adrian Desmond's *Politics of Evolution* shows how the evolutionary theories of Lamarck and Geoffroy Saint-Hilaire were taken up by reformers in the medical profession, especially in Edinburgh, where the structure of medical education was looser than in London. Poorer students shared the radicalism of political activists determined to challenge the existing structures of social and professional activity. While evolutionary speculations appeared occasionally in the more adventurous natural history journals, illegal news-sheets and magazines promoted any idea that seemed to discredit orthodox politics and orthodox science. There were also those who wanted reform rather than revolution, middle-class dissenters willing to tolerate at least some of the new ideas.

D'Holbach's vision of material nature having creative powers was an inspiration after the publication of the cheap translation of his book in 1820–1, but he proposed no mechanism of change. The first new initiative came from the followers of Lamarck who extended this world-view by invoking two transformative processes, a progressive trend driving successive generations to ascend a hierarchy of complexity, distorted by a secondary process of adaptation to the environment. The latter worked via the inheritance of acquired characteristics, later identified as the primary mechanism of 'Lamarckism'. In the 1830s attention switched to a different approach proposed by Geoffroy Saint-Hilaire, which

35 Secord, *Victorian Sensation*, pp. 69–76.

36 Wallace, *My Life*, 1: 232–3 and 247–62; Spencer, *Autobiography*, 1: 200–1, which gives the actual details of his reading.

focused on abrupt changes to the process of embryological development known as saltations. British zoologists used all of these elements to attack the image of a universe designed by God, but without pulling them together to form a coherent alternative position.

Lamarck's transmutationism was expounded in his *Philosophie zoologique* of 1807, although this was not translated into English until much later. His position was in no way an anticipation of Darwinism: he postulated parallel lines of development advancing up the chain of being and did not believe in extinction. The process of the inheritance of acquired characteristics merely distorted the lines of advance, providing adaptation without natural selection.[37] It was once thought that Lamarck had been marginalized by Cuvier's dominance of the French scientific community, but we now know that his ideas were taken up by others resisting Cuvier's influence. They interacted with British naturalists, many linked to the Edinburgh medical community.[38]

The most active Lamarckian was Robert Edmond Grant, who spent his long vacations in Paris during the 1820s. Darwin later recollected his astonishment at hearing Grant's support for Lamarck when he (Darwin) had been a medical student in Edinburgh in 1826–7.[39] Grant lectured and wrote on evolutionary themes both in Edinburgh and after his move to take up the chair in zoology at University College, London, in 1827. Others shared his enthusiasm, including Robert Knox, the anatomist subsequently discredited for buying bodies for dissection from the murderers Burke and Hare. An anonymous article in favour of Lamarckism once thought to be by Grant is now known to be one of several translations from French sources. There is evidence that the idea began to spread among other Scottish naturalists who were not associated with the political radicals.[40]

By 1830, attention was switching to Geoffroy's 'transcendental anatomy', which saw an underlying unity in the structures of living things and proposed that new species arose from sudden distortions of

[37] See M. J. S. Hodge, 'Lamarck's Science of Living Bodies', and Ludmilla Jordanova, *Lamarck*. For details of how the inheritance of acquired characteristics works, see Chapter 1.

[38] See Pietro Corsi, *The Age of Lamarck*, and on later developments Toby Appel, *The Cuvier-Geoffroy Debate*. On the British links, see Desmond, *The Politics of Evolution*.

[39] Darwin was far more influenced by Grant than he later admitted; see Phillip Sloan, 'Darwin's Invertebrate Program, 1826–1836'.

[40] James Secord challenged the attribution to Grant, suggesting Jameson as the true author, but Pietro Corsi has now traced the French sources. See Secord, "Edinburgh Lamarckians: Robert Jameson and Robert E. Grant," and Corsi, "Edinburgh Lamarckians? The Authorship of Three Anonymous Articles (1826–1829)." See also Bill Jenkins, *Evolution before Darwin*, especially chaps. 5 and 6.

embryological development. Grant and others began to promote this new version of transmutationism, but interest in Edinburgh lapsed as catastrophism became dominant in geology and evangelicalism grew in religious circles. The topic went underground, although Grant included it in a series of lectures to the Zoological Society of London in 1833–4. He attempted to cover up the more radical aspects of his thinking, but was gradually marginalized within the English scientific community. The topic simmered on, however, with others from both radical and reformist backgrounds publishing books on the new science, including John Fletcher and William B. Carpenter. It would come to the surface with a vengeance in 1844 with the publication of Chamber's *Vestiges*.

If evolutionary ideas were circulating within the medical community, how far did they diffuse into the wider public? Lectures by Grant and others to both local and national societies would have attracted non-specialist audiences from all levels of society. There were also networks of amateur naturalists who might encounter articles by Grant and Jameson in the *Edinburgh New Philosophical Journal* in libraries attached to local Mechanics Institutes and similar educational forums. Anatomists with radical political views were in contact with working-class activists calling for reform, allowing Lamarckian ideas to appear in the cheap and often illegal literature distributed by these figures. There might be little technical detail, but the idea of natural change was incorporated into the campaign to discredit the conservative view of nature and society. The idea of a self-creative nature leading up to humanity formed the scientific background to an ideology of social progress.[41]

There were many ephemeral publications of which the illegal penny-magazine *The Oracle of Reason* was the most successful. It published a forty-eight-part 'Theory of Regular Gradation' from 1841 to 1843. The first six parts were by Charles Southwell, but when he was imprisoned for publishing material banned by the government, it was taken over by William Chiltern. They attacked the image of nature as a system designed by God and adopted a materialist viewpoint in which the first living things were generated by purely natural forces. Orthodox publications on the fossil record were cited with their message reworked to bring out an element of continuous progress in the history of life: 'the simplest forms of animals were contemporaneous with the lowest, or earliest strata, and as we proceed upwards from the granites to the tertiary, the number of complex forms increase, and at last, upon the tertiary strata we find man'. The implication that humans had originated from an ape-

[41] In addition to the above, see Secord's *Victorian Sensation*, especially pp. 308–20.

like ancestor was driven home with an image of 'Fossil Man' borrowed from a French publication – a savage-looking figure carrying a club. The progress of the animal kingdom was linked to hopes for the future, even to the extent of imagining developments beyond the current state of humanity.[42]

Less radical groups seeking to challenge the status quo also saw natural history as a valuable component of the education that would help improve individuals and society at large. The school at Robert Owen's co-operative centre at New Lanarck used the variety of species to stress the adaptability of living organisms, a position associated with the Lamarckian principle of the inheritance of acquired characters and the progress of life on earth.[43] Grant lectured to Mechanics' Institutes around the country, his Lamarckism carefully toned down to conceal his materialist sympathies. Literary and philosophical societies promoted science's role in improving the world and were open to transformist ideas, provided they were not presented in an openly atheistic context.[44]

Herbert Spencer's experiences at the Derby Philosophical Society shaped his move toward evolutionism. Although aware of Erasmus Darwin's work through the local connection, he was also open to Lamarckism, which he encountered through reading the critique in Charles Lyell's *Principles of Geology* in 1840. The attack had the opposite effect to that intended because Spencer was immediately attracted to the theory. Even negative publicity was helping to spread information about the transmutationist alternative in moderate circles. Spencer's first article on the topic, 'The Development Hypothesis', published anonymously in the socialist magazine *The Leader* in 1852, criticized 'those who cavalierly reject the theory of Lamarck and his followers' (the phase was altered to 'theory of Evolution' in reprinted versions of the article).[45]

The Lamarckian alternative to special creation was now being circulated in a sanitized form, but it was still risking ostracism for anyone in these less-radical circles to suggest openly that humanity was the summit of a continuous progress through geological time. Eventually, the claim that the ascent of life on earth was a prelude to the future perfection of

[42] Southwell and Chiltern, 'Theory of Regular Gradation', quotations from parts 17 and 7. The first page with the image of 'Fossil Man' is reproduced in Secord, *Victorian Sensation*, p. 312,

[43] See Desmond, 'Artisan Resistance and Evolution', pp. 92–3.

[44] See, for instance, Secord, *Victorian Sensation*, pp. 65 and 192–9.

[45] Spencer, *Autobiography*, 1: 176, and 'The Development Hypothesis', *The Leader*, 20 March 1852, p. 280. See my 'Herbert Spencer and "Evolution" – An Additional Note'. For the reprinted version, see Spencer, *Essays Scientific, Political and Speculative*, 1: 381–7, and for Lyell's attack on Lamarck, see his *Principles of Geology*, vol. 2, chap. 1.

humanity would be promoted more widely, but to gain acceptance in liberal circles it would have to be made not in the context of the radicals' materialistic challenge to orthodoxy but in a reconciling spirit suggesting that the whole process unfolded in accordance with a divine plan.

Vestiges of Creation

The challenge was taken up by Robert Chambers, who with his brother had founded an Edinburgh publishing house to promote the interests of those hoping to benefit from the opportunities offered by industrialization.[46] Interested in phrenology and natural history, he promoted an ideology of science-based progress in a world governed by natural laws. To placate those with religious sensibilities, he argued that the laws were instituted by the Creator, so the advance in biological and mental complexity leading to humanity's moral and spiritual nature is part of a divine plan. The evolutionism advanced in his *Vestiges of the Natural History of Creation* of 1844 offered no naturalistic explanation of the process; the pattern of development that unfolded was built into nature by God.

Vestiges popularized the basic idea of evolution and thus took at least some of the flak that would have been aimed at the *Origin of Species*. It helped to dissociate evolutionism from radical politics and thus made it socially acceptable to the middle classes. But it did not convince the scientists, nor did it anticipate the Darwinian approach, proposing predetermined hierarchies of development rather than an open-ended 'tree of life'. James Secord's detailed study of the *Vestiges* affair suggests that the book may even have skewed how the public subsequently interpreted Darwin's theory by conditioning them to expect a theory based on a progressive trend in nature.[47]

Chambers' publishing house produced material aimed at the rising middle classes. *Chambers's Edinburgh Journal* disseminated uplifting information, including articles on geology and natural history, with occasional allusions to the gradual progress of life through geological time.[48] The firm also issued cheap editions of books on phrenology and related topics. Chambers was not a scientific expert, but he had wide interests and was willing to accept what were regarded as highly controversial theories. His breadth of knowledge allowed him to spot

[46] See Aileen Fyfe, *Steam-Powered Knowledge*.
[47] Hodge, 'The Universal Gestation of Nature', and the conclusion of Secord, *Victorian Sensation*.
[48] Secord, *Victorian Sensation*, p. 94.

generalizations missed by narrow specialists, so where most studies of the fossil record stressed its discontinuity, he saw the overall trend of gradual progress.

Chambers knew the dangers of negative publicity and had *Vestiges* issued anonymously by another publisher. His book was the first 'evolutionary epic', anticipating the surveys of the post-Darwinian period that used the broad sweep of life's development to create a sense of awe and wonder at nature's achievements. It surveyed cosmic progress from the origin of the solar system through to the emergence of humankind. Its narrative style resembled that of novels and popular science texts intended to guide the reader through their subject matter. The story was told in a way designed to show how the laws of nature established by the Creator ensured a gradual but inevitable progress toward higher levels of organization. The planetary system condensed from a rotating cloud of dust under the law of gravity. Primitive forms of life were generated through the action of electricity in a process that is still active today – here Chambers drew on experimental work that most authorities greeted with suspicion.

Several chapters then surveyed the ascent of life through the fossil record, giving the impression of a continuous path toward humanity. The extinct monsters of the past were presented as analogous to living forms – the first edition didn't mention dinosaurs and later versions described them as crocodile-like creatures with affinities to the lizards.[49] Chambers then presents his 'Hypothesis of the Development of the Vegetable and Animal Kingdoms', which involved progressive extensions of individual development. He argued '*that the simplest and most primitive type, under a law to which that of like-production is subordinate, gave birth to the type next above it, that this again produced the next higher, and so on to the very highest*, the stages of advance being in all cases very small, namely from one species to another; so that the phenomenon has always been of a simple and modest character'.[50] The main classes of vertebrate life branched off one by one from an ascending linear sequence defining the main path of development (see Fig. 2.3). A chapter titled 'The Mental Constitution of Animals' invoked phrenology to explain how the expansion of the brain in the course of evolution produced an equivalent increase in mental powers, culminating in those of humanity. Chambers also argued for the gradual progress of

[49] Chambers, *Vestiges*, 5th ed., p. 121. The first edition is reprinted edited by Secord in Chambers, *Vestiges of the Natural History of Creation and Other Evolutionary Writings*.
[50] *Vestiges*, 1st ed., p. 222, and for similar wording, see 5th ed., p. 213.

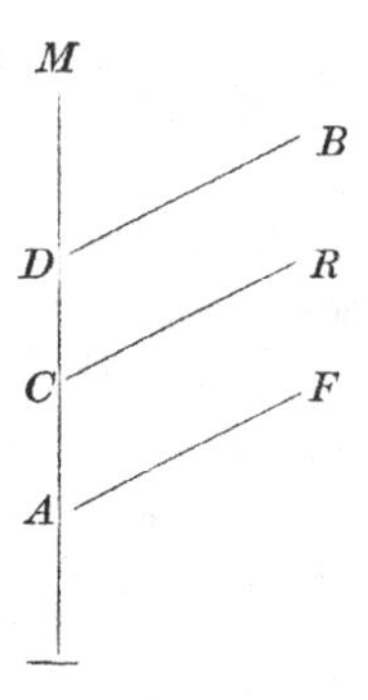

Fig. 2.3 Diagram representing the development of the vertebrates through time. M indicates mammals, B birds, R reptiles and F fishes. From Robert Chambers, *Vestiges of the Natural History of Creation*, 5th ed. (1846), p. 220.

civilization and emphasized that even our social life is governed by laws, citing the regularities being found in studies of crime and other activities.

The account of biological evolution paid little attention to adaptation and proclaimed the inadequacy of Lamarckism.[51] Changes in the environment could only trigger the upward steps that produced new species, the actual direction of change being predetermined by a law of development that was closer to Lamarck's primary mechanism of change focused on the progression of life. Chambers postulated successive extensions of the growth process but provided no naturalistic explanation of the new characters produced – evolution was the unfolding of a predetermined pattern of creation. An argument borrowed from Charles Babbage, who had invented a mechanical computer, implied that there might be hidden laws programmed into nature by the Creator to generate occasional 'jumps' that would seem like miracles to a human observer.[52] The progressive trend could thus be seen as the unfolding of a divine plan. The first edition also contained a chapter outlining W. S. MacLeay's circular system of classification in which the species could all be fitted into an orderly pattern built up of five elements arranged in a circle. This too implied a highly structured plan of nature, although it didn't fit well with the linear hierarchy of progress and was dropped from later editions.

In a sequel entitled *Explanations* and in later editions of *Vestiges*, Chambers began to argue for multiple lines of development called *stirpes*,

[51] *Vestiges*, 1st ed., pp. 230–1 (this critique does not appear in the 5th ed.).
[52] *Vestiges*, 1st ed., pp. 206–11; 5th ed., pp. 213–19.

all ascending from simple to complex forms in parallel. The lines could independently transit from one class up to the next, although each major upward leap had to begin with an aquatic form that then developed into terrestrial ones. A number of genealogies were proposed, mostly based on living rather than fossil forms and all quite unlike the phylogenies that would be proposed by post-Darwinian naturalists. The lines of development unfolded in different geographical areas, leading Chambers to suggest that the absence of mammals in the Galápagos Islands indicated that evolution here had not been active long enough to reach that grade of organization. In very late editions he postulated three parallel lines of development running through the animal kingdom.[53]

The notion of parallelism would re-emerge in some non-Darwinian theories later in the century, but no one reading *Vestiges* would anticipate a model based on a branching tree of life constantly divided by environmental pressures. Nor was there any recognition of the role played by extinction. There was little discussion of how our species might have evolved from an ape-like origin, although an account of the origin of human races endorsed the view that the black races are inferior to the whites. Later editions of *Vestiges* offered conflicting views on whether this involved a lower level of development from an ape ancestry (see Chapter 4 for details). Given the role played by the issue of an ape origin in the Darwinian debates, its relative absence in this earlier phase is striking – it didn't even feature strongly in the widespread reaction to Chambers' book. He may have deliberately avoided a topic that had already raised some hackles, and he was lucky that the huge public interest in the discovery of the gorilla would not come until the 1860s, when it helped to shape the reaction to Darwinism.

Even so, Secord's extensive survey shows that the reaction to Chambers' more generalized account of humanity as the high point of animal development was intense, although it varied depending on the interests and opinions of the readers. The initial impact was on well-to-do readers – the book appeared at the relatively expensive price of 7/6d and sales of the early editions were limited. It was the talk of the town for a season and there was much speculation on the identity of the author. Those who did not have strong religious opinions were willing to take the possibility of progress through law in their stride, establishing the

[53] The later reference to MacLeay is in *Vestiges*, 11th ed., p. 220. On *stirpes*, see Chambers, *Explanations: A Sequel to 'Vestiges of the Natural History of Creation'*, pp. 68–75 (facsimile in *Vestiges … and Other Evolutionary Writings*) and *Vestiges*, 5th ed., pp. 280–2. Secord discusses this aspect of the theory in *Victorian Sensation*, pp. 388–9. On the Galapagos, see *Explanations*, pp. 161–2, and *Vestiges*, 5th ed., p. 292, and 11th ed., p. 229.

presumption that what would later be called 'evolution' must follow a predetermined path. Benjamin Disraeli's novel *Tancred* of 1847 mocked the vague notions of development then circulating – the hero talks with a lady who has read the latest sensation:

> You know, all is development.... First there was nothing, then there was something, then, I forget the next, I think there were shells, then fishes; Then we came, let me see, did we come next? Never mind that, we came at last. And the next change there will be something very superior to us, something with wings.[54]

Those who did have conservative religious views were, of course, horrified. The dean of York, William Cockburn, wrote a strong letter to the *Times*.[55] Richard Owen refused calls to get involved and the task of rebuttal fell to Adam Sedgwick, who published a vitriolic eighty-five-page critique in the *Edinburgh Review*. He complained about ladies being told 'they are the children of apes and the mothers of monsters', but there were only isolated comments on the ape connection – a bigger target was phrenology.[56] Even those with more liberal opinions took issue with the book for its scientific blunders. T. H. Huxley wrote a vicious review of a later edition – like Darwin, he distrusted the implication that natural law could be equated with design.[57]

Soon *Vestiges* began to extend its reach beyond the leisured classes and the scientific community. A cheap edition was published in 1847 and sold over 10,000 copies. By the time the eleventh edition appeared in 1860, sales had reached 20,000, and it was the end of the century before sales of Darwin's *Origin* overtook those of *Vestiges*.[58] The *Atlas* newspaper published a summary in a supplement, soon reprinted as a pamphlet priced at only 4d.[59] The book found its way into libraries where the working classes could read it. Herbert Spencer recommended it for the Derby Mechanics' Institute, and A. R. Wallace purchased a copy in 1846. He thought it opened up important new perspectives, although his future travelling companion H. W. Bates was less impressed.[60] There

[54] Quoted in Secord, *Victorian Sensation*, p. 189, in his discussion of the early reaction.
[55] For the *Times* correspondence, see ibid., p. 177, and for Cockburn's dispute with Sedgwick, pp. 232–47.
[56] Sedgwick, 'Vestiges of the Natural History of Creation', quotation from p. 3, other comments on pp. 1, 65 and 82.
[57] On Darwin and Huxley's responses, see Secord, *Victorian Sensation*, pp. 426–33 and 500–6.
[58] Ibid., p. 526. [59] Ibid., p. 136.
[60] Spencer, letter to Edward Lott, 18 March 1845, quoted in his *Autobiography*, 1: 267; Wallace, letters to H. W. Bates, 4 November and 28 December 1846, quoted in *My Life*, 1: 254–5.

were acrimonious debates in Liverpool, where the book became a focus for disputes between conservatives and their political opponents. Working-class radicals were generally sympathetic although suspicious of the hints at design. William Chiltern praised it as a vindication of his articles in the *Oracle of Reason*.[61]

In Edinburgh *Vestiges* was used to boost phrenology, to the disgust of those with strong religious views. Hugh Miller, already a popular writer on fossils and a strong supporter of miraculous creations, attacked *Vestiges* in the pages of the Free Church magazine *The Witness*. When cheap editions threatened to spread the book's infidelity, he brought out his *Footprints of the Creator* to refute it – the book soon sold over 5,000 copies.[62]

Vestiges was also published in the United States just as Louis Agassiz was arriving to great acclaim. It sold for 75 cents and was initially advertised as the work of Sir Richard Vyvyan, one of the suspects frequently named in Britain. A response by the clergyman George B. Cleever was included, later replaced by David Brewster's critique from the *North British Review*. Over twenty editions were published in the United States, and the eventual sales outstripped those in Britain.[63]

Agassiz lectures to large audiences at the Lowell Institute in the winter of 1846–7. His vision of a divine plan unfolding toward humankind resembled that of *Vestiges* apart from his insistence that the sequence represented a series of distinct creations. Like Hugh Miller he saw discontinuity as necessary to maintain the gulf between humanity and the brutes that perish. He attacked *Vestiges*, insisting that apart from its infidelity it was also bad science.[64] Harvard's professor of botany, Asa Gray, also criticized the book in his own Lowell lectures, dismissing it as a rehash of Lamarckism and Geoffroy's developmental saltations. He hoped to respond in the *North American Review*, but the job went to Francis Bowen, who was equally critical. Gray got his chance when *Explanations* appeared, giving him the opportunity for a forty-page critique. Gray conceded that if the books' thesis were accepted, it would produce a revolution in religious thinking – a prophetic statement in view

61 On Liverpool, see Secord, *Victorian Sensation*, chap. 6, and on Chiltern, p. 313.

62 Ibid., pp. 279–82.

63 Secord admits that there is no detailed analysis of the American response; see ibid., p. 382, where there is a reproduction of the advertisement for the second American edition giving details of the price and newspaper notices. He notes that there is a limited discussion of American responses in Milton Millhauser, *Just before Darwin*, chap. 5. American reviews are included in the list appended to Secord's edition of *Vestiges*, pp. 222–9.

64 The Lowell lectures are covered in Lurie, *Louis Agassiz*, pp. 126–8.

of his later endorsement of Darwinism.[65] Other scientists and theologians joined in with critical reviews, and Edward Hitchcock attacked it in his *The Religion of Geology* in 1846.

Despite the critiques, many found Chambers' book a revelation, including Abraham Lincoln.[66] Here as in Britain, *Vestiges* seems to have made the idea of development through law more acceptable to middle-class opinion. By the time Darwin published, there was a more open mind on the topic except in those who still felt the need to preserve the distinct nature of human spirituality. In this respect, *Vestiges* paved the way for Darwinism – yet it also shaped public opinion in ways that influenced how people read the *Origin of Species*. Darwin did not encourage the belief that the pattern of evolution is governed a preconceived divine plan his was a purely naturalistic explanation. But many non-scientific readers saw the *Origin* as another attempt to show how humans had appeared as the high point of creation.

There were alternatives to Chambers' model of linear development. Owen and W. B. Carpenter had begun to appreciate how the development of each class in the fossil record could be seen as the branching out of divergent trends of adaptive specialization. Darwin would interpret this as evidence for the action of natural selection on different populations derived from a common ancestor. This was hinted at in the more popular second edition of his *Journal of Researches* from the *Beagle* voyage, which mentioned the species of the Galápagos Islands and included an image comparing the beaks of the different finches adapted to their various ways of life (see Fig. 2.4). He speculated that 'one might really fancy that from an original paucity of birds in this archipelago, one species had been taken and modified for different ends'.[67] Here was a hint at the new model for understanding the history of life on earth he was developing for eventual publication.

Herbert Spencer was also beginning to expand his own view of how life had both specialized and progressed, exploiting the Lamarckian explanation of adaptation he would continue to defend even after acknowledging the power of Darwin's alternative theory. His *Principles of Psychology* of 1855 attracted attention and criticism for its attempt to explain how our mental powers had developed through exercise over many generations. His 1857 article 'Progress: Its Law and Cause' in the *Westminster*

[65] See A. Hunter Dupree, *Asa Gray*, pp. 144–9. [66] Secord, *Victorian Sensation*, p. 38.

[67] Darwin, *The Voyage of the Beagle*, p. 361. The edition cited is a modern reprint of the second edition of the *Journal of Researches into the Natural History and Geology of the Countries Visited during the Voyage of H.M.S. Beagle round the World, under the Command of Captain Robert FitzRoy, R.N.* On the popularity of this edition, see Desmond and Moore, *Darwin*, pp. 327–8, and Janet Browne, *Charles Darwin: Voyaging*, pp. 466–8.

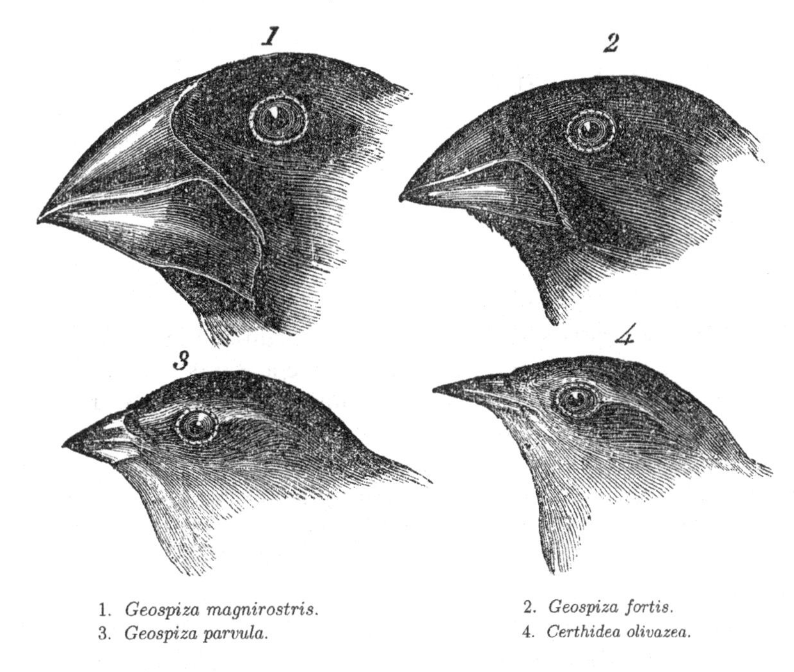

Fig. 2.4 Beaks of Galapagos finches adapted to different modes of feeding. From Charles Darwin, *Journal of Researches* ([1845], new edition 1860, this image from the 1873 printing), p. 379.

Review outlined the evolutionary philosophy he would develop through the 1860s.[68] That decade would see the start of a process that would establish a very different and much less goal-directed model of evolution – but the process would take time to coalesce into anything resembling a coherent alternative to the more developmental way of thinking popularized by Chambers.

[68] 'Progress: Its Law and Cause' is reprinted in Spencer, *Essays*, 1: 1–60.

3 Reacting to the *Origin*

The *Origin of Species* was published toward the end of 1859, initiating a debate that fluctuated through the next decade and was renewed when the *Descent of Man* appeared in 1871. This initial phase of the 'Darwinian Revolution' was successful to the extent that the general idea of evolution became widely accepted by scientists and the educated public by the early 1870s. But this did not mean the triumph of Darwinism in anything like its modern form. Alternatives to natural selection flourished, some so different they compromised Darwin's vision of the history of life as a divergent tree. The basic notion of common descent did gain some traction, although the situation remained fluid, and it would not be until the later 1870s that coherent efforts to visualize the history of life in evolutionary terms would proliferate.

How was this confused situation perceived by the cultural elite, and how did the issues make their way from the highbrow periodicals into media aimed at a wider readership? Ellegård's classic survey of how Darwin's theory was conveyed to the general reader suggests that it was some time before the topic began to gain the attention of the lower-middle and working classes.[1] Chapter 2 showed that in fact the basic idea of evolution had already begun to circulate in liberal and radical circles, but in a form very different to that proposed in Darwin's theory. His new model had to make its way into a climate of opinion already shaped by very different ideas about how the process might work.

The means of dissemination were becoming ever more fluid. Although books like the *Origin* were expensive, they were available in libraries, and if successful would eventually be reprinted in cheaper editions. Reviews in the highbrow magazines were now paralleled by comments in a much wider range of periodicals thanks to less stringent government restrictions on the press. There was an expansion of magazines aimed at the well-educated middle classes and of popular science journals seeking to attract

[1] See the discussion of Ellegård's *Darwin and the General Reader* in Chapter 1 above.

a wider interest in new discoveries. Much of the material in these organs was written by authors who were not active scientists, although many had amateur interests or were concerned about the implications of new ideas. Some scientists did get involved, either to garner support for their work or to promote particular theories. Active supporters of Darwinism such as Thomas Henry Huxley pressed their case in publications aimed at a variety of readerships and gave public lectures, some aimed at the working classes. Comments proliferated in sermons and religious publications and in magazines focused on natural history. Some aspects of the debate made their way into satirical magazines, although seldom in a form that the scientific community would feel comfortable with.[2]

The pre-Darwinian ideas of evolution had already become identified with religious and ideological positions. Conservatives were horrified, but Chambers had enabled the respectable middle classes to accept the basic idea of progress by presenting it as the unfolding of a divine plan. This compromise was already being challenged by Herbert Spencer's efforts to link Lamarckism to free-enterprise individualism, and inserting Darwin's theory into this mix was inevitably going to be a complex process. The response to his theory was also influenced by contingent factors within the scientific and cultural developments of the time.

There were changes at work within the scientific community. Expertise was becoming a crucial source of authority, replacing that of the aristocracy and the Church. Expert non-professionals (such as Darwin) still had a role, but influence was passing to paid teachers and researchers who felt that talent, not birth, should open the way to advancement. Evolutionary ideas, especially those that did not compromise with the old idea of a divinely created cosmos, were a useful weapon in their struggle against conservative forces. Huxley, a champion of the fight for professional recognition, supported Darwin even though he doubted that natural selection was an adequate explanation of evolution. His willingness to spread the word to the working classes was tempered by the need to present it in a way that did not imply revolution.

Controversy over the theory was temporarily marginalized in theological circles by concerns over the liberal views on the interpretation of the Bible advanced in a collected volume, *Essays and Reviews*, published in the following year. Another distorting factor was the sudden emergence into public consciousness of the one issue that Darwin had hoped to avoid: human origins. Popular fascination with accounts of the gorilla,

[2] On the expansion of publications, see Lightman, *Victorian Popularizers of Science*; Cantor et al., eds., *Science in the Nineteenth-Century Periodical*; and Ruth Barton, 'Just before *Nature*'.

newly discovered in Africa, prompted immediate speculations about the apes as human ancestors. At the same time, archaeologists – quite independently of developments in biology – confirmed the huge antiquity of the human species and highlighted its primitive origins, while anthropologists sought to portray 'savages' as relics of the past. The possibility that the 'missing link' might be found not in the fossil record but in some remote corner of the world today would shape the way many perceived Darwin's initiative. Here, at least, Chambers' vision of linear development would resist the impact of the Darwinian model of evolution as an ever-branching tree.

Getting Published

Darwin had a huge correspondence network, but only in the late 1850s did he allow friends, including the geologist Charles Lyell and the botanist Joseph Dalton Hooker, to know the details of his theory. He was planning an extended publication, but this was interrupted by the arrival in 1858 of a letter from Alfred Russel Wallace, then collecting in the Malay Archipelago (modern Indonesia), containing a theory with close resemblances to his own. The story of how arrangements were made to read Wallace's paper and extracts from Darwin's own work to the Linnaean Society is well known, as is the fact that there was only limited reaction to this event and the publication of the papers. Darwin now set to work on the shorter account of his theory that became – to give the full title – *On the Origin of Species by Natural Selection, or The Preservation of Favoured Races in the Struggle for Life*. Wallace did not return to Britain until 1862, and thus took no part in the initial debate sparked by the book.[3]

The *Origin* was a very different book to *Vestiges* – and to anything that Herbert Spencer would write on the topic. There was no introductory account of the origin of life and no comprehensive survey of the progress recorded in the fossil record. Darwin launched straight into a description of his proposed mechanism of natural selection, including his views on the still poorly understood phenomena of variation and heredity (which allowed for both the inheritance of acquired characters and the appearance of undirected variations stimulated by environmental change). He appealed to the artificial selection used by horticulturalists and animal breeders to give his readers a model by which to appreciate how environmental pressures could also select which variants can reproduce. This was an important strategy because a significant number of lay readers would

[3] For more details on the development of the theory, see my *Evolution*, chap. 6.

have some familiarity with these practical applications of selection (see Fig. 3.1). It also drove home the point that change was open-ended and driven by external factors because the variations involved were multifaceted and hence imposed no direction. The power of the environment to determine which variations would flourish was emphasized by noting Thomas Malthus' claim that in every generation more are born than can survive (and breed), so the resulting 'struggle for existence' will weed out all but those best adapted to the prevailing conditions.

The later chapters showed that a wide range of areas studied by naturalists could be illuminated by applications of the theory, which also gave Darwin the opportunity to head off some of the objections he anticipated. Biogeography figured strongly, and embryology was brought in to point out the significance of the process of specialization seen in the developing organism. A chapter on palaeontology was mainly concerned to show that the available fossils gave an imperfect record of the history of life, so little evidence of continuous evolutionary transitions could be expected. Darwin had limited interest in trying to reconstruct detailed ancestries for particular groups, and his diagram showing the pattern of evolution as a branching tree was meant only to illustrate the general principles of his system.

The book ended with an appeal to readers concerned about the theory's wider implications. Darwin admitted that 'Light will be thrown on the origin of man and his history' even though the topic had not been addressed in the main text. He also argued that to imagine the Creator operating through the laws the Creator had established, rather than by miracle, did nothing to diminish our sense of his power. The complexity of the whole system was emphasized via the analogy of the 'tangled bank' to give an impression of what we would now call the ecological forces bearing on every species. Finally, Darwin tied his theory to the prevailing view that the history of life is a story of progress. Since natural selection works only for the good of each being, 'all corporeal and mental endowments will tend to progress toward perfection'. Despite the apparent harshness of the mechanism, the end result has been the production of the higher animals (and, by implication, humanity). The final sentence pulls it all together:

> There is grandeur in this view of life, with its several powers, having been originally breathed into a few forms or into one; and that, whilst this planet has gone cycling on according to the fixed law of gravity, from so simple a beginning endless forms most beautiful and most wonderful have been, and are being, evolved.[4]

[4] Darwin, *Origin of Species*, 1st ed., p. 490. In later editions 'breathed' was changed to 'breathed by the Creator'.

Fig. 3.1 Varieties of domestic pigeons used to illustrate Darwin's views on the power of artificial selection, one of many different images created to make this point. From C. M. Beadnell, *A Picture Book of Evolution* (1934 ed.), p. 145; original edition by Dennis Hird (1906/7), part 1, p. 191.

The suggestion that life might have been 'breathed' into the earliest forms was intended to show that the theory did not necessarily displace the Creator altogether – there was no connection with the idea of spontaneous generation. There is progress here, but it was a more complex and less predictable version of that ideology than that offered by *Vestiges*. There was also diversity, so there can be no single goal, and not all branches of the tree of life will advance to higher states. The harshness of the system means that many of the forms produced will not represent an advance as judged by human moral standards.

The *Origin of Species* was published by John Murray on 24 November 1859. Of the first printing, approximately 1,100 copies were available to the trade, of which 500 were taken by Mudie's Subscription Library. The remainder were sold to book dealers on the first day, generating the myth that it was snapped up immediately by the public – although it must have sold quickly because a new edition was immediately planned. The price was fourteen shillings per copy, making it a relatively expensive book, and cheap editions did not become available until it went out of copyright at the end of the century. Darwin made numerous changes in the subsequent editions and added an extra chapter responding to criticisms to the sixth and final version of 1872. Sales remained fairly modest, and it was 1890 before they exceeded those of *Vestiges* at a total of 40,000. In the United States, Asa Gray headed off the publication of a pirated edition and arranged for Appleton of New York to publish an authorized edition with a print run of 2,500 copies, which rapidly sold out. For the next printing Gray persuaded Darwin to add a 'Historical Sketch' to show that he did not claim to be the first to suggest the basic idea of evolution. Reprints of the revised editions were published with the cooperation of Murray and sold well, although it was not until 1891, when a new copyright act was introduced, that a variety of new editions were published.[5]

First Reactions: Britain

Darwin hoped to convert younger naturalists with more flexible opinions, and the *Origin* certainly triggered a debate in scientific circles. Charting its influence beyond the scientific community is a complex process, however. There were supporters determined to promote the

[5] For details of the publication, see Michèle Kohler and Chris Kohler, 'The *Origin of Species* as a Book', and Janet Browne, *Charles Darwin: The Power of Place*, pp. 88–90 and 132–3. *The Correspondence of Charles Darwin*, vol. 8 (1860), gives details of the American editions in appendix 4.

theory, but they disagreed over the exact nature of the 'Darwinism' they were trying to get into circulation. The differing perceptions were soon transmitted to media aimed at the lower-middle and working classes, but external factors and preconceived opinions would shape the wider reaction. Despite Darwin's efforts to avoid the topic, his theory was immediately associated with the claim that humanity had originated from the apes, which in turn fed into debates over the relationship between the human races. In the United States this included the issue of slavery and the threat of civil war. Here, though, the initial debate was less active than in Britain, with the real excitement beginning in the next decade.[6]

Huxley introduced the term 'Darwinism' in 1860[7] and it soon came into general use, but what did it mean? Darwin adopted a purely naturalistic approach to the topic with natural selection as the main (but not the sole) agent of change. He saw adaptation to changing environments as the ultimate driving force, but others favoured trends predetermined by internal forces. Hooker, most concerned with the implications for biogeography, took on board Darwin's more open-ended, adaptationist position. Huxley was a morphologist who liked the naturalistic slant of Darwin's theory but was tempted by the idea of internally directed variation. He was suspicious of Lamarckism – yet this was the alternative favoured by Herbert Spencer. Asa Gray argued that Darwinism was compatible with belief in a Creator who directs the laws of nature toward beneficial ends. Initially, then, 'Darwinism' meant little more than 'evolutionism', and only in subsequent decades did it come to mean 'evolution by natural selection'.

Historians have mostly focused on reviews in the elite monthly and quarterly journals, which must be mentioned here if only because they helped to define issues that would be commented on by authors of more popular material. There was certainly much opposition from conservative quarters, the two most notable reviews being by Richard Owen and the bishop of Oxford, Samuel Wilberforce. Owen's in the *Edinburgh Review* was a nuanced critique, since whatever his subsequent reputation, he was by no means opposed to the basic idea of transmutation. Like many others, he felt that if the process occurred it would have to be

[6] For details of the media coverage, see Ellegård, *Darwin and the General Reader*; Browne, *Charles Darwin: The Power of Place*, chap. 3; and Desmond and Moore, *Darwin*, chaps. 33–8. Ronald L. Numbers, *Darwinism Comes to America*, notes the muted initial debate here, see p. 31, while Cynthia Eagle Russett's *Darwin in America* explicitly focuses on the period after 1865. The detailed list of reviews and comments assembled by John Van Wyhe for the Darwin online website lists 292 responses to the *Origin* and 262 to the *Descent of Man*.

[7] In his *Westminster Review* article; see Huxley, *Collected Essays*, 2: 78.

directed by something more coherent than a combination of purposeless variation and brutal struggle. He also primed Wilberforce for his assault in the *Quarterly Review*, which identified many of the objections that would dog the theory over subsequent decades. Everyday variations seem to be confined within well-defined limits, thus guaranteeing the fixity of species, and the fossil record of the time revealed no continuous transformations. Darwin's theory was just another wild speculation to be taken no more seriously than earlier efforts by his own grandfather and Lamarck. Darwin was accused of implying that humans are nothing more than improved apes, a charge indicative of what was to come.[8]

For the defence, Huxley was first off the mark with a review in the *Times* and a short notice in the December 1859 issue of the new monthly *Macmillan's Magazine*. The next month saw a more substantial and very positive review by William B. Carpenter in the *National Review*. But it was Huxley's extended and far more pungent piece in the *Westminster Review* for April 1860 that has become famous for its eminently quotable statements on the theory's wider implications: Darwinism is 'a veritable Whitworth gun in the armoury of liberalism' and 'Extinguished theologians lie about the cradle of any science as the strangled snakes beside that of Hercules.' As one of the newly emerging breed of professional scientists, Huxley wanted to see expertise in technical fields displace the authority traditionally claimed by churchmen. Nevertheless his enthusiasm was qualified in a way that disturbed Darwin and reveals the limits of his conversion. Here was a thoroughly naturalistic theory, but was it enough to explain the deeper structures that define the main forms of animal life, as opposed to their superficial adaptations? Doubts would remain until the breeders had formed a genuine new species by artificial selection. Huxley also cautioned Darwin against ruling out saltations, sudden jumps prompted by internal factors that might have no adaptive significance.[9]

Darwin had used little technical language, so his book could be read by any well-educated person, but at 490 pages in the first edition it was a heavy read. George Eliot (Mary Ann Evans) thought it interesting but 'not impressive, from want of luminous and orderly presentation'. She did accept that it 'marks an epoch' and moved the world a step toward

[8] Owen's review is reprinted in Hull, *Darwin and His Critics*, pp. 175–213. See Wilberforce, 'On the Origin of Species', pp. 254–5 on Lamarck and Erasmus Darwin; also pp. 263–4 on Lyell and 257–8 on human origins.

[9] Carpenter's review is reprinted in Hull, *Darwin and His Critics*, pp. 87–117. Huxley's is reprinted in his *Collected Essays*, 2: 22–79, quotations on pp. 23 and 52. Huxley's limited conversion to Darwinism was first stressed by Michael Bartholomew, 'Huxley's Defence of Darwinism', and is now widely acknowledged.

'brave clearness and honesty'.[10] Her partner George Henry Lewes claimed in the *Cornhill Magazine* that it was 'in every body's hands', but this meant the middle class. For many it would be the reports in newspapers and magazines that gave them their first impressions. There were several newspaper reports (only the *Daily News* was hostile), but the most effective was Huxley's substantial review in the *Times* on Boxing Day 1859 (although published anonymously, his authorship soon became widely known). He stressed Darwin's focus on divergence, comparing the development of life to the congregation leaving church, with everyone making their way along the aisle together and then going their separate ways outside. His description of natural selection brings out the severity of the struggle for existence: a population under pressure from the environment is 'like the crew of a foundered ship, and none but the good swimmers have a chance of reaching land'.[11] He was soon recognized as the leading campaigner for the theory, although the term 'Darwin's bulldog' was used only after his death later in the century.[12]

The *Cornhill* in which Lewes wrote was one of several newly founded weekly and monthlies including *Macmillan's Magazine* and the *Fortnightly Review* that provided a format for debate. Robert Chambers gave the book a short but supportive notice in his popular *Chambers' Journal*, and the *English Churchman* was surprisingly positive. Charles Dickens' *All the Year Round*, with a circulation of 80,000, published a review insisting that if natural selection was supported by the facts, no opposition from religious forces could block its acceptance. The author was David Ansted, who also included an account of the theory in a popular book on geology.[13] The notices were not all formal reviews; contrary to Ellegård's count of two articles the *Cornhill* actually mentioned Darwin's ideas at least a dozen times (Lewes' comments were in a general series on animal life).[14] Joseph Hooker did his bit by getting a favourable review into the weekly *Gardener's Chronicle*, reaching a

[10] J. W. Cross, ed., *George Eliot's Life*, 2: 143 and 148.
[11] The review is reprinted in Huxley's *Collected Essays*, 2: 1–21, for the references see pp. 5 and 18.
[12] See John Van Wyhe, 'Why There Was No "Darwin's Bulldog"'.
[13] Browne, *Darwin: Power of Place*, p. 101, and on Lewes' comment p. 128; for the original quotation, see Lewes, 'Studies in Animal Life', chap. 5, p. 603, and for more details of his views, see below. There is also a list of reviews in the *Darwin Correspondence* volume (1860), appendix 7, while Ellegård gives the fullest analysis. On the response in Dickens' *All the Year Round*, see George Levine, *Darwin and the Novelists*, pp. 127–9, and Ruse, *Darwinism as Religion*, pp. 60–1, the latter identifying the author.
[14] Gowan Dawson, 'The *Cornhill Magazine* and Shilling Monthlies in Mid-Victorian Britain', pp. 129–30. For Lewes' articles, see his 'Studies in Animal Life', of which chaps. 4 and 5 deal with Darwin.

readership that would become familiar with Darwin's later botanical research. Hooker concentrated on the analogy with artificial selection and noted that the alternative of special creation had no empirical support. The magazine later reprinted Huxley's *Times* review and published Hooker's response to a letter arguing for saltations.[15]

Huxley and a group of like-minded thinkers, later known informally as the X Club, founded new periodicals with the aim of spreading ideas such as Darwinism. Their efforts encountered the perennial problem of how to appeal to readers with different levels of interest. In 1860 the *Natural History Review*, originally founded in 1854, was relaunched with Huxley and a group of fellow scientists as co-editors. It published a number of articles on the question of human origins. Darwin and Hooker had warned Huxley not to get involved, and there were endless problems of production and only limited sales. Huxley bailed out in 1863 and the journal closed a year later. The X clubbers then got involved with the weekly *Reader*, originally founded by Christian Socialists with J. Norman Lockyer as science editor. Again there were financial losses, with half of the 4,000 copies printed having to be given away. It closed down in 1867, and thereafter the group confined themselves to reaching the public through general magazines. Lockyer, however, would go on to persuade the publisher Macmillan's to launch *Nature* in 1869 – this would experience similar problems but survived to become a leading organ for the wider scientific community.[16]

Printed accounts were not the only means of dissemination. In February 1860 Huxley lectured on Darwin's theory at the Royal Institution in London. He used a display of pigeons to illustrate the diversity produced by artificial selection, but again indicating his reservations about the power of the natural equivalent, much to Darwin's annoyance. Toward the end of 1862 he was lecturing to working men on the phenomena of organic nature, with the sixth and final lecture dealing with Darwinism. The lectures were taken down in shorthand and published in cheap pamphlets, thus reaching an even wider lower-class audience. Having outlined the hypothesis of natural selection, he declared that it 'will be found competent to explain the majority of the phenomena exhibited by species in nature'. He did, however, repeat his

[15] The first part of Hooker's review is reprinted in Hull, *Darwin and His Critics*, pp. 81–5; on the reprint of the Huxley review and Hooker's response to W. H. Harvey's letter on saltations, see *Correspondence of Charles Darwin*, 8: 120 and 94.

[16] On these publications, see, for instance, Ruth Barton, *The X Club*, pp. 177–83, 202–14 and 406–8. On the *Natural History Review*, see Geoffrey Belknap, 'Natural History Periodicals and Changing Perceptions of the Naturalist Community'. The story of *Nature* will be covered in later chapters.

warning about the failure to produce a new species by selection. He also alluded to the question of human origins, while conceding that Darwin himself had not addressed this topic. He was, in fact, already deeply involved in the controversy over the degree of our relationship to the apes and had already lectured on the theme to a working-class audience (see Chapter 4 below).[17]

The theory was also debated at public meetings, the most famous of which was that of the British Association for the Advancement of Science held at Oxford in July 1860. This has become legendary, largely on the strength of Huxley's sharp response to Wilberforce's sneering remark about claiming descent from an ape. Contemporary sources disagree on the exact words uttered, but the legend has it that the huge audience thought that Huxley had demolished the bishop. Historians have recognized for some time now that this interpretation was a later construct by the supporters of scientific naturalism. In fact, both public and private descriptions of the debate at the time suggest little agreement over who won, and some note that Hooker, who spoke after Huxley, gave a better defence of the theory. The most recent analysis notes a number of external factors that influenced the audience, including the fact that Wilberforce, as bishop of Oxford, was not popular with the local clergy.[18]

Fourteen notices of the meeting were published in newspapers and magazines, the most substantial in the weekly *Athenaeum*. They focused mainly on the debate over the general theory of natural selection. The *Athenaeum* account does not mention the spat over human origins, but the magazine did note an earlier session of the meeting in which Huxley had clashed with Owen over the degree of similarity between the structures of the ape and human brain. The two rivals were already at loggerheads over the issue, and their antagonism soon spilled over into material aimed at a wide audience. From February to May 1861 Huxley lectured to working men on 'the relation of man to the rest of the animal kingdom'. His case for a close relationship with the apes would subsequently be published in his *Man's Place in Nature*.

Hooker's response to Wilberforce at the Oxford meeting specifically criticized the claim that Darwin's theory involved the descent of one living species from another. The model of a branching tree for evolution implied that related species are derived from a common ancestor that would not

[17] See Huxley's *Collected Essays*, 2: 447–75, for the sixth lecture, quotation from p. 463. The preface, p. vi, notes the circumstances of the pamphlet publication, and on their impact, see Desmond, *Huxley: The Devil's Disciple*, 309–11.

[18] Frank A. L. James, 'An "Open Clash between Science and the Church"?' The article lists the contemporary press accounts. For the text of the most detailed, in the *Athenaeum*, see the *Correspondence of Charles Darwin*, vol. 8, appendix 6.

possess the specialized characters developed by any of its descendants. In 1868 Hooker got another chance to promote Darwinism when he was appointed president of the British Association's meeting in Norwich. In his widely reported address, he presented the general case for evolutionism, stressing aspects of Darwin's biogeographical work and his botanical studies. He also drew attention to the work of Lubbock and others who were establishing an extended view of human prehistory.[19]

Aspects of the theory Darwin had proposed certainly got into general circulation. As early as November 1859 the satirical *Punch* included a skit about marriage partners entitled 'Unnatural Selection and the Improvement of Species', and in the following decades it would routinely parody various aspects of the theory.[20] Darwin began to receive letters from ordinary people who had heard about his theory. In March 1862, George Harris, 'a poor working man with a wife and four children to support', asked for a copy of the *Origin*, offering to do tailoring work in return.[21] The basic model of natural selection had been described often enough to allow a wide audience to appreciate its operations, although the reviews were often less clear on the implications of a theory in which this was the main mechanism of change. It is hard to determine how many readers would have realized that the branching-tree model of the development of life that Darwin proposed was open-ended, because each branch was shaped solely by the pressures of the local environment. We shall see how the overall impact of the theory was shaped to a significant extent by popular interest in the possibility that the human race had evolved from the gorilla, which encouraged a more linear model of evolution.

First Reactions: The United States

The response in the United States paralleled that in Britain to some extent, but there were also significant differences. The level of debate seems to have been lower at first – the *New York Times* chided the American Association for the Advancement of Science for avoiding the issue at its 1860 meeting.[22] In part this was because the country had

[19] On the lectures, see Adrian Desmond, *Huxley: The Devil's Disciple*, pp. 292–5. For Hooker's remark at the Oxford meeting, see the *Correspondence of Charles Darwin*, 8: 596, and for his 1868 address, see his 'Address of the President'.

[20] See Richard Noakes, '*Punch* and Comic Journalism in Mid-Victorian Britain'.

[21] George Harris to Darwin, 3 March 1862, in *Correspondence of Charles Darwin*, 10: 99. In 1864 Darwin also received a manuscript on his theory from an Edinburgh baker, well-argued but with poor spelling; see *Correspondence*, 12: 142–3 and 148. For the *Punch* skit, see Browne, *Charles Darwin: The Power of Place*, p. 110.

[22] Noted in Numbers, *Darwinism Comes to America*, p. 31.

more pressing matters to worry about – John Brown was hanged for his involvement in the raid on Harper's Ferry only eight days after the *Origin* was published.[23] Human origins and the race question were becoming a matter of life and death in a country that was tearing itself apart on these issues. Wider arguments over the materialistic implications of the new theory would inevitably be sucked into the controversy over the nature of humanity.

The theory's impact on religious belief could not be ignored, however. Among the scientists, Darwinism was rejected by those who clung to the hope of seeing a rational or purposeful plan in nature and supported by those who favoured empiricism. But empiricism did not necessarily mean the kind of scientific naturalism espoused by Huxley. Louis Agassiz may have played a role similar to that of Owen in Britain, but Asa Gray certainly did not share Huxley's views on the theory's religious implications. As yet the United States did not have the organized middle- and working-class movements that provided Huxley with an audience anxious to overthrow the traditional structure of society. Americans regarded themselves as already liberated from the British hierarchy (apart from enslaved people, of course). Religion played a more active role in the American psyche, and there was much opposition to the theory in traditionalist quarters. But there were also liberals willing to move with the times: Gray was an abolitionist and a deeply religious believer anxious to minimize the disparities between Darwin's theory and the vision of a universe designed by a benevolent Creator. This approach would eventually consolidate around alternatives to natural selection that made the process seem less harsh and more directed.

Debates erupted soon enough within the scientific community, with Agassiz as the leading critic and Gray as the defender, aided by William Barton Rogers. The students at Harvard were already discussing the topic ahead of formal debates at their Natural History Society in mid-1860. Agassiz dismissed the theory at the Boston Society of Natural History on 15 February and was answered by Rogers, leading to a series of four meetings on the theme over the following months. The topics of religion and human origins did not come up, and the events were ignored by the local press.[24] Gray and Rogers also clashed with Agassiz at the January meeting of the American Academy of Arts and Sciences, again leading to a series of further discussions in which Agassiz was supported

[23] See Desmond and Moore, *Darwin's Sacred Cause*, p. 317.

[24] Numbers, *Darwinism Comes to America*, p. 31. Other accounts include Browne, *Charles Darwin: The Power of Place*, pp. 132–5; Dupree, *Asa Gray*, chap. 14; and Lurie, *Louis Agassiz*, chap. 7.

by the philosopher Francis Bowen and the philanthropist John Amory Lowell. Gray addressed the theological implications of the theory directly, knowing that with a lay audience these might count for more than scientific details. By now the issues were getting into the local press, and Gray sent Darwin cuttings from the newspapers reporting on his efforts.

Reviews were also appearing in both the scientific journals and more general magazines. Gray persuaded the *American Journal of Science* to reprint Hooker's survey of the flora of Tasmania, which supported Darwin on the topic of geographical distribution. His own review of the *Origin* soon appeared and a follow-up on the wider issue of 'Design versus Necessity'. He then reached out to a wider audience in the *Atlantic Monthly* to reassure the readers that 'Natural Selection [is] not Inconsistent with Natural Theology.' He wanted to understand the theory in a way that allowed it to be interpreted as an expression of the Creator's purpose. Where Darwin saw the variations on which selection works as random or undirected, Gray advised him to assume 'that variation has been led along certain beneficial lines'. Darwin was not happy with the suggestion – if variation pushed species in the right direction anyway, what need was there for selection? But he realized that Gray's arguments would head off some of the objections felt by religious believers and made arrangements for the article to be published as a pamphlet in Britain.[25]

Over the next decade the idea of evolution would become widely accepted, but Gray's preference for directed variation would find expression in various non-selectionist explanations, including Lamarckism. The wider debate would become embroiled in the issues raised by the question of how the 'lower' races could be fitted into the sequence of human development.

Defending Darwinism

If even Huxley had doubts about the efficacy of natural selection, hostile critics raised more serious objections. Many naturalists still believed that species are defined by rigid boundaries beyond which no variation is possible. How new variations would be transmitted to future generations was also problematic. Darwin's own theory of heredity, 'pangenesis', was eventually published in 1868 but gained little support (and is based on principles very different to those of modern genetics). It allowed for both

[25] Gray's reviews are reprinted in his *Darwiniana*: the first *American Journal of Science* review is pp. 9–61, the follow-up 'Design versus Necessity' is pp. 62–86 and the *Atlantic Monthly* pieces, pp. 87–128 and 129–77, quotation from p. 148.

Lamarckism and undirected variation, but did nothing to undermine the popular assumption that the characteristics of the parents are blended together in the offspring. Pangenesis seldom featured in the more popular literature, but an influential critique of natural selection by the engineer Fleeming Jenkin in the *North British Review* used the concept of blending heredity to argue that a favourable variant would be swamped by interbreeding with unchanged members of the population. Only saltations that established a different centre of variation could produce a new species. Darwin was forced to rethink how his theory was presented, and the issue subsequently gave rise to the belief that his theory could not become convincing until supplemented by the new interpretation of heredity introduced by genetics. In fact, a plausible version of the theory *could* be based on blending heredity – as A. R. Wallace would later point out, most characters show a range of variation so there is always a significant number of individuals at the favoured extreme of the range.[26]

Wallace returned to Britain from the tropics in 1862 and published a series of papers based on his observations to support Darwin's claim that there is no clear distinction between local varieties and true species. His travelling companion from an earlier expedition to South America, Henry Walter Bates, also supported Darwin with his discovery of mimicry in insects. Some species are able to camouflage themselves for protection, but Bates showed that others actually mimic the warning colours of inedible species. Since insects have no voluntary control of their colours, the adaptations cannot be due to the Lamarckian effect and are easily accounted for by natural selection. Wallace published an account of mimicry in the *Westminster Review* of 1867, later reprinted in a collection of his papers. He also reached a wider readership with his account of his travels, *The Malay Archipelago*.[27]

Perhaps the most well-established argument against any form of evolutionism was the belief that the fossil record showed only discontinuous transitions from one geological formation to the next. The Darwinists needed to counter this claim, but they needed to do it in a way that headed off rival interpretations of how life has developed. Darwin's theory was based on the image of the diverging tree of life, but alternative images were already in play. Chambers' *Vestiges*, still in print, proposed a model of predetermined development based on the ascent of a linear

[26] The various objections are listed in Ellegård's *Darwin and the General Reader*, chap. 13. Jenkin's 1867 review is reprinted in Hull, *Darwin and His Critics*, pp. 302–50. For more information on the resulting debates, see Jean Gayon, *Darwinism's Struggle for Survival*, and for a brief account my *Evolution*, chap. 6. On Wallace's later views on variation, see Chapter 6 below.

[27] Wallace's paper on mimicry is reprinted in his *On Natural Selection*, pp. 45–125.

hierarchy of classes toward the mammals and, eventually, to humanity as its goal. As the following chapters will show, this model continued to have influence through the later nineteenth century, partly in the form of a modified version of the tree image that gave the tree a central trunk rising up to humanity, with adaptive divergences dismissed as side branches of lesser importance.

A more linear model was also represented in the popular view that our immediate ancestry lies in the great apes, popularized as a result of Paul Du Chaillu's 1862 account of his discovery of the gorilla. Although Darwin had avoided the topic of human origins in the *Origin*, the image of the ape as ancestor inspired a host of cartoons and caricatures in the 1860s and was retained in many later efforts to present the 'lower' races as survivals of intermediate stages in the ascent. The later editions of *Vestiges* complicated the issue further by introducing the concept of what became known as parallel evolution. Instead of a single line of ascent, this model proposed multiple lines of development, each rising through a similar predetermined sequence but at different rates. The various living races had independently reached the human level from different ape origins. Pushing the model further back into the past, multiple lines of reptilian evolution might have evolved in parallel toward the avian and mammalian forms. The birds and mammals were not classes unified by divergence, each from a single common ancestor – they were grades of development that could be achieved independently by several different lineages. Darwin supported his rival view by appealing to Owen's work on the fossil mammals, which had provided evidence of diversification from more generalized earlier forms. His position needed more detailed examples of adaptive specialization and, even more crucially, evidence of how the ancestral form of a new class had emerged from a single lineage within the previous class.[28]

Reconstructing the evolution of life on earth would dominate perceptions of evolution through the later nineteenth century, but the process was slow to get underway. Surveys of the fossil record very similar to those popular in earlier decades continued to appear, including John Phillips' *Life on Earth*, which included a diagram showing the rise and fall of the vertebrate classes. John Page's popular *Past and Present Life of the Globe* of 1861and *The Earth's Crust* of 1864 offered more substantial comments, but like Phillips, Page thought along the same lines as Agassiz. The fossils revealed coherent patterns of development indicating an underlying divine plan. Louis Figuier's *The World before the Deluge*

[28] See Adrian Desmond, *Archetypes and Ancestors*, chap. 6, and my own *Life's Splendid Drama*.

(translated in 1865) offered a pictorial survey of the inhabitants of the successive geological periods but also presented the process as the unfolding of a divine plan. Some of its images were reminiscent of those published decades earlier, including those of the dinosaurs.[29]

Darwin made no effort to speculate about ancestries in the *Origin*, although he did address the issue in private.[30] He was acutely aware of what Lyell had called the 'imperfection of the fossil record' and appealed to this to counter claims that the apparently sudden appearance of each major group indicates divine creation. Although Owen had provided limited evidence of continuous trends within groups, Darwin had little hope that further discoveries would throw light on the major innovations in the history of life. He even seems to play down the prospect that substantial changes in the earth's populations could have occurred in the course of geological time.[31] This would imply diversification before the first fossil-bearing rocks were laid down. Huxley too was convinced of what he called the 'persistence of type' – he was now spending more of his time on fossils, but he thought the record so imperfect that the first examples of any type found in the rocks were almost certainly preceded by others of which we have no knowledge. This perception explains why the Darwinists were initially reluctant to get involved with what could only be speculative reconstructions of ancestries, at least in public. They were also reluctant to admit the possibility of what we now call episodes of mass extinction, because this would give too much ground to the catastrophists.

Darwin's supporters, and at least some reviewers, realized the significance of his image of the tree of life. We have seen that Hooker made the point that one living species should not be assigned as the ancestor of another in his response to Wilberforce at the 1860 British Association meeting. G. H. Lewes' account of the theory in *Macmillan's Magazine* also stressed that the similar characteristics in a group of related species

[29] See Phillips, *Life on Earth*, pp. 175–204 and the diagram on p. 66. For a detailed description of Page's work and its reception, see Lightman, *Victorian Popularizers of Science*, pp. 223–38. On Figuier, see Rudwick, *Scenes from Deep Time*, chap. 6; Figuier's image of dinosaurs was reminiscent of Mantell's; see *The World before the Deluge*, facing p. 257 in the first edition and facing p. 297 in the cheaper edition of 1872.

[30] See, for instance, his thoughts on the origin of the mammals in a letter to Lyell, 23 September 1860, *Correspondence of Charles Darwin*, 8: 377–80.

[31] See Richard Delisle, 'Natural Selection as a Mere Auxiliary Hypothesis (sensu stricto I. Lakatos) in Charles Darwin's *Origin of Species*'. I have to say that I find Delisle's further suggestion that Darwin was really thinking of network-like relationships rather than divergent ones rather unconvincing, especially in light of his interest in the origin of mammals shown in the 1860 letter to Lyell.

are inherited from a common ancestor and used the example of the great cat species to explain the principle of divergence. He warned against those who ridiculed the theory by claiming it implied absurd relationships. The point still had to be made, however, when the *Descent of Man* appeared: the London *Examiner* noted that Darwin's theory implied humanity 'is related to the gorilla, not as a grandson or great-grandson, but as a grand-nephew or great grand-nephew'.[32] Some branches of the tree may diverge further from the common ancestor than others, but so-called living fossils will be very rare.

The situation with respect to the fossil record soon improved, creating a foundation on which the evolutionists could begin to reshape the popular vision of how life on earth had developed. The first discovery of a creature that seemed to bridge the gap between two supposedly distinct vertebrate classes came in 1862 when Owen secured the fossil *Archaeopteryx* for the British Museum. Here was a bird with several reptilian features, undermining the claims of those who insisted that each class had been created fully formed. It precipitated a flurry of scientific and popular interest in the project to work out how the first bird might have evolved.

The remains of *Archaeopteryx* had been found in the fine-grained Solnhofen limestone of Germany, which preserved even the impression of its feathers. Owen went to considerable expense to buy it for the Museum, where it was displayed to fascinated crowds and prompted a flood of articles in the popular press. A skull with teeth (hence unlike any modern bird) was described in 1865, enhancing the view that it must throw some light on the origin of the class. Huxley too admitted that it was a true, if primitive, bird but insisted that it came too late in the geological sequence to throw direct light on the origin of the class. He had already begun to develop a hypothesis tracing the birds back to a newly discovered dinosaur named *Compsognathus*, a small creature that ran on hind limbs almost identical to those of birds. In a lecture to the Royal Institution in 1868 he proposed that birds had first emerged as ground-dwelling forms analogous to the modern Ratitae (ostriches etc.) and only later took to the trees and began to fly. On this model, *Archaeopteryx* was an offshoot from this later stage in the class's development.[33]

These proposals had ramification in the United States, where Edward Hitchcock had been championing fossil footprints found in Connecticut

[32] Lewes, 'Studies in Animal Life', chap. 4, pp. 441 and 445–7; the *Examiner* quotation is from Desmond and Moore, *Darwin's Sacred Cause*, pp. 373–4.

[33] On the early ideas about bird evolution, see Dawson, *Show Me the Bone*, pp. 303–16; Bowler, *Life's Splendid Drama*, pp. 261–80; and Desmond, *Huxley: Evolution's High Priest*, p. 359.

as evidence that birds had existed long before any skeletal remains were known. Now he was beginning to admit the possibility that the tracks had been made by reptiles of the kind Huxley had postulated for the ancestors of true birds. In 1872 Othniel Charles March reported the discovery in Kansas of toothed birds he called the Odontornithes, adding to the evidence that the class had reptilian ancestry.[34]

Huxley's decision to begin reconstructing the history of life in evolutionary terms was to a large extent inspired by his reading of Ernst Haeckel's *Generelle Morphologie* of 1866. Haeckel was emerging as Germany's leading Darwinist, but it was his enthusiasm for the use of embryological rather than fossil evidence that inspired his efforts to uncover clues about ancestral stages. Darwin (whose German was far less fluent than Huxley's) also hailed the *Generelle Morphologie* and its more popular successor the *Naturliche Schöpfungsgeschichte* as pioneering efforts to uncover the actual course of evolution that had led eventually to the emergence of humankind, suggesting that the later book made parts of his *Descent of Man* more or less superfluous. The *Descent*'s section identifying the key stages in the ascent toward humanity was pure Haeckel.[35] It would be 1876, though, before the latter's more popular work was translated as the *History of Creation.*

The Rise of Spencer

The 1860s also saw the emergence of another contribution to the naturalistic interpretation of evolution, one that some commentators in the later nineteenth century would regard as even more important than Darwin's. Herbert Spencer popularized the term 'evolution' (Darwin hardly ever used it), and he did so in the context of an all-embracing philosophy of cosmic progress that operated at the physical, biological and social levels. In biology at least the Darwinian and Spencerian programmes were complementary. Both presented evolution as a process governed solely by natural law, and both appealed to natural selection and Lamarckism to explain the process – although Spencer was by far the more enthusiastic Lamarckian. Historians of the social sciences find it hard to accept that Spencer was a follower of Lamarckism, a reluctance that helps to explain the popular misconception that he promoted a

[34] On Hitchcock and the bird tracks, see Dawson, *Show Me the Bone*, pp. 316–29, and on Marsh's discoveries Bowler, *Life's Splendid Drama*, pp. 267–8.

[35] See the introduction to the *Descent of Man*, 1st ed., 1: 4; 2nd ed., p. 3. On Huxley's debt to Haeckel, see his letter to the latter, 21 January 1868, in L. Huxley, ed., *Life and Letters of Huxley*, 2: 303.

'social Darwinism' based purely on natural selection. But the term 'Lamarckism' came to denote acceptance of the inheritance of acquired characteristics as an evolutionary mechanism, and Spencer's focus on self-improvement as the driving force of progress proclaims his Lamarckian credentials. This is confirmed by his later opposition to biologists who began to question the mechanism's validity.[36]

The relationship between Darwin's theory and Spencer's philosophy is complex, and the parallels and differences would shape how evolutionism affected the world-views of the later nineteenth century. They had different purposes: Darwin's was a theory intended primarily to resolve scientific questions, while Spencer used evolution as the foundation for a cosmic ideology of progress in all areas of the natural and social sciences. The contrast was noted by Chauncey Wright in *The Nation* for 1866: he saw Darwin's theory as objective science, while Spencer's 'synthetic philosophy' was a cosmological system charged with a mission.[37] Darwin's views were not without ideological implications, but for Spencer such implications were the whole point of the exercise.

Both accepted that evolution was progressive in the long run, and neither endorsed the linear model of development aimed at a particular goal (although both could be perceived as slipping into its orbit on issues related to human origins). Evolution was inherently divergent, producing not only more complex structures but also more complex assemblies of differing structures. The products could be ranked on a scale of complexity, and many different forms could reach the same level on the scale, although only one had reached the degree of mental and social activity humans value so much. For Spencer, though, progress was the central feature of evolution: even though it had no single goal it was more or less inevitable in every branch of the tree of life. Darwin was far more aware of the evidence from natural history showing that adaptive specialization did not necessarily result in increased complexity of structure and might even lead to degeneration. Episodes of progress toward higher-level structures and functions were rare and opportunistic, not the central driving force. To the extent that Spencer popularized the idea of evolution, his system ensured that it would be associated with a more intensive element of progressionism.[38]

Even for Spencer, progress worked through the constant interaction between organisms and their environment, a process equivalent to

[36] On this point, see Piers J. Hale, *Political Descent*, chap. 2, and Bowler, 'Herbert Spencer and Lamarckism'.

[37] This was in Wright's review of Spencer's *Principles of Biology*; see his 'Spencer's Biology'; see also Fichman, 'Ideological Factors in the Dissemination of Darwinism in England'.

[38] In have explored these issues more fully in my *Progress Unchained*.

Darwin's focus on local adaptation. But Spencer saw this as a far more abstract tendency for the organism to come into equilibrium with its surroundings. He accepted natural selection as a valid mechanism and coined the phrase 'survival of the fittest' to denote its action. In human affairs he had little sympathy with those who couldn't measure up to environmental challenges. On this latter point, both Spencer and Darwin can be seen as tending toward what was later called 'social Darwinism', especially as both saw evolution as a process driven by competition between individuals. Theirs was a way of thinking imbued with the logic of free-enterprise individualism and laissez-faire economics, in which personal success generates social progress. This emphasis would be applied with ever-greater force in applications of evolutionism to the development of human societies.

In addition, though, Darwin and Spencer both accepted a role for the Lamarckian process of the inheritance of acquired characteristics in which an organism's positive responses to environmental challenge are passed on to its offspring to shape the future of the whole population. For Darwin this played a minor role, even after he had been forced by his critics to emphasize that he had never ruled it out. For Spencer it was the dominant mode of progressive evolution, representing an even more effective consequence of the struggle for existence. Competition stimulates all but the weakest to seek ways of adapting themselves to an ever-more complex environment, and at the same time tends to improve the efficiency of their mental processes. Assuming that such improvements can be passed on to later generations was crucial for Spencer's thinking and explains why he saw progress as a far more pervasive consequence of the interactions. In this respect he was more a social Lamarckian than a social Darwinist, although his focus on struggle may have misled many of his readers. This was not, however, an anticipation of the kind of Lamarckism later promoted by Samuel Butler and those who invoked a creative life-force that transcended the material world.[39]

Spencer had worked out his philosophy in the 1850s. From the start he was committed to the Lamarckian mechanism, as in his 1852 article on 'the development hypothesis'. His overall vision of cosmic evolution was outlined in 'Progress: Its Law and Cause' in 1857 and extended in *First Principles* in 1862. His *Principles of Psychology* had already explained the structure of the mind in Lamarckian terms. In 1864 his *Principles of Biology* extended his thinking on evolution by incorporating Darwin's theory alongside his existing support for Lamarckism. The book had little

[39] On their relationship, see Lightman, 'The "Greatest Living Philosopher" and the Useful Biologist'.

on the history of life on earth and no evolutionary trees, but the non-linear aspect of his position was made clear in a diagram adapted from Huxley's views on classification. The main animal phyla were shown radiating out from the simplest organisms at the centre, with the vertebrates having moved further away from the starting point than the other phyla.[40] A chapter on Darwin's hypothesis of natural selection introduced the term 'survival of the fittest' that would soon come to define it in the public mind. Spencer accepted that selection could explain some phenomena that Lamarckism could not, and that during the earliest phases of evolution it must have been the dominant mechanism. But once animals became more complex and better coordinated, the Lamarckian process would become steadily more significant.[41]

Spencer's books and collected essays were published in the United States by Appleton of New York, and in 1864 the firm compiled a separate collection of his essays under the title *Illustrations of Universal Progress*. An editorial preface noted that much of the attention had been aroused by Spencer's volume on *Education*, which criticized academic book-learning and called for intellectual and moral training that would promote successful interaction with the real world.

The *Illustrations* volume opened with an 'American Notice of the New System of Philosophy by Herbert Spencer', warning that many would ask how the author 'stands related to the problem of Religion'. The publishers realized that to gain wide acceptance they would have to defuse any suggestion that the system was hostile to Christianity. They admitted that Spencer called the Absolute Cause of things the 'Unknowable', acknowledging the limitations of our senses. But they insisted that this was not a negative expression of ignorance, because the laws of thought impose a *consciousness* of this 'Incomprehensible, Omnipotent Power'. Spencer was thus acknowledging that there is a 'Supreme but inscrutable Cause of which the universe is but a manifestation and which has an ever-present disclosure in human consciousness'.[42] Along with the emphasis on self-improvement, this assumed link with religion meant that Spencer's free-enterprise ideology of progress could be seen as little different from the kind of 'muscular Christianity' popularized by Charles Kingsley (discussed below).[43]

[40] Spencer, *Principles of Biology*, 1: 303. [41] Ibid., chaps. 11 and 12.

[42] Spencer, *Illustrations of Universal Progress*; the editorial preface is pp. xxiii–xxiv and the publisher's announcement pp. v–xxii, citations from pp. vi–vii and xvii–xx.

[43] See James Moore, 'Herbert Spencer's Henchmen', and more generally the same author's *The Post-Darwinian Controversies*.

In the 1870s Spencer's evolutionary philosophy would become hugely influential, especially in the United States. His ideology of unrestrained free-enterprise later became known by the pejorative term 'social Darwinism', giving the impression that it promoted a ruthless struggle in which the losers went to the wall. Historians now recognize the one-sidedness of this interpretation: Spencer realized that in a social environment even self-reliant individuals must work with others and learn the need for cooperative and sometimes altruistic behaviour (a point that Darwin himself accepted).[44] The harsher image of social Darwinism emerged as the result of a two-way interaction between scientific theories and contemporary values. Darwin's concept of natural selection inspired a number of rival ideologies promoting the need for competition, but at the level of individualism Spencer's self-help Lamarckism was more influential than the image of the 'struggle for existence'. Its influence was forgotten by later commentators because anything involving struggle was assumed to be derived from Darwin's theory. The resurgence of Darwinism in the twentieth century led a generation of historians to assume that the Spencerian ideology was a mere extension of Darwin's theory of natural selection. In fact, they were complementary, but by no means identical.

Defending Design

For those with religious beliefs liberal enough to allow creation to be regarded as a process rather than an event, evolution was acceptable only if it could be seen as the expression of a divine purpose. The naturalistic approach taken by Darwin and Spencer eliminated any hope of seeing such a purpose in the results because the interactions that drive the process are the result of natural laws operating in a mechanistic fashion. To retain some element of design in nature, one would have to show that the results made the Creator's intentions obvious. In effect, the laws governing evolution would have the Creator's intentions built into them in a way that achieved ends impossible for laws operating in the normal way. To complicate matters, there were two very different views on what the imposed design was intended to produce. Chambers' *Vestiges* had paved the way for one approach by arguing that the laws of development forced living things to progress in a predetermined direction, ending with

[44] On social Darwinism, see Chapter 8 below. For reinterpretations of Spencer, see Robert J. Richards, *Darwin and the Emergence of Evolutionary Theories of Mind and Behavior*, chap. 6; Hale, *Political Descent*, chap. 2; Thomas Dixon, *The Invention of Altruism*, esp. chap. 8; and Mark Francis, *Herbert Spencer and the Invention of Modern Life*.

the production of humanity. There was a structured pattern built into nature that proclaimed its ultimate purpose. But Chambers ignored the other focus of the argument from design: Paley's claim that the adaptation of each species to its environment confirmed the Creator's benevolence. Darwin had deliberately subverted this argument by making adaptation a natural process, but Asa Gray had counted that it would be better to think of variation being slanted to provide each species with the new characteristics it would need.

Some of the many objections raised against natural selection were intended to bring out its incompatibility with any form of design in nature. Gray's position was perhaps the weakest because his focus on adaptation made it hard to see how natural laws could foresee all the environmental challenges that would be faced by the vast number of species. More fundamental critiques argued that the history of life could not be reduced to an endless sequence of local adaptations – something far more structured must be involved. This was Agassiz' view, adapted to an evolutionary format by Chambers. The Duke of Argyll's *Reign of Law*, reprinted five times in 1868, proclaimed that God's purpose in the development of life is expressed in trends designed to achieve meaningful goals, including the creation of beautiful forms such as the plumage of birds. For these critics, natural selection – indeed, any theory in which new forms were produced solely in response to environmental pressure – was, in the words of Sir John Herschel, 'the law of higgledy-piggledy'. Darwin simply made this point more visible by basing his theory on undirected variation. Later idealists were less committed to theistic evolutionism, but the appeal of a natural philosophy based on order and predictability remained a potent source of opposition to the Darwinian perspective.[45]

The most comprehensive survey of the anti-Darwinian position was St George Mivart's *Genesis of Species* of 1871. Although not published as a mass-market book, it became widely recognized as a source of objections to be raised against Darwinism, some of which are still in use by opponents today. It was soon issued in America by Appleton (who were also responsible for the works of Huxley, Lubbock and Darwin himself). The book was criticized in the *North American Review* by Chauncey Wright – so effectively that Darwin commissioned a pamphlet version to be

[45] On this early phase of theistic anti-Darwinism, see Bowler, *The Eclipse of Darwinism*, chap. 3, and *The Non-Darwinian Revolution*, pp. 90–104. Herschel's comment was made in private and subsequently reported to Darwin, but he made his preference for predesigned laws of development clear in his *Physical Geography* of 1861.

published in Britain.[46] Mivart postulated innate powers within living things that controlled variation to drive evolution in predictable directions whatever the environmental constraints – hence his willingness to envisage multiple lines of development ascending in parallel through the fossil record. Like Owen, Argyll and many others at this early point in the debate, he was willing to suggest that these innate forces represented the will of God imprinted on nature. Owen's 'derivative creation' (as opposed to direct supernatural intervention or miracles) was a form of theistic evolutionism in which the changes predesigned by the Creator are imprinted onto nature – just as Chambers had suggested. Mivart pointed out that even A. R. Wallace, the co-discoverer of natural selection, conceded that supernatural agency had been needed to create the human mind.[47]

Mivart's world-view envisioned patterns and goals built into the laws of development that were incompatible with the Darwinian claim that the similarities between related forms are derived from the common ancestor of their group. Some patterns were supposed to operate entirely independently of a form's affinities. Mivart pointed to cases where similar structures had emerged in entirely unrelated animal types, as in the eyes of vertebrates and in cephalopods such as the squid. (The Darwinians countered by arguing that there are very few ways in which a functional visual apparatus can be constructed.) The critics also assigned significance to connections that the Darwinians saw as superficial and hence no clue to affinity. Mivart and Harry Govier Seeley linked the pterosaurs (flying reptiles) and the birds, although their wings are constructed on very different anatomical foundations. The most persistent challenge to the Darwinian perspective came from theories in which multiple lines within one class were thought to advance independently to the status of the next highest, as when Mivart subsequently argued that the mammals are polyphletic, with monotremes, marsupials and placentals having different reptilian ancestors. This model was very similar to the parallel lines of development suggested in the later editions of *Vestiges*.[48]

[46] Wright's response to Mivart is reprinted in Hull, *Darwin and His Critics*, pp. 384–408, and was reprinted as his *Darwinism*. Wright's reactions are discussed in Martin Fichman, 'Ideological Factors in the Dissemination of Darwinism in England'.

[47] Mivart, *Genesis of Species*, chap. 12; on Wallace, see pp. 301–3.

[48] Seeley gave an 1875 Royal Institution lecture on the pterosaurs, reported in the *Illustrated London News*; see Dawson, *Show Me the Bone*, pp. 325–7. On Mivart's views and the similar idea adopted by Richard Owen, see Desmond, *Archetypes and Ancestors*, chap. 6, and Bowler, *Life's Splendid Drama*, e.g., pp. 285–93, on claims that the mammals are polyphyletic.

In the United States, Agassiz was promoting the belief that the history of life represented the unfolding of a divine plan, also visible in the development of the human embryo. His younger followers were more willing to accept that such a plan might unfold gradually and began to endorse the possibility of evolution guided by predetermined trends. In 1867 the palaeontologist Edward Drinker Cope – who would later make his name describing fossils from the US West – presented a paper to the Philadelphia Academy of Natural Sciences arguing for non-adaptive patterns of development 'conceived by the Creator according to a plan of His own, according to His pleasure'. Like his friend Alpheus Hyatt, an Agassiz student, he saw a parallel between the patterns seen in the fossil record and the development of the embryo. They argued that evolution is driven by an acceleration of growth in which the pattern built into embryological development is gradually extended in the course of successive generations. As in Agassiz's world-view, this allowed the evolutionary history of the species to be recorded in the development of the modern embryo.[49]

In the 1870s Cope and Hyatt would turn their attention to the focus on adaptation that had led Asa Gray to suggest beneficial lines of variation. They were now willing to accept that the parallel trends they saw in the evolution of many groups were initiated by the founders of the group adapting themselves to a particular lifestyle. This choice of lifestyle then imposed a constraint that drove all subsequent generations to specialize further, sometimes to an extent that made them vulnerable to extinction. They would not accept natural selection as the process responsible for the initial adaptation, turning instead to the Lamarckian process of the inheritance of acquired characteristics. Unlike Spencer, though, Cope rejected the mechanistic interpretation of how an organism could adjust itself to the environment, suggesting instead that the process was driven by a life-force derived from the Creator. He became a founder member of what became known as the American school of neo-Lamarckism (discussed in Chapter 6 below).

Cope's move from theistic evolutionism to Lamarckism was anticipated by others with liberal religious views who preferred a more purposeful alternative to natural selection. Bernard Lightman has shown that in Britain the tradition of religious believers writing about nature continued, with some being willing to adopt an evolutionary interpretation of God's creative power.[50] Lamarckism offered them an

[49] Cope, 'On the Origin of Genera', quotation from p. 269. The paper was reprinted in his *On the Origin of the Fittest* of 1887, and his views on the divine source of the creative activity were expressed in his *Theology of Evolution* in the same year.

[50] See Lightman, *Victorian Popularizers of Science*, chap. 2.

increasingly attractive way of seeing how that power could be delegated to living things by giving them the ability to respond positively to environmental challenge – a process so multi-faceted that it was impossible to imagine it being imposed by a predetermined pattern of development. Long vilified by religious thinkers, the Lamarckian process was now welcomed as a less materialistic alternative to natural selection.

Liberal theologians on both sides of the Atlantic began to recognize this point, an early example being that of Charles Kingsley, who welcomed the *Origin* (Darwin had sent him a presentation copy) and was quoted in later editions as the 'celebrated author and divine' who conceded that evolution unfolding according to divinely established laws gave us 'just as noble a conception of the Deity' as special creation.[51] Kingsley initially accepted the open-endedness of Darwin's vision, but he soon began to indicate his preference for what was, in essence, the Lamarckian mechanism of change. This can be seen in his much-loved children's story *The Water Babies*, originally serialized in *Macmillan's Magazine* in 1862. At one point Tom, the water baby, encounters Mother Carey – a figure from legend personifying the sea – who is 'making new beasts out of old all the year round'. But she actually does nothing herself: 'I sit here and make them make themselves.' Tom has already been told the story of the Doasyoulikes, who dislike hard work and eventually degenerate over many generations to end up as apes, the largest of which is a gorilla shot by the explorer Paul Du Chaillu. The moral of the story is made clear in the conclusion: if you work hard and keep yourself clean, you will become a better person and contribute to the improvement of your race; if you don't, you and your offspring will gradually degenerate.[52]

This was what came to be known as 'muscular Christianity' – the emphasis was on the individual's moral responsibility for themselves and for their race. Kingsley had no time for the inferior races who had not kept up the struggle to improve themselves, and the suggestion that the Doasyoulikes end up as gorillas shows that he had not escaped a simple, linear model of human origins. Lip-service is paid to Darwinism when he argues that humans can be changed into beasts (or vice versa) by 'circumstance, and selection, and competition', but the focus on self-improvement to ensure racial progress is pure Lamarckism. He would later express his support for Mivart in the reprinted version of a lecture

[51] Darwin, *Origin of Species*, 6th ed., p. 422.

[52] Kingsley, *The Water Babies*; on Mother Carey pp. 270–3, on the Doasyoulikes pp. 229–39 and for the conclusion pp. 328–30. See Lightman, *Victorian Popularizers of Science*, pp. 71–81, and Piers Hale, 'Monkeys into Men and Men into Monkeys'.

entitled 'The Natural Theology of the Future'.[53] Darwin himself allowed a limited role for Lamarckism, and Kingsley's writings may have helped to deflect the opposition of conservative theologians. But with supporters like this it is hardly surprising that readers with religious leanings would come to associate the theory with the implication that progress is achieved through effort and initiative, not the selection of random variations.

The positions that emerged in the debates of the 1860s would continue to play a role into the early twentieth century. There would be no consensus on the actual process of change even when the concept of progressive evolution became a dominant force in the ideologies of the period. Whatever the scientific and social factors helping to convert the world to evolutionism, wholesale acceptance of natural selection did not play a primary role at this stage – even though catchphrases such as the 'struggle for existence' and the 'survival of the fittest' were widely deployed.

The issues raised by Spencer and the opponents of any naturalistic evolutionism would unfold over the following decades in the form of powerful alternatives to the Darwinian selection theory. Evolutionism would become a central theme in Western culture, but the divisions over how the process actually worked were not resolved. Mivart's 'groves' of parallel developments were replaced by the tree-like model of divergent evolution, although often with a central trunk defining a main line of development toward humankind. The concept of parallelism was mostly applied to the side branches of the tree of life. The 'evolutionary epics' of the late nineteenth century would be told in language implying that the history of life on earth displayed both progress and a degree of inevitability. At the same time, evolutionism would routinely be applied to social issues, especially by the followers of Spencer's system. Social evolution depended on the assumption that humanity had originated in a primitive state, presumably from an ape ancestry, and thus had to take on board a series of assumptions about this last and in some respects most important stage in the overall ascent.

[53] Reprinted in Kingsley, *Scientific Lectures and Essays*, pp. 313–36, see the footnote on p. 313.

4 Human Ancestry

Darwin may have hoped to avoid the controversial topic of human origins when the *Origin of Species* was published, but *Vestiges* had already ensured that it could not be ignored. The popular press almost immediately associated his theory with the belief that we have originated from a great ape ancestor. The publicity given to Paul Du Chaillu's discovery of the gorilla highlighted the incongruity of such an apparently ferocious beast giving rise to our superior mental and moral powers. Huxley demonstrated our close relationship to the apes, which was widely (but incorrectly) assumed to imply that a living ape was our direct ancestor. Darwin himself did not enter the arena until his *Descent of Man* appeared in 1871, although he had to respond to a whole series of developments that had taken place during the preceding decade. He was anxious to undermine the assumption that a living ape such as the gorilla could be a surviving relic of our ancestral form. He also responded to the widespread use of a hierarchical model that depicted non-European cultures and races as surviving relics of the ascent from the apes. In the absence of the 'missing link', this model had been constructed by archaeologists and anthropologists who had vastly extended the antiquity of the human species and then looked to living peoples as examples of earlier stages in the advance.

The idea of a simple ape-to-human transition, coupled with the broader image of a progressive cultural development, tended to encourage the impression that evolution is a predetermined ascent toward an inevitable goal. Both deflected attention from Darwin's image of a branching tree rather than a ladder, with the branches being driven apart by different adaptive responses to environmental challenge. Indeed, Darwin himself found it difficult to throw off some aspects of the old way of thinking, but the *Descent of Man* does include what we would now call an adaptive scenario to explain why the human line acquired faculties that the ancestors of the modern apes did not. This suggestion was ignored by the ideologues who saw evolution as a vision of inevitable progress toward the white races and modern Western civilization.

This chapter charts the emergence of the widespread fascination with the living apes as potential ancestors and the exploitation of this model in literature and images designed to show the non-white races as equivalent to the intermediate links missing from the fossil record. For physical anthropologists this was a straightforward issue of morphological transformation: the expansion in the size of the cranium and the gradual loss of ape-like features such as heavy brow-ridges. Thanks to the legacy of phrenology, cerebral capacity was seen as a surrogate for a race's level of intelligence. The appeal to living examples was all the more important because at the time it was impossible to demonstrate the advance from the fossil record. The human remains found at Neanderthal in Germany remained controversial, and it was not until the end of the century that the popular image of the Neanderthal race as shambling ape-men emerged.

To fill in the details not supplied by fossil evidence, the new discipline of prehistoric archaeology used stone tools to confirm the high antiquity of the human race. It also allowed the construction of a hierarchical sequence of cultural stages leading from the most primitive form of stone tools to the ages of bronze and iron that formed the starting point of the historical record. Cultural anthropologists looked to the various levels of civilization enjoyed by living races and mapped them on to the prehistoric sequence, with the most primitive 'savages' at the bottom. Psychologists would eventually use the increasingly popular belief that the development of the modern individual recapitulates the evolutionary history of its race to equate the mind of the savage with that of the European child. By stressing the supposedly unpleasant aspects of savage life, the model of steady mental and cultural progress could be made to appear consistent with the popular image of an ascent from a ferocious ape ancestry.

Vestiges of Humanity

Darwin's initial attempt to downplay the issue of human origins was doomed to failure because the public was already alerted to this implication of evolutionism. Chambers' *Vestiges* had not, however, stressed the claim that the apes represent a clue to our immediate ancestry. The issue emerged only peripherally in the book's account of the origin of the human races, which endorsed the claim that the black races are inferior to the whites. Given the major role the issue of an ape origin would play in the Darwinian debates, its relative absence in this earlier phase is striking – it had already been identified as a sensitive issue. Lyell's discussion of Lamarck's theory had attacked the suggestion of the 'conversion of the orang-outang into the human species'. Charles Southwell's

1841 article in the radical *Oracle of Reason* had been introduced by an image of 'Fossil Man' with remarkably ape-like features.[1] Chambers may well have decided to play down any link with so controversial a position.

On the issue of a racial hierarchy, Chambers could afford to be a little more open, since the suggestion that the 'lower' human races formed intermediates between the apes and Europeans had already emerged in discussions based on the chain of being. One did not have to be an evolutionist to see the great apes as lying immediately below humanity in the chain and hence to imagine that the 'lower' races were intermediates.[2] The first edition of *Vestiges* offered confused opinions on the status of the human species because Chambers used the entomologist W. S. MacLeay's system of classification in which species, genera and so on were arranged in circles, which did not fit in with the claim that development followed a linear sequence. Chambers suggested that our species concentrated qualities from all the Primate family and could be placed at its head as the 'type of all types'.[3] He then argued that most of the human races had originated in north-east Asia, although the 'Negro' might have had a separate origin in Africa. Both areas were where great apes are also found. The races were ranked in a hierarchy defined by the development of the brain in an individual Caucasian (European), with the 'Negro' brain being equivalent to that of a child just before birth. The Caucasian race was supposed to have advanced beyond the status of the original human type, while the others had degenerated to various degrees. The differences were presented as stages in an intrinsic pattern of development.[4]

In later editions Chambers accepted that the 'Negro' had a separate origin and again noted the geographical proximity to the apes. The racial hierarchy was maintained, but he now accepted the possibility of parallel evolution, several lines of development having advanced up the same scale but at different rates, so 'separate developments may have attained various points in the scale of human organization'.[5] In the last editions he returned to the idea of a single origin for all the races but reinforced the implication that the pattern defining the human line of development was

[1] Lyell, *Principles of Geology*, 2: 14–17; see Lamarck, *Zoological Philosophy*, pp. 169–73; the image from Southwell and Chiltern's 'Theory of Regular Gradation' is reproduced in Secord, *Victorian Sensation*, p. 312.

[2] See, for instance, Nancy Stepan, *The Idea of Race in Science*, especially chap. 1, 'Race and the Return of the Great Chain of Being'.

[3] *Vestiges*, 1st ed., p. 266 on the orangutan and pp. 271–3 on the position of humans in the Macleay system.

[4] Ibid., p. 296 on the apes and pp. 306–9 on the racial hierarchy.

[5] *Vestiges*, 5th ed., pp. 311–12 on the apes, quotation from p. 324.

preordained and had a predictable outcome. 'The signal superiority of the human species is thus prepared for and betokened in the immediately preceding portions of the line: it might have been seen, ere man existed, that a remarkable creature was coming upon the earth.'[6] *Vestiges* thus moved ever more clearly toward a position in which evolution is the unfolding of a coherent plan of development with a preordained goal, without stressing the potentially controversial link with the apes.

The issue of an ape-like origin for humanity seems not to have become a major focus for opponents in the 1840s. Sedgwick complained about ladies being told 'they are the children of apes and the mothers of monsters', but there were only a few isolated comments on the topic in his review – the real target was phrenology.[7] Sedgwick had seen a chimpanzee at the London Zoo, but as yet, the image of the ape as a monster to be feared rather than ridiculed had hardly entered the public consciousness. Secord's detailed study of the reactions records few references to the topic, although in 1846 P. T. Barnum displayed an orangutan at his American Museum in New York, proclaiming it to be 'the Grand Connecting Link between the two great families, the Human and the Brute'.[8]

Conservatives in the scientific community preferred to stress the gulf between apes and humans. Richard Owen had begun studies of primate anatomy in the 1830s, and in 1854 his lecture to the British Association on the topic – in which he dismissed the possibility of a link between apes and humanity – attracted wide attention.[9] It was from this date onwards that the level of popular interest expanded, thanks mainly to the discovery of the gorilla and descriptions of its allegedly ferocious nature. Specimens eventually arrived in Britain, and a much-publicized exhibition toured the country in 1855.[10] The issue of a potential link to the apes thus came to play a prominent role in the debate sparked by the *Origin of Species*, having been muted through the period in which *Vestiges* served as the primary example of evolutionary speculation. Had Du Chaillu's observations of the gorilla hit the headlines a decade earlier, the situation might have been very different.

Racial Origins

Chambers' reflections on the different races of humanity represent views that we now find offensive but which were common at the time.

[6] *Vestiges*, 11th ed., p. 223.
[7] Sedgwick, 'Vestiges of the Natural History of Creation', quotation from p. 3, other comments on pp. 1, 65 and 82.
[8] John Betts, 'P. T. Barnum and the Popularization of Natural History', pp. 353–4.
[9] Rupke, *Richard Owen*, pp. 260–6. [10] Desmond and Moore, *Darwin*, p. 452.

Evolutionism absorbed the hierarchical model and gave it a new explanatory basis, but the explanation was more often based on non-Darwinian versions of the theory. By modern standards Darwin and most of his followers would be counted as racist. Yet by the standards of the Victorian period their hatred of slavery put them in the liberal camp, even if they suspected that the black races were not equal to the white. The hard-core racists of the period were polygenists who saw the various races as distinct biological species separate for so long that any point of divergence was beyond the realm of speculation. For obvious reasons this position was popular among British imperialists and in the slave-owning states of the American South. The claim that black Africans were a separate form of humanity with a permanently lower level of mental development made it easier to argue that they should be kept in a state of servitude.

Such claims had become commonplace before the emergence of Darwinism – Chambers was echoing the view of Louis Agassiz, for instance, who recognized parallel sequences of creation leading toward the modern races and claimed there were similarities to the appearance of the various ape species. Agassiz endorsed the work of American physical anthropologists who were determined to show both that the races were quite distinct and that some retained a more ape-like character. The work of Samuel George Morton established what was seen to be a hierarchy of races based on their cerebral capacities, assuming that brain size was an indication of intelligence – a view that would have seemed natural to any supporter of phrenology.

In the 1850s Josiah C. Nott and George Gliddon emerged as America's leading polygenists. Their *Types of Mankind* appeared in 1854 with an introduction by Agassiz and went through seven editions. Their *Indigenous Races of the Earth* followed in 1857. *Types of Mankind* included depictions of black Africans with ape-like physical features and described research intended to show that the various non-white races had smaller cranial capacities than Europeans. The claim that the substantial mental and physical differences between the races required them to be treated as distinct species would soon be reconciled with evolutionism and the belief in an ape origin for humanity, but only with an element of parallelism incompatible with the Darwinian focus on divergence. For polygenists, the retention of ape-like characteristics by the 'lower' human species meant that they were independent lineages that had not been able to climb as far up the scale defining the advance from ape to full humanity. In the theory of parallel evolution, lower types visible in the world today are not surviving relics of earlier stages in the progress of the highest,

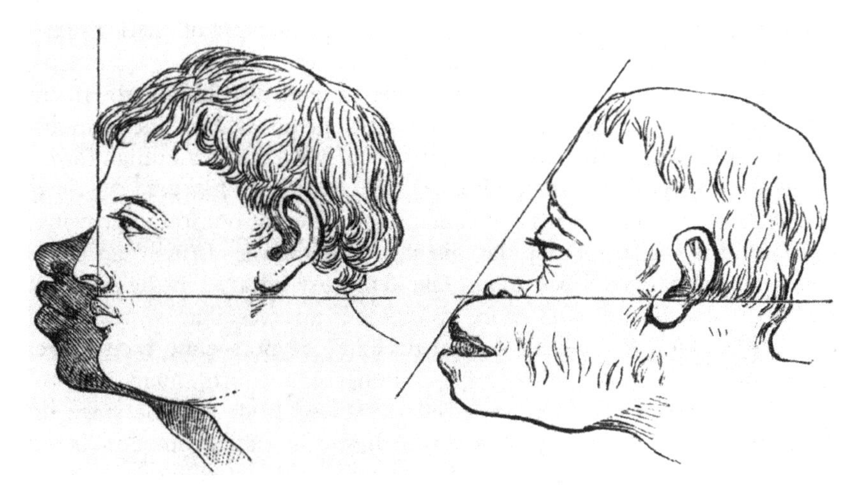

Fig. 4.1 Profiles of racial types with comparison to ape. From Robert Knox, *The Races of Men* (1850), p. 404.

although they do illustrate what those earlier stages would have looked like.[11]

Polygenism flourished in the slave-owning states and also gained support in Britain, where it was actively promoted in Robert Knox's *The Races of Men* of 1850 (Fig. 4.1). James Hunt became a leading activist and founder of the Anthropological Society of London in 1863 – a breakaway group from the Ethnological Society that had become a centre for the monogenist view that all the races belong to a single species. By the following year the Society had 271 fellows, and in 1865 it campaigned to defend Governor Edward Eyre of Jamaica, who had suppressed a revolt with much brutality. The affair split the country along the same lines as the slavery issue in the United States, although the British sense of racial superiority tended to be expressed as much in anti-Irish literature and caricatures. The level of support for outright polygenism is, however, less certain – only half of the 500 printed copies of Hunt's pamphlet 'The Negro's Place in Nature' were actually sold. The most extreme version of the theory of parallelism emerged in a

[11] There is an enormous literature on the history of race in science; see, for instance, Stanton, *The Leopard's Spots*, pp. 155–96 on America, and Stepan, *The Idea of Race in Science* on Britain. See also Livingstone, *Adam's Ancestors*, esp. pp. 176–9 on Nott and Gliddon, and Desmond and Moore, *Darwin's Sacred Cause*, chap. 9 on Agassiz. On the physical anthropologists' efforts to create a hierarchy based on cerebral capacity, see Stephen Jay Gould, *The Mismeasure of Man*.

translation published by the Anthropological Society of Carl Vogt's *Lectures on Man* in 1864. Here the image of separate lines of development ascending a scale upward from the ape was extended by linking the main racial types of modern humans to different ape ancestors. The orangutan, the chimpanzee and the gorilla had each evolved into a human form, some more successfully than others.[12]

The racial hierarchy with ape characters at the bottom found growing support in a variety of popular displays and images. British cartoons depicting the native Irish with ape-like features appeared in the satirical magazine *Punch*, although there were other sources that avoided the temptation.[13] Over the next several decades the non-white races were all too often portrayed with sloping foreheads and prominent jaws to convey the impression they were less advanced from the ape stage of development. The message was driven home in exhibitions and freak shows in which individuals from remote parts of the world and unfortunates with excessive body hair were displayed as 'missing links'. An exhibition of a 'bushman' family in 1847 led British newspapers to comment on their allegedly ape-like appearance. In 1860 Barnum's Museum in New York advertised a black individual exhibited as the 'man monkey', a 'connecting link between the wild native African and the orang utang' (Americans were less interested in the gorilla than the British).[14]

For the general reading public, fed on a diet of these cartoons and displays, it may have seemed obvious that human evolution had followed a predetermined path toward its final goal, with some populations lagging behind others in the drive toward higher levels of mental and moral development. The scientists too tended to assume that there must have been an inherent tendency for the brain (and hence intelligence and the moral sense) to expand toward the fully human level. Later efforts to

[12] On Knox and Hunt, see Efram Sera-Shriar, *The Making of British Anthropology, 1813–1871*. On Hunt's pamphlet, see Douglas A. Lorimer, 'Science and the Secularization of Victorian Images of Race', p. 218. The figure for the membership of the Anthropological Society is derived from the material printed in the conclusion of their translation of Paul Broca's *On the Phenomena of Hybridity*, pp. 91–100 (Broca insisted that the races could not, in fact, interbreed successfully).

[13] On Nott and the polygenists, see, for instance, David Livingstone, *Adam's Ancestors*, pp. 176–9; on the later appearance of anti-Irish cartoons, see Lewis P. Curtis, Jr., *Apes and Angels*, p. 31, but for a more cautious interpretation, see Roy Foster, *Paddy and Mr Punch*, chap. 9. William Stanton's account of early race science, *The Leopard's Spots*, says little about ape links.

[14] For the 1847 display, see Secord, *Victorian Sensation*, pp. 357–8; more generally, see Rebecca Bishop, 'Evolution im Zoo' (the Barnum advertisement is on p. 177), and Browne, *Charles Darwin: The Power of Place*, pp. 339–50.

explain human evolution took it for granted that the force driving our separation from the apes was the expansion of the brain.

The Antiquity of Man

Other sources of information also became available for those who wanted to visualize what the most primitive type of humanity had been like. New factors unrelated to Darwin's initiative entered the public consciousness and focused attention onto our prehistoric origins. Archaeologists were uncovering relics of the stone-age past, fuelling the assumption that the first humans had been – to use a term then popular – primitive 'savages'. Anthropologists assumed that those populations in the modern world still using low levels of technology were little more than surviving relics of early stages in cultural evolution. The question of whether they should also be seen as relics of biological evolution was all too easily answered by those who erected the hierarchy of races.

Whatever the ancestral form of humanity, the process of evolution must have been spread over a considerable period of time so that something resembling modern humanity should have appeared long before the timescale traditionally ascribed to human history. In the early decades of the century, the age of the earth had been extended by the geologists, but the appearance of humanity was still thought to have occurred only a few thousand years ago, more or less in line with the Genesis story. By another coincidence that had a major impact on the reception of Darwinism, this assumption was challenged just as the *Origin* was published. Geologists and archaeologists now began to extend the antiquity of humans back into the last phases of the geological record. There were no fossils apart from the highly controversial Neanderthal remains (discussed below), but evidence of stone tools found alongside the remains of extinct animals began to be taken seriously, showing that a creature capable of making the tools, and hence presumably human, must have already been in existence at the time. By extending the timescale and simultaneously showing that the earliest humans had developed only a primitive level of technology, these discoveries helped to eliminate at least one argument used against theories of gradual transmutation.

Stone tools had been reported from deposits containing the remains of extinct animals earlier in the century but had been dismissed as fakes or later intrusions. In 1858–9 British geologists visited the French sites where these discoveries were being made and became convinced that some, at least, were genuine. This meant that tool-making people had already been in existence tens of thousands of years ago, greatly extending the timespan of human history and introducing the concept of pre-history – a long period before events were recorded in monuments or writing. Reports by eminent figures including William Pengelly and Hugh Falconer were delivered to learned societies and widely reported in the press – Falconer wrote letters to the *Times* giving details of his changing opinions. The whole episode was summed up by Charles Lyell in his *Geological Evidences of the Antiquity of Man* in 1863. This was reprinted several times in that year and was reviewed by many periodicals. Darwin's niece Julia Wedgwood reviewed it for *Macmillan's Magazine*, although Darwin himself was disappointed by Lyell's reluctance to accept that humans could have appeared by a process of gradual evolution. Lyell still believed in a sudden jump by which humanity's mental and moral powers lifted us above the animal kingdom.[15]

Excavations continued apace, and it soon became apparent that there were several stages in the development of stone technologies over a considerable period of time. In 1865 Sir John Lubbock's *Pre-historic Times* introduced the distinction between the Palaeolithic and Neolithic, the old and the new stones ages. Soon French experts were distinguishing several developmental stages within the Palaeolithic, providing evidence of gradual technological progress over time. But what kind of cultural lives had these distant ancestors enjoyed? Opponents of evolutionism such as the Duke of Argyll insisted that a primitive level of technology was not necessarily a sign of a low level of cultural or moral development. To most archaeologists and the general public, however, it seemed obvious that these early humans had lived in a state of savagery, presumably equivalent to that of the most primitive races still existing today. The sub-title of Lubbock's book said it all: *Pre-historic Times: As Illustrated by Ancient Remains and the Manners and Customs of Modern Savages*. It was as though some races, perhaps held back by unfavourable conditions, had failed to progress toward higher levels of cultural development and had thus preserved earlier stages in the process through to the present.

This assumption was shared by the anthropologist Edward Tylor, who proposed a theory of cultural evolution in which all races ascended the same scale of cultural progress, but at different rates. His *Researches into the Early History of Mankind* of 1865 was in fact a survey of the different stages of development shown by living races. At this point Tylor did not assume that a lower culture equated with a lower level of intelligence – a

[15] These developments are explored in Donald K. Grayson, *The Establishment of Human Antiquity*, especially chaps. 8 and 9, and A. Bowdoin Van Riper, *Men among the Mammoths*. See Julia Wedgwood, 'Sir Charles Lyell on the Antiquity of Man'.

'psychic unity' ensured that they would progress in the same way, although not at the same rate. Eventually, though, even he would succumb to the prevailing assumption that 'savages' were held back by their inferior mental powers.[16]

Europeans both at home and in North America took it for granted that their own society represented the high point of cultural development. The human species had dispersed around the globe and had diversified physically into the various races, but each population had then tended to progress toward higher levels of culture at a rate determined by the conditions to which it was exposed. Those trapped in really unfavourable environments have remained stuck at the level of stone-age savagery. Others have made it to various intermediate levels of civilization. The great empires of modern Asia were seen as equivalent to Europe in the age of Rome; they were not directly related to the Romans but had independently (but belatedly) reached an equivalent level of cultural development. Only Northern Europeans – exposed to the most stimulating environment – have achieved the highest level. This was a theory of evolution, but one based on very different principles to Darwinism. It was a classic expression of the idea of evolutionary parallelism driven by a tendency to advance in a predetermined direction.

To a public trying to come to grips with the implications of Darwinism, this independent suggestion of a linear model of cultural evolution made it harder to appreciate the logic of his theory's model of divergence. This, in turn, helped to promote a much simpler idea of how our species had evolved from the ancestral ape form. If the lower stages of human cultural evolution were still preserved in the world today, perhaps those involved were also stuck at an earlier stage in our biological evolution. The suggestion that the gorilla, or perhaps the orangutan – and not some hypothetical generalized ape lost in the fossil record – was the real ancestor primed the public to imagine that the 'savages' discovered in remote areas of Africa or Asia were also biological relics of past. The races of humanity that still lived with primitive levels of technology preserved not only past levels of culture but also earlier stages in the physical and mental progress upward from the ancestral ape.

Man's Place in Nature

It was hard for many to accept that humanity's superior mental powers did not require special divine intervention, and even harder to concede

[16] See, for instance, George W. Stocking, *Victorian Anthropology*, and my own *The Invention of Progress*. Tyler's later viewpoint is noted in Chapter 8 below.

that our moral sense could emerge from animal origins. The theory of natural selection simply added to the problem – how could our sense of responsibility for others have been created by the essentially selfish mechanism of the 'survival of the fittest'? Critics such as Bishop Wilberforce immediately seized on this point to attack the new theory, and the issue would remain one of concern for conservative religious believers. Even Alfred Russel Wallace came to believe that divine guidance would have been needed to produce some human faculties.[17]

The situation was not helped by the publicity given to the discovery of the gorilla and to its alleged ferocity. Scientists now had specimens for anatomical study, prompting the much-publicized debate between Owen and Huxley over the degree of similarity between the ape and human brains. Even more significant for the public were the reports of explorers such as Paul du Chaillu, who provided Owen with specimens and lectured to huge crowds at the Royal Geographical Society. His *Explorations and Adventures in Equatorial Africa* appeared in 1861, containing accounts of gorilla-hunting and images of the animal as a dangerous adversary. R. M. Balantyne's adventure story *The Gorilla Hunters* popularized Du Chaillu's discoveries, and there were hundreds of references to the gorilla in newspapers through 1861 and 1862 (Fig. 4.2). Some began to suspect that the gorilla's ferocity had been exaggerated, but the damage had been done in terms of the popular perception. A frisson of excitement was generated by hints of sexual liaisons between apes and human females. If religious thinkers found it hard to believe that our mental and moral powers had emerged from such a bestial foundation, the general public seemed to be fascinated by the possibility, and the claim that we are descended from an ape, often identified with the gorilla, was seen as an integral part of the evolutionists' thesis.[18]

If humans had evolved from apes, what had the intermediate stages looked like? The supposedly 'lower' races of today might be interpreted as equivalent to the later steps in the sequences, but only the most bigoted polygenists would take them to be a halfway house. The term 'missing link' came to symbolize the earlier gap in the assumed process by which the gorilla had been transformed into a modern human. The

[17] See Malcolm Kottler, 'Alfred Russel Wallace, the Origin of Man, and Spiritualism', and Frank Miller Turner, *Between Science and Religion*, of which chap. 4 is on Wallace.

[18] James G. Paradis notes over twenty significant items on the gorilla published in 1861; see his 'Science and Satire in Victorian Culture', pp. 157–8, and for the figures on newspaper references, see Jochen Petzold Regensburg, 'How like Us Is That Ugly Brute, the Ape'. See also Richard Noakes, '*Punch* and Comic Journalism in Mid-Victorian Britain', and Browne, *Darwin: The Power of Place*, pp. 156–62; on the sexual implications, see Gowan Dawson, *Darwin, Literature and Victorian Respectability*, chap. 2.

Fig. 4.2 Frontispiece to R. M. Ballantine, *The Gorilla Hunters* (new ed. 1897).

fossil record ought to reveal even more primitive ape-human intermediates than anything alive today. Imagining a complete series of intermediates would become part of the popular iconography of evolutionism, and many of Darwin's contemporaries thought that this kind of transformation was exactly what his theory implied. In fact, the claim that one modern species could morph into another was more compatible with the linear model of the chain of being than with the image of evolution as a branching tree. For those who appreciated the logic of the tree image, it would be necessary to work out what the common ancestor of the great apes and modern humans would have looked like and then trace the sequences from that to its various later descendants. Unfortunately, the fossil record at the time provided virtually no relevant evidence.

The issue was clouded by the much-publicized debate between Richard Owen and T. H. Huxley on the degree of difference in the structure of the ape and human brain. Huxley claimed victory and his verdict had generally been accepted by historians, although an interpretation more sympathetic to Owen is certainly possible.[19] Owen was by no means completely opposed to the idea of transmutation, but his conservative inclinations led him to insist that humanity's mental and moral faculties required a major difference in brain structure. He had defended this position through the 1850s and had reinforced it in papers published in 1857 and 1859. At an earlier session in the 1860 meeting of the British Association in which he subsequently clashed with Wilberforce, Huxley had challenged Owen. He insisted that despite the much greater size of the human brain, it contained no unique structures. He rejected Owen's claim that the apes lacked an organ in the human brain known as the hippocampus minor and backed up his position in an article published in the *Natural History Review*. From February to May 1861 he lectured to working men on the topic of 'The Relation of Man to the Rest of the Animal Kingdom' and confided to his wife that 'By next Friday evening they will all be convinced they are monkeys.'[20] There was another clash with Owen at the 1862 meeting of the British Association, and Huxley then went on to publish his *Evidence as to Man's Place in Nature* the following year.

The first part of Huxley's book surveyed the growth of information about the great apes. The second section was based on the 1860 lectures

[19] See, for instance, Rupke, *Richard Owen*, chaps. 6 and 7.

[20] Huxley to his wife, 22 March 1861, in *Life and Letters of Huxley*, 1: 190. See Desmond, *Huxley: The Devil's Disciple*, p. 292, and more generally Sherrie L. Lyons, 'Convincing Men They Are Monkeys', and Alison Bashford, *An Intimate History of Evolution*, pp. 238–50.

in which he had emphasized the close relationship between humans and the apes and included his detailed refutation of the claim that the hippocampus minor was unique to the human brain. Huxley also argued that the attempt to base our superior status on the possession of a physical structure in the brain was hardly likely to increase respect for our moral faculties. The third and final part evaluated the limited evidence of fossil human remains, focusing especially on the Neanderthal skull discovered in Germany in 1857. This had provoked much controversy, but Huxley accepted that it was genuinely ancient and potentially threw light on the nature of primitive humanity. Despite having some ape-like features, including prominent brow-ridges, he concluded that it could not be counted as a link between the apes and humanity because its cranial capacity was fully as large as a modern human's. To make sense as a 'missing link' a fossil skull would have to have a capacity intermediate between the ape and human levels.

Man's Place in Nature confirmed our anatomical relationship with the apes and argued that this was congruent with the aim of applying Darwin's theory of transmutation to the question of human origins. But it was surprisingly limited in its efforts to explore exactly how and why some hypothetical ape-like ancestor might have acquired an expanded mental capacity while other ape species had merely specialized for their life in the forests. The frontispiece, actually designed by Waterhouse Hawkins and by no means to Huxley's satisfaction, depicted a sequence of ape and human skeletons arranged so that they could easily be seen as a linear transition, anticipating cartoons still in circulation today (Fig. 4.3). The image seemed to endorse the popular view that a species such as the gorilla represented the ancestor that had evolved into the human form.[21]

Huxley's book was a sensation, selling a thousand copies in the first week and provoking a series of lampoons in the popular press. The debate with Owen was the source of much hilarity: *Punch* had a cartoon emphasizing the link to the gorilla accompanied by a piece of doggerel verse under the title 'Am I a Man and a Brother?', and an anonymous pamphlet ridiculed the affair as 'A Sad Case'. Darwin himself was caricatured as an ape, and in 1871 *Harper's Weekly* ran a cartoon in which a gorilla complains that he has stolen his pedigree. The Huxley-Owen debate was caricatured in Kingsley's *The Water Babies*, where a professor 'had even got up once at the British Association, and declared that apes had hippopotamus majors in their brains just as men have. Which was a

[21] On the frontispiece, see Gowan Dawson, 'A Monkey into a Man'.

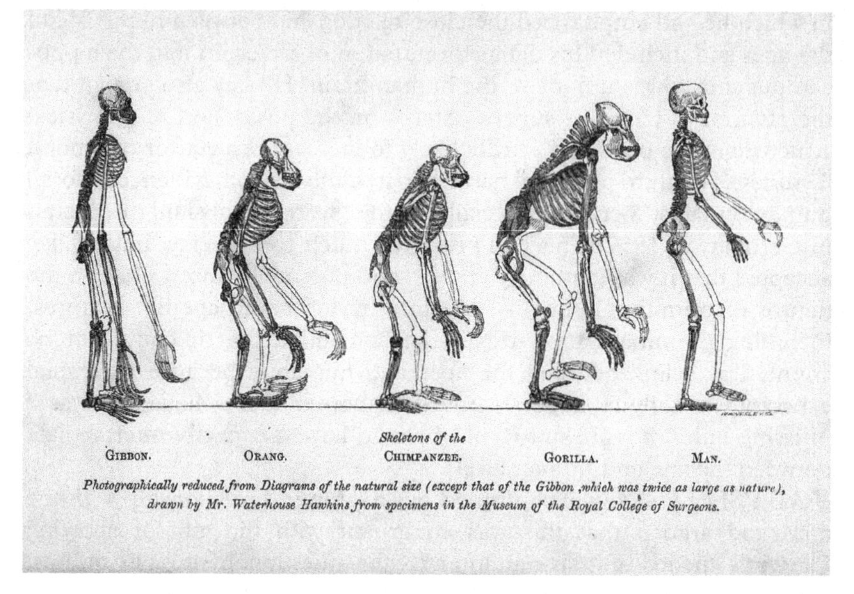

Fig. 4.3 Frontispiece to T. H. Huxley, *Man's Place in Nature* (1863).

shocking thing to say; for if it were so, what would become of the faith, hope, and charity of immortal millions?' The book became a children's favourite and was much reprinted, editions from 1886 onwards containing an image of Owen and Huxley peering at a water baby in a glass jar.[22]

The assumption that humans had evolved from a great ape ancestor interacted with the anthropologists' theory of a linear pattern of technological and cultural progress, generating poisonous implications for European attitudes toward other races. Everyone wondered what the creatures who made the earliest, most primitive, stone tools had actually looked like. Louis Figuier's *The World before the Deluge* (translated from the French in 1865) depicted the first humans in a pastoral landscape reminiscent of the Garden of Eden. In later editions this was replaced by one in which they had stone-age tools and were dressed in animal skins, but at least they appeared to be fully human.[23] The fossils unfortunately could not be used to clarify the situation. Thanks to the caution of Huxley and other experts, the Neanderthals were not yet seen as

[22] Quotation from pp. 152–3 of the edition of *The Water Babies* cited in the bibliography; the image of Owen and Huxley is on p. 69. For the other examples cited, see, for instance, Lyons, 'Convincing Men They Are Monkeys'.

[23] On Figuier's book, see Martin Rudwick, *Scenes from Deep Time*, pp. 206–10.

candidates for the 'missing link'. In 1873 an anonymous article in *Harper's Weekly* did attempt a visual reconstruction of 'Neanderthal Man', describing him as ferocious-looking and gorilla-like, but conceding that he was human.[24] It was only after better specimens were discovered in the 1880s that the Neanderthals began to acquire their eventually iconic (but misleading) status as shambling ape-men. In the meantime, there were other all-too-obvious candidates to be found in the allegedly 'lower' human races still alive today.

Darwin Enters the Fray

In 1871 all of these issues came to a head in Darwin's own belated contribution to the debate, *The Descent of Man and Selection in Relation to Sex*. This was not what the public might have expected – two thirds of the book was devoted to sexual selection, relevant because it was the basis for Darwin's explanation of how the human races had developed their distinctive physical characteristics. Only the early chapters addressed the more general issues that had emerged in the debate over human origins, and even here there were some omissions. Darwin provided plenty of evidence to minimize the gulf between humans and animals, and his *Expression of the Emotions in Man and the Animals* the following year added more. But the debate over the structure of the ape and human brains was ignored, although a passage written by Huxley was inserted into later editions. There was little about the gorilla and the other great apes, although Darwin did not conceal the fact that his theory assumed a common ancestor that had probably been a more generalized ape inhabiting a forested environment. He outlined the earlier stages in the ascent of vertebrate life toward the primates, following the explanatory technique pioneered in Germany by Ernst Haeckel. There was little on prehistoric archaeology and only a passing reference to the Neanderthal skull. Darwin did, however, include Lubbock's claim that modern savages could be seen as illustrating the habits of our distant ancestors and quoted figures from the physical anthropology of J. Barnard Davies claiming that non-white races had smaller brains than Europeans.

Darwin thus endorsed some aspects of the linear model of development, but he also exploited the logic of his own theory of divergent evolution. If getting bigger brains has been the driving force of the ascent,

[24] This article is noted in Marianne Sommer, 'The Neanderthals', pp. 147–9, and the image is reproduced in Bårbel Auffermann and Gerd-Christian Weniger, 'Von Wildern Månnern und Frauen', p. 144.

why have the great apes themselves not participated? Darwin believed that a more challenging environment stimulated the development of intelligence – in the *Origin* he had argued that the highest mammals had emerged in 'the more efficient workshops of the north'.[25] To explain why only one branch of the great ape stock had embarked on a process of brain expansion, he suggested that the population representing our ancestors had moved out of the trees and stood fully upright as an adaptation to an open environment. This had indirectly favoured intelligence, perhaps because our hands were free to become better adapted to tool-making. The apes remained apes because they had continued to specialize for living in the trees.[26] Darwin's insight that the adoption of bipedalism had preceded and probably stimulated the expansion of the brain was shared by Ernst Haeckel but largely ignored until the twentieth century in favour of the assumption that the development of the brain must be the driving force of the whole process.

Darwin believed that once the human family had begun to take up tool-making, its physical characteristics would become less bestial and its intelligence would gradually be enhanced. But in the eyes of his religious critics, the most important human faculty was our moral sense, assumed to be lacking in the animals. Darwin showed that animals did sometimes show affection for others, citing the example of a zoo monkey that defended its keeper from an attack by a baboon. He also accepted that individualistic natural selection would not enhance such primitive feelings into the moral sense of modern humanity. He argued instead that our social instincts would be enhanced when our ancestors began to live together in families and then in larger tribes. In the competition for resources, the groups that cooperated more effectively would be more successful than those that were less cohesive. Selection operating between groups, not individuals, would become the driving force promoting the instinct for self-sacrifice and care for others. This would initially apply only to other members of one's own group, but as intelligence expanded these feelings would be applied more widely.[27]

Darwin's view on the formation of the human races also used the model of branching evolution from a common ancestor, although he did not believe that the physical characters defining the races could be explained in terms of adaptations to their differing environments. His

[25] Darwin, *Origin of Species*, pp. 379–80.

[26] Darwin, *Descent of Man*, 1: 142–4, 2nd ed., pp. 52–3; the Barnard Davies figures are at 1: 146, and 2nd ed., p. 54. On the lack of interest Darwin's views on bipedalism at the time, see my *Theories of Human Evolution*.

[27] The development of the moral sense is outlined in the *Descent of Man*, chap. 3, in the 2nd ed., chap. 4. For the monkey and the zoo keeper, see 1: 78, and in the 2nd ed., p. 103.

theory of sexual selection depended on the important point that the crucial factor is not the *survival* of the fittest but their enhanced reproductive potential. Any characteristic that improves an individual's ability to obtain a mate will be improved even though it does not aid survival and in some cases may even be a hindrance. The peacock's ungainly tail has become so large because the females accept it as an indication of a male's fitness, while the antlers of the male stag are used primarily to fight off rivals during the rutting season. Darwin argued that the physical features of the human races might have evolved because geographically separated cultures acquired different views on which characteristics were sexually attractive.

Female choice was sometimes a factor in sexual selection, reflecting the Victorian vision of women as coy and choosy. Darwin was more interested in the parallel view of males competing for access to the most favoured women, which could drive the development of useful characteristics as well as those that were seen as sexually attractive. The theory certainly reflects the sexual stereotypes of the age, although Darwin can hardly be accused of creating them. There were cartoons exploiting his idea, but sexual selection gained very few adherents in his own time and had to wait until the twentieth century before scientists began to take it seriously.[28]

Some 2,500 copies of the first edition of the *Descent of Man* were sold, and the book was widely reviewed on both sides of the Atlantic. A second edition in one volume and at half-price was published in 1874. American publishers issued the book in unauthorized editions which made it more widely available. Curiously, Darwin seems to have received little abuse from the reviewers, apart from those publishing in conservative religious periodicals who refused to accept any position that would rob humanity of its spiritual status. As far as those with more liberal views were concerned, the battle had already been fought when the *Origin of Species* had been linked to the issue of human origins, and our animal ancestry was now taken more or less for granted. Some reviewers were already endorsing the view that a theory of progressive evolution leading up to humankind was, in the words of the Liverpool Leader, 'perfectly consistent with the belief in the Creator'.[29]

[28] For a recent study of sexual selection's relationship to contemporary sensibilities, see Gowan Dawson, *Darwin, Literature and Victorian Respectability*, chap. 2; on the theory's later history, see, for instance, Helena Cronin, *The Ant and the Peacock*.

[29] On the reception of the *Descent of Man*, see Bowne, *Darwin: The Power of Place*, pp. 350–6, quotation from the *Liverpool Leader* on p. 351; also Desmond and Moore, *Darwin's Sacred Cause*, pp. 364–75.

Toward 'Social Darwinism'

As the controversy over the *Descent of Man* subsided, evolutionism began to gain increasing influence on the public imagination – although its manifestations were complex, sometimes contradictory, and by no means always Darwinian. It was Herbert Spencer's philosophy that came to be associated most closely with Darwin's views on social evolution, even though his thinking was based more on the inheritance of acquired characteristics than on natural selection. From around 1870 Spencer's philosophy increased in popularity, and in the United States especially it appealed to the industrialists who thrived on individualism and laissez-faire economics. The suggestion that individual competition was the main driving force eventually led to his ideology being identified as a ruthless 'social Darwinism', even though Spencer himself believed that self-improvement was the main driving force. Historians now recognize that the businessmen who took Spencer as an advocate of unrestrained competition were missing the real point of his philosophy. To live in a social environment, the self-reliant individual has to learn how to get on with others, in effect acquiring the foundations of morality.[30]

Spencer had little interest in prehistory, and his views on the progress from savagery to civilization did not follow the cultural anthropologists' predetermined hierarchy. But he did see the move from militaristic societies to those based on individualism as the most significant consequence of the historical trend. In his popular treatise *Education: Intellectual, Moral and Physical* he also endorsed what would become an important feature of later ideas about mental and cultural evolution, the assumed parallel between the individual's psychological development and the cultural evolution of the race. The growing child recapitulates the stages that society has passed through in the course of history:

> Do not expect from a child any great amount of moral goodness. During the early years every civilized man passes through that phase of character exhibited by the barbarous races from which he is descended. As the child's features – flat nose, forward-opening nostrils, large lips, wide apart eyes, absent frontal sinus, etc. – resemble for a time those of the savage, so, too, do his instincts.[31]

[30] See, for instance, Thomas Dixon, *The Invention of Altruism*, esp. chap. 8, and Mark Francis, *Herbert Spencer and the Invention of Modern Life*.

[31] Spencer, *Education*, pp. 220–1. In general, though, Spencer adopted a much less recapitulationist form of the relationship between embryology and evolutionary progress – his essay 'Progress: Its Law and Cause' opens with a reference to the link, but in the context of von Baer's interpretation of the development of the embryo as a process of differentiation; see Spencer, *Essays Scientific, Political and Speculative*, 1: 2–3.

The assumed link between the child and the savage would become a major feature of late nineteenth-century psychological thinking, although more often linked to the linear model of development preferred by the anthropologists.

Darwin and Spencer agreed that society would be better off without its least productive members, although neither suggested that they should be actively prevented from breeding. The complexities arising when biological theories were translated into social terms were, however, illustrated by Walter Bagehot in articles published in the *Fortnightly Review* from 1867 and compiled in his *Physics and Politics* of 1872, an early contribution to the International Scientific Series. The book claimed to apply the principle of natural selection to social evolution, but where Spencer focused on individual competition, Bagehot argued that the most important form of struggle was between societies and nations. It was social coherence that ensured success, based on a centralized state that used religion to boost cultural unity. He conceded that in the modern world the state needs to encourage individual enterprise to allow a flexible response to external challenges, but his focus on group competition would resonate with the growing influence of imperialism in late nineteenth-century Europe. Even in America, Bagehot was profiled in *Popular Science Monthly*, which relentlessly publicized Spencer's ideas.[32] The stage was thus set for the age of social Darwinism discussed in Chapter 8 below.

[32] *Popular Science Monthly*, 12 (February 1878), pp. 489–90; on Bagehot's ideas, see Hale, *Political Descent*, pp. 125–7.

5 Evolutionary Epics

By the time Darwin died in 1882, he had become a cultural icon, often caricatured yet regarded as someone who had made a major contribution to science. He was buried with much publicity in Westminster Abbey. Five years later there was considerable press attention when his statue was unveiled in the Natural History Museum. In 1887 the *Illustrated London News* included retrospectives on his life and work prompted by the publication of his *Life and Letters*. By this time popular biographies had already begun to appear in which he was treated as a 'hero of science' – some were written for children, a sure sign that his work was no longer seen as threatening.[1] It was in a *Cornhill Magazine* article published in 1888 that the popular science writer Grant Allen claimed that evolutionism had at last become a dominant theme in Victorian culture.

Allen suggests that this was a relatively recent development, which would imply that the debates of the 1860s had not embedded the idea in the public mind.[2] Those debates had identified elements that could attract popular interest, but some of these – as Allen pointed out – did not reflect how the theory was understood by scientists. The experts still did not agree on how the process of evolution worked – all the alternatives to natural selection were still in play in the 1880s. But the very different model of the ascent of life based on independent parallel lines, as suggested by Chambers and Mivart, had been overtaken by a consensus based on the tree of life, although there were disagreements over the tree's shape.

[1] On Darwin as a public figure, see Browne, *Charles Darwin: Voyaging*, chap. 10. On his funeral, see Desmond and Moore, *Darwin*, chap. 44, and Browne, pp. 495–7. Desmond and Moore's image 90 shows the unveiling of his state as pictured in the *Graphic*. On popular biographies, see Lightman, 'The Many Lives of Charles Darwin'; the article in the *Illustrated London News* for 10 December 1887 is reproduced in Schwartz, ed., *Streitfall Evolution*, p. 19.

[2] Lightman, 'The Popularization of Evolution and Victorian Culture', pp. 286–7; See Grant Allen, 'Evolution'.

The period from the 1870s to the end of the century was the heyday of evolutionism's influence, although in the 1890s the situation changed in response to new ideas about heredity and new discoveries in the fossil record. Allen was one of the most prolific authors using evolutionary science for a wider purpose, in his case to promote Herbert Spencer's naturalistic philosophy, although he had a Darwinian sense of the harshness embedded in the process. He used familiar animals and plants to explore how evolution had shaped the world around us. His friend Edward Clodd, also a Spencerian, wrote popular books surveying the whole evolution of life on earth – these were perhaps better examples of a genre that has been called the 'evolutionary epic'. This tactic was pioneered in Chambers' *Vestiges* and was adapted to the post-Darwinian world in the German biologist Ernst Haeckel's *History of Creation*, translated in 1876. It too had an ideological agenda – Haeckel's naturalistic 'monism' was a Germanic equivalent of Spencer's synthetic philosophy. Yet the same technique could also be used by writers such as Arabella Buckley who promoted the view that the progress of life was the unfolding of a divine purpose.

The evolutionary epics highlighted the ascent of life and assumed a strong element of continuity as displayed in the development of the embryo (the recapitulation theory). Like Darwin and Spencer, their authors adopted Lyell's uniformitarian model of earth history, playing down anything that would resemble the old catastrophism. The transition from the Age of Reptiles to the Age of Mammals, for instance, was assumed to be gradual, any apparent discontinuity being due to the imperfection of the record. Despite growing evidence of a huge diversity of dinosaurs that had disappeared relatively abruptly, the possibility that evolution was an episodic process interrupted by mass extinctions and explosions of diversity was not recognized. The Darwinian emphasis on the continuity of change was so effective that to be an evolutionist meant to believe that all transitions are slow and gradual. This assumption dominated popular surveys of evolutionism until new fossil evidence became more widely publicized at the end of the century.

The authors of the evolutionary epics celebrated the diversity of modern life, but the extent of divergence earlier in life's history was played down. By picking a few living examples of each class as examples, they could give the impression of a series of upward moves, minimizing the significance of any side branches. Focusing on fossils leading to dead-ends would hardly imply a progressive trend running through the ascent of life. The transition from the apes to humanity was included as the last step in the process, but these surveys seldom presented the gorilla as our ancestor, and (with some exceptions) they did not invoke the hierarchy of racial types. Accounts of the nastiness of our ape or savage ancestors did not fit the model of an overall progress toward higher mental and moral faculties. There were thus two types of evolutionary epics. Some were broad visions of progress that could equally well be given a Spencerian or a theistic emphasis. Another type, explored in Chapter 8 below, confined itself to human origins and admitted both the bestiality of the apes and the limited extent to which 'primitive' races had advanced beyond this state.

One issue that remained open was the actual mechanism of change. Most popular authors still felt it necessary to explain how natural selection works, and some – Grant Allen is a good example – made their readers aware of the harshness implicit in the mechanism's activity. There was a general tendency to assume that the struggle for existence must allow the 'best' to come to the fore, and this muted the worst implications of Darwinism, especially as high-profile figures such as Haeckel and Spencer emphasized a Lamarckian as well as a selectionist component. Use inheritance was more congruent with the recapitulation model of development and the widespread tendency to assume that each step must involve a gradual progress toward higher things. This consensus soon began to unravel, however, as some Lamarckians became hostile to the selection theory. In the 1880s Samuel Butler's assault on Darwinism ensured that readers outside the scientific community would become aware that the synthesis of Darwinism and Lamarckism was now under threat.

Popular literature thus reflected in a simplified form the compromises and tensions within the evolutionists' camp. How important was natural selection, and what were the alternatives? Did evolution inevitably push toward some moral or spiritual goal, or was it an amoral process whose outcomes were diverse and sometimes nasty? Was there a main line of development leading toward humankind and some bright future, and if so, why were there so many different forms of life, each with its own specialist adaptations? Popular writers presented their interpretation of these issues without necessarily going into the details of new evidence or the technical debates of the scientists. In the 1890s, however, new factors would emerge to drive home the challenges faced by the progressionist vision, including the discovery of new fossils and a growing belief in the power of heredity.

Disseminating the Theory

Writers seeking to promote evolutionism had an expanding array of outlets for their work. They wrote articles for general magazines like the *Cornhill* and for popular science periodicals such as *Science Gossip*, a monthly priced at only four pence to promote natural history and microscopy. *Science Gossip* was aimed at amateurs, but it also published extracts from more technical works. Allen and Clodd represented a group in touch with elite scientists but seeking to popularize their research as a means of gaining a living. Tensions could emerge: the astronomy writer Richard Proctor founded his magazine *Knowledge* in 1881 because *Nature* had become too closely identified with the elite scientific community. Clodd became its assistant editor. Some popularizers were far more radical – Edward Aveling wrote for atheist magazines and lectured to secularist groups, his promotion of Darwinism causing embarrassment to Huxley and others anxious to preserve the movement's respectability. By the 1890s cheaper magazines such as *Science Siftings* catered for an even wider readership, publishing sensationalized material the scientific community dismissed as trivial.[3]

In the United States newspapers such as the *New York Times* now had weekly science columns, and general magazines such as *Harper's Monthly* and *Atlantic Monthly* included frequent pieces on developments in science. When both magazines declined to serialize Spencer's work on sociology, E. L. Youmans founded *Popular Science Monthly* in 1872 to take on the task; by the end of the year it had 12,000 subscribers. *Scientific American* had a wider remit that included natural history, but retained much of its original focus on applied science. There was considerable interaction between British and American editors, and the same article often appeared on both sides of the Atlantic.[4]

This was also the case for books: Haeckel's *History of Creation* was published in London by Henry S. King and in New York by Appleton. Youmans created the International Scientific Series with the same publishers to issue semi-popular texts by experts, including Spencer's *Study of Sociology*. By the 1890s the series was in difficulties, the books being still too technical for the average reader – although rivals such as the Contemporary Science Series issued by Walter Scott took over a similar role. There were also series of cheaper books, again mostly issued

[3] Lightman's *Victorian Popularizers of Science* provides details on all these developments; see chap. 5 on evolutionary epics, chap. 6 on *Knowledge* and chap. 7 on Huxley's contributions. *Science Siftings* is profiled in Erin McLaughlin-Jenkins, 'Common Knowledge'; see also Susan Paylor, 'Edward B. Aveling'. Gowan Dawson highlights the problems created by the radicals in his *Darwin, Literature and Victorian Respectability*. On the growing popularism of the media, see Peter Broks, 'Science, Media and Culture'.

[4] There is some information on newspaper and magazine coverage in R. V. Bruce, *The Launching of Modern American Science, 1844–1876*, pp. 151–2 and 354–5, the latter reference including the suggestion that coverage dropped in the 1880s. On the dinosaur discoveries, see below.

simultaneously in Britain and the United States. Clodd's *Story of Creation* of 1887 was published by Longmans, Green in both London and New York, with the shorter and even cheaper *Primer of Evolution* appearing in 1895.[5]

Some experts were still willing to publicize their work, often with a view to promoting their profession or a particular ideological position. Huxley's writing for the more respectable magazines increased in the 1870s, and he also wrote the article on evolution for the 1878 edition of the *Encyclopaedia Britannica*. He still lectured to huge audiences, including those attracted during a tour of America in 1876. Evolution could also generate publicity via the annual meetings of the British and the American Associations for the Advancement of Science – John Tyndall's controversial address at the Belfast meeting of 1874 included references to both Darwin's and Spencer's theories. O. C. Marsh's survey of evolutionary evidence shown by the American fossil record, delivered to the Nashville meeting of 1877, was widely reported and reprinted in *Popular Science Monthly*. Coverage of evolution in the American media actually decreased in the 1880s – the discovery of dinosaurs in the American West, for instance, did not attract much attention until they went on display in the late 1890s. This seems at variance with Allen's claim that evolutionism had achieved cultural dominance, but perhaps there was more interest in the social applications of the idea than in the biological theory. *Popular Science Monthly*'s coverage of Spencer certainly focused more on his ideas about social evolution.

Curators at natural history museums were at first reluctant to incorporate evolutionary, let alone Darwinian, themes into their displays, although fossils were increasingly presented in terms of temporal sequences. Marsh established the Peabody Museum at Yale for the display of his fossils, indicative of an expansion that would soon match that taking place in Europe. In London the new Natural History Museum opened in 1881. It became a popular centre for the display of zoological specimens and fossils, although there were complaints at first about inadequate lighting.[6] Washington, DC, and Philadelphia already had equivalent museums, but it was only toward the end of the century

[5] See Charles Haar, 'E. L. Youmans'; also Roy MacLeod, 'Evolutionism, Internationalism and Commercial Enterprise', and Leslie Howson, 'An Experiment with Science for the Nineteenth-Century Book Trade'. Prices for the Clodd books are from contemporary advertisements.

[6] On the poor lighting at the museum, see H. N. Hutchinson, *Creatures of Other Days*, p. xiv. On the gradual incorporation of evolutionary themes, see Arthur MacGregor, 'Exhibiting Evolutionism'; Juliana Adelman, 'Evolution on Display'; and more generally Rachel Poliquin, *The Breathless Zoo*.

that other American cities were endowed with similar institutions, mainly through the munificence of the capitalists of the Gilded Age. They provided the public with the opportunity to see newly discovered fossils at first hand, and those who commissioned and mounted the displays were becoming ever more adept at using them to convey messages about the process of evolution and its implications.

Finally, use of the term 'evolutionary epic' reminds us that the theory could be expressed in formats that paralleled those of fictional depictions of the drama of human life. Victorian novels of adventure on a global scale can be seen as expressions of an imperialist ideology linked to the progressionism of the evolutionary epics. A few novels addressed evolutionary themes directly, not always in sympathy with the Darwinian theory. Sir Henry Bulwer-Lytton's *The Coming Race* (1871) imagined a superior version of humanity with psychic powers living beneath the earth's surface. H. G. Wells' *The Time Machine* explored the theme of evolutionary degeneration. Usually seen as early forms of science fiction, such accounts may well have had significant impact on how their readers viewed the theory and are thus relevant to this survey. Rudyard Kipling's stories make occasional references to evolutionary themes, sometimes with Lamarckian implications as in the *Just So Stories*.

There are, of course, many less direct connections between evolutionary thinking and contemporary fiction and poetry. Chapter 8 below notes that accounts of how humanity evolved from its ape-like ancestors had a structure that can be compared to that of folk-tales and adventure stories.[7] Literary scholars including Gillian Beer and George Levine find numerous references to evolutionary themes, along with parallels between the ways in which the language is used. The parallels are suggestive of broad literary and cultural developments, but this may be an appropriate point to remind the reader of an argument made in the introduction to this study. When the links to evolutionism are only implicit, they are less relevant if we are seeking to understand how the actual theories of evolution were disseminated. Darwin may have helped to impose the belief that nature is a scene of constant struggle on the public imagination, but he was not the only source of an idea that was widely exploited in ways not always related to the mechanism of natural selection. Other components of his theory were also representative of ways of thinking common to other naturalists and intellectuals. Aspects of his approach to nature and his language may resonate with changing styles of thought and representation, but here he is best seen not as a

[7] See Misia Landau, *Narratives of Human Evolution*. More generally, see, for instance, Beer, *Darwin's Plots*, and Levine, *Darwin and the Novelists*.

direct influence but as someone participating in these wider cultural developments. No one looking for information about how natural selection – or any other evolutionary mechanism – operates would have been advised to start by reading the novels we now recognize as classics of Victorian prose.

To give an example: Joseph Conrad's *Heart of Darkness* (1899) is a vivid expression of the horrors of imperialism and can be seen as the expression of a wider sense that the world is driven by chance rather than any purposeful historical process. It challenged the ideology of progress by exposing the potentially nihilistic implications of the selection theory noted by some of its earliest critics. Some readers would recognize the resonances with Darwinism, but only if they were already aware of how the mechanism operates. If those unfamiliar with the science of the time wanted to understand the connection with Darwinism, they would have to be directed to a work of popular science for information. The study of how the theory was disseminated operates at a different level to the resonances noted by literary scholars because it focuses on sources that directly address the theory and its implications.

Expanding Horizons

Although Darwin had little interest in reconstructing the course of life's history, a broadly defined Darwinian synthesis was one of the factors inspiring the next generation of biologists to work out how each major group of species had acquired its new set of structures and functions. They accepted the limitations imposed by the imperfection of the fossil record (while always hoping for important new discoveries) because they also had indirect lines of evidence such as embryology. Reconstructing the history of life on earth in evolutionary terms became a major scientific project and provided a model for the idea of progress in human society.

Their inspiration for this initiative was provided by the leading German Darwinist, Ernst Haeckel. He used a panoply of biological techniques to construct hypotheses about the main steps in evolution even where fossil evidence was lacking. Like Huxley, he was a morphologist interested in how animal forms were related, and he was particularly adventurous in using embryology to throw light on the evolutionary past. Darwin realized that similarities of embryological development could indicate evolutionary relationships, but Haeckel went beyond this to argue that the main steps in a species' evolutionary past were often repeated in the development of the modern individual. This was the recapitulation theory, a progressionist model with only loose links to Darwin's theory.

Haeckel inspired a generation of evolutionary biologists to study key stages in the history of life on earth. Michael Ruse has dismissed this whole episode as second-rate science, and the problems encountered were certainly far more complex than had been anticipated. By the end of the century, younger biologists were bailing out to concentrate on other areas such as the study of heredity. But the early evolutionists did not have the benefit of hindsight, and their enthusiasm inspired the popular epics charting the ascent of life. New discoveries did occasionally supply hard evidence – Huxley's appeal to Marsh's horse fossils on his American lecture tour is a classic example. But to a large extent the extended story of 'life's splendid drama' was initially told using other forms of evidence to fill in the gaps.[8]

Globally, Haeckel was probably the most influential popularizer of evolutionism. The first introduction to his work for English-speakers was the translation of his *History of Creation*, supervised by Huxley's protégé E. Ray Lankester. A more detailed account focused on the development of the human species was translated in 1879 as *The Evolution of Man*. Both were expensive two-volume works (thirty shillings for the British editions) so they would not reach a mass market, although they would be required reading for anyone hoping to write for a wider audience. The print runs were modest – 1,000 for the first edition in Britain and 500 for later reprints. *The Evolution of Man* was eventually issued as a cheap paperback for only one shilling, but this was not until the turn of the century when Haeckel's *Riddle of the Universe* hit the headlines.[9]

Haeckel's naturalistic philosophy saw everything as governed by law rather than design by God, which in principle should have ruled out any suggestion that the human species is the predestined goal of creation. He was an enthusiastic Darwinian, firmly committed to the theories of common descent and natural selection. Yet historians argue about the extent to which his vision of evolution still implied a predetermined goal. Robert Richards sees him as a committed Darwinian, but others including Michael Ruse and the present author suspect that his commitment to progress reflects an element of idealist thought.[10] The sub-title of the

[8] See Michael Ruse, *Monad to Man*; my own *Life's Splendid Drama* is an extended reply to Ruse's position, the title borrowed from a phrase coined later by the palaeontologist W. D. Matthew.

[9] Parts of Huxley's *Academy* review are reprinted in his *Collected Essays*, 2: 107–19. Details of Haeckel's publications are given in Nick Hopwood, *Haeckel's Embryos*, pp. 163–7 and 194 for the *New York Times* review.

[10] See Robert J. Richards, *The Tragic Sense of Life*, and Michael Ruse, *Monad to Man*, pp. 179–81. The situation is complicated by the fact that Richards thinks that Darwin too was influenced by Romanticism; see Richards and Ruse, *Debating Darwin*.

History of Creation proclaims it to be an exposition of the views of Darwin, Goethe and Lamarck, implying at least some non-Darwinian influences.

Haeckel's vision would be familiar to anyone who had read *Vestiges* or was aware of the scope of Spencer's philosophy. *The History of Creation* covers everything from the formation of the planets through the origin of life and its evolution up to and including modern humanity. The main focus was the development of life and the reconstruction of the steps by which new forms were generated. Haeckel coined the term 'phylogeny' to denote the sequence of ancestral forms identified for each species. His approach was Darwinian in that he regarded all groups of related species as monophyletic, formed by divergence from a single common ancestor, so the overall process could be represented as a branching tree. The challenge was to recognize the source of that ancestral form, which must have emerged from a single branch of a preceding group. Haeckel assumed that this process must have occurred in a particular location, so that a study of geographical distribution might locate the ancestral 'home' of each group.

The fossil record was obviously relevant, but Haeckel was well aware of its imperfection and routinely used anatomical structures and their development in the embryo to determine relationships. Darwin argued that relationships can be inferred from homologies, cases where superficially different forms can be seen as modifications of a single underlying structure inherited from a common ancestor. He also knew that embryology could extend the scope of the argument when different forms have similar patterns in the early stages of individual development. This was a major component of Haeckel's technique, reinforced by images showing the close similarities between the early embryos of a number of vertebrate species, including humans, which he attributed to their common descent (see Fig. 5.1). As Nick Hopwood has shown, these images were copied by many later evolutionists but also were criticized as deliberately overemphasizing the similarities. They remain controversial today.[11]

Haeckel's main focus was the vertebrates and the line leading up to humankind. The origin of the vertebrate type itself – which occurred before the first fossil-bearing rocks were laid down – provided a good example of the power of embryology. Haeckel adopted the theory of Alexandr Kovalevskii, who found that the larvae of the tunicates, or sea-squirts, had vertebrate characteristics even though the adult animal is a sessile form that passes water through its body cavity. It was the larvae that represented the ancestral form of the first vertebrates. The classes

[11] This is the main theme of Hopwood's *Haeckel's Embryos.*

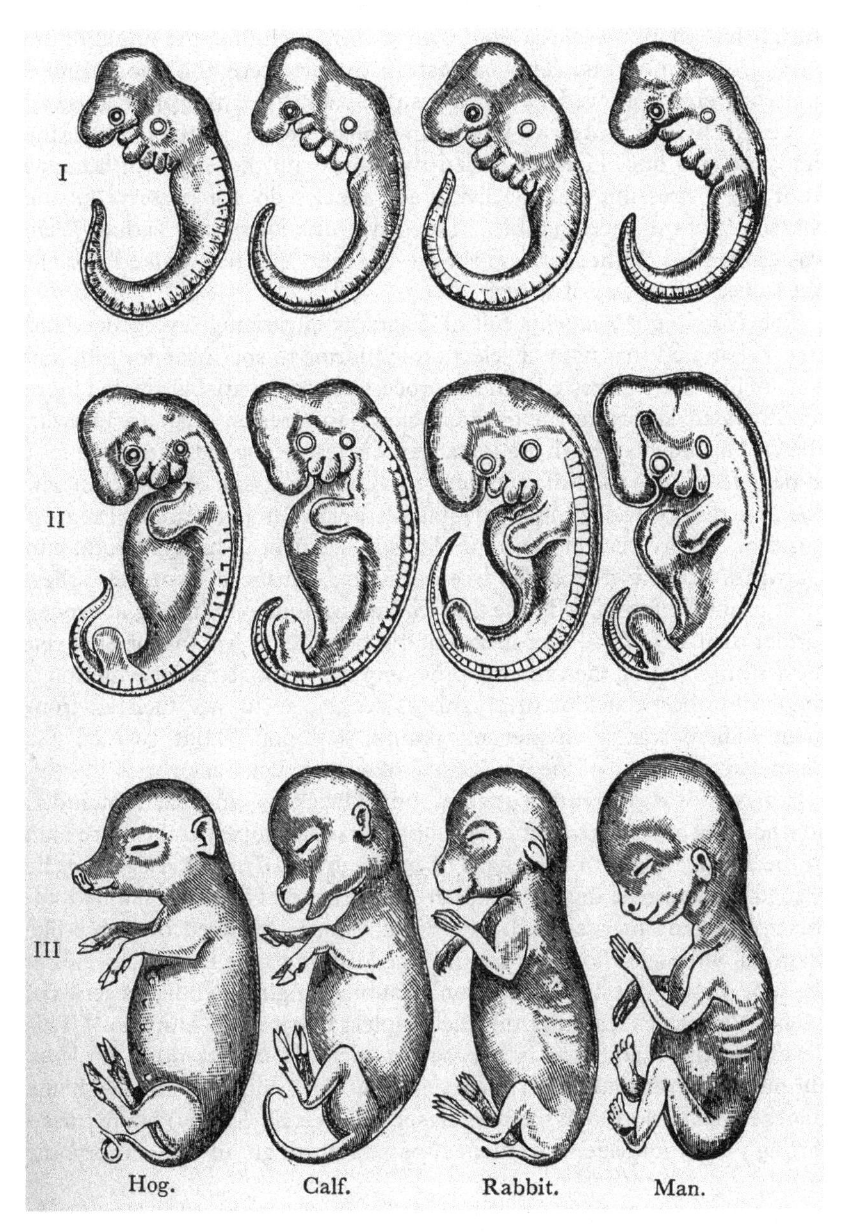

Fig. 5.1 The development of different embryos from a common starting point. This is one of many versions that were adapted from those published by Ernst Haeckel, in this case C. M. Beadnell, *A Picture Book of Evolution* (1934 ed.), p. 209, adapted from Ernst Haeckel, *The Evolution of Man* (1879), vol. 2, plate 7, facing p. 362.

from fish to mammals were dealt with in turn, including the origin of the birds from dinosaurs. The earliest mammals were the monotremes (platypus etc.) followed by the marsupials and then the placentals, all preceded by an unknown 'promammalian' form distinct from the dinosaur-bird line. Haeckel ended the story with the origin of humans from apes, stressing that the living ape species do not preserve all the characters of the ancestral form. The as-yet-unknown intermediate stage was christened 'Pithecanthropus', an ape-man that had walked upright but lacked the power of speech.[12]

The *History of Creation* is full of diagrams illustrating divergence, and Haeckel argued that natural selection led forms to specialize for different ways of life. But he insisted that the process almost invariably created more sophisticated structures, a position closer to Spencer than to Darwin. Haeckel acknowledged that adaptation to a less active lifestyle could lead to degeneration – the tunicates offered a good example of this – but saw this as a rarity because most adaptations involve improvement. Progress was thus a more central, more predictable, element in his system than in Darwin's. All branches of the tree of life had advanced from the earliest most primitive forms, with the line leading to humanity having advanced further than any other. The source of this confidence was, as for Spencer, the assumption that the variation providing the raw material of evolution is shaped by the behaviour of organisms seeking to master their environment. There was a chapter on natural selection – but two on the Lamarckian process of the inheritance of acquired characteristics.[13]

Progress was a central theme, but Haeckel's approach included another way of visualizing the development of life superficially more akin to the linear model of the chain of being. In the *History of Creation* he argued that because the mammals are at the head of the animal kingdom, their phylogeny merits special attention. This is followed by a tree-like diagram, although there is a central line leading to the human species at the top. A more detailed discussion of human origins then lists a series of twenty-two stages leading from the simplest forms up to humans.[14] This developmental sequence is the sole topic of *The Evolution of Man*, although the recapitulation theory is now the central theme, with the stages in the evolution of the human species (our phylogeny) being traced through the equivalent sequence passed through in the embryonic

[12] Haeckel, *The History of Creation*, vol. 2, chaps. 17–23; for details of Haeckel' views and subsequent debates, see Bowler, *Life's Splendid Drama*.

[13] On divergence and progress, see Haeckel, *The History of Creation*, 1: 277–9; on natural selection, see chap. 7 and on Lamarckism chaps. 8 and 9.

[14] Ibid., 2: 231–2, p. 241 for the diagram, and pp. 278–95.

development of every modern individual (our ontogeny). The main stages are reduced to ten, and the implication that this sequence should be given a central status in the overall picture of evolution is driven home with a much-reproduced image depicting the process as a gnarled tree with an obvious central trunk leading up to humankind at its head (see Fig. 5.2). All other forms are demoted to side branches as though they had stepped off the main escalator of progress.[15]

Haeckel may not have intended his readers to imagine that evolution has a preferred direction built into it. Seeing ontogeny as a speeded-up version of the history of life on earth does not mean that evolution itself has the same goal-directed quality. Haeckel notes that giving a linear sequence of antecedents for one form does not detract from the need to see the whole sweep of evolution as a branching tree (one could construct a linear phylogeny for any individual species).[16] Recapitulation is simply a consequence of the Lamarckian mode of evolution in which new characters acquired by the organism have to be added on at the end of their offspring's ontogeny in order to be inherited, thus leaving the previous adult form intact as the penultimate stage. (Darwin's model of undirected variation sees it more as a distortion of individual development, thus eliminating the previous adult form.) The logic of recapitulation need not imply that evolution is as goal-directed as the development of the embryo.[17]

The presentation of human phylogeny as the ascent of a linear sequence may thus be only a rhetorical technique used to inform readers about our own evolutionary past. Since we are the only species that has attained full intelligence, we are entitled to focus on our own development, but this does not imply that nature itself has humans as its predetermined goal. This point is valid for the *History of Creation*, although the same cannot be said for the *Evolution of Man*, which shows much less awareness of the diversity of life. Some historians suspect that Haeckel's world-view contains a submerged relic of the old idealist vision of creation, and one cannot imagine Darwin or Huxley using the development of the human embryo to depict the whole sweep of evolution in the animal kingdom.

The question at hand is not what Haeckel believed but what impression the lay readers of his books (in translation) would have gained about the overall nature of evolution. It is worth noting that the appeal to

[15] Haeckel, *The Evolution of Man*, 2: 184–6; the diagram is facing p. 188.
[16] Haeckel, *The History of Creation*, 1: 314.
[17] The classic survey of the recapitulation theory and its influence is Stephen Jay Gould's *Ontogeny and Phylogeny*; on Haeckel, see pp. 76–85.

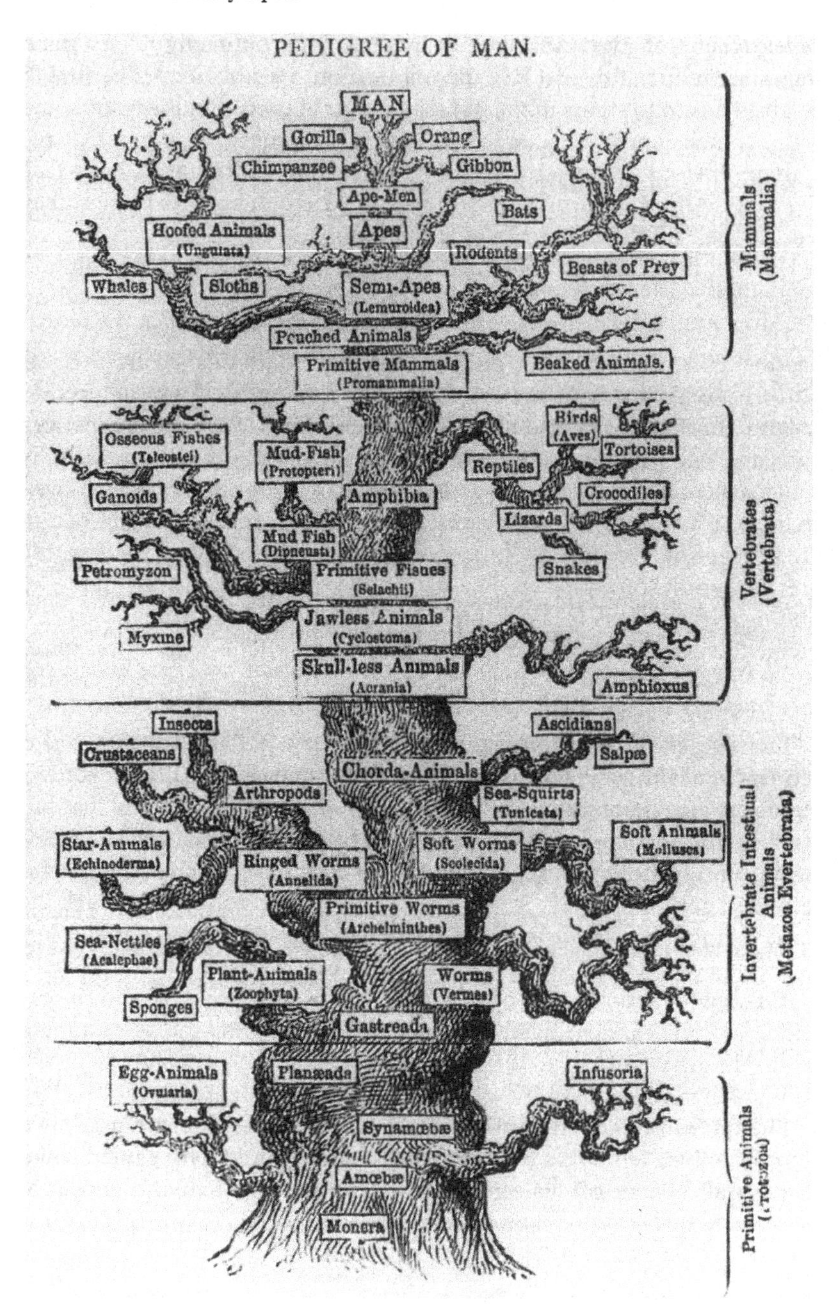

Fig. 5.2 Ernst Haeckel's 'tree of life' with a central trunk running up to humanity at the top. Haeckel, *The History of Creation* (1879), vol. 2, facing p. 188.

embryology as a model might have upset some readers in that more prudish age – the *New York Times*' review of the *Evolution of Man* made this point.[18] Some of those who did consult the book may well have been left with the impression that the parallel between ontogeny and phylogeny implied that both processes share the common element of being goal-directed. This impression is certainly given by some of those defending evolutionism against the charge it undermines religion. Thus, for instance, we find John William Draper (who had recently published his controversial *History of the Conflict between Science and Religion*) addressing the Unitarian Church Institute of Springfield, Massachusetts, on 11 October 1877, providing an overview of the progress of life on earth and concluding:

> Man passes now through the same series of transmutations which his animal predecessors passed through in immense spaces of time, long ago. The progress he makes in the lapse of a few days in the darkness of the womb is the same that has been followed by the procession of animated Nature in the lapse of myriads of centuries in the daylight of the world.[19]

Draper's hearers may well have assumed that both processes were equally directed toward a preconceived goal.

Most books and articles written by authors seeking to popularize evolutionism also assumed that the process is inherently progressive, and some imply that humankind is the intended goal. The recapitulation theory fitted neatly with this developmental hierarchy. As the prototype of the Darwinian evolutionary epic, certain aspects of Haeckel's work thus promoted the belief that humans have a cosmic significance, traditionally enhanced by Christianity but now an integral part of a Western culture convinced that it was the acme of history. The parallel between individual and racial development also featured strongly in efforts to seek a biological understanding of human behaviour, via the claim that children and 'savages' exhibit characteristics typical of earlier phases in the ascent of life (see Chapter 8 below).

Haeckel's work was controversial even among scientists, but his influence on Huxley suggests that his approach helped to transform what counted as Darwinism in science and beyond. Morphological studies expanded in the 1870s with the reconstruction of phylogenies as the underlying agenda. E. Ray Lankester gave courses on comparative anatomy at University College London, attended by a large number of

[18] Noted by Hopwood, *Haeckel's Embryos*, p. 194.

[19] Draper, 'Dr. Draper's Lectures on Evolution', p. 187. His *History of the Conflict between Science and Religion* was a contribution to the International Science Series.

medical students. Huxley eventually taught at the Normal School of Science in London, where the young H. G. Wells studied in 1884–5 and later recalled that the goal of the programme was the 'determination of the relationship of groups by the acutest possible criticism of structure' using anatomy, embryology and the available fossils. In the United States, William Keith Brooks' work at the new Johns Hopkins University in Baltimore introduced a cohort of students to morphology, some of whom went on to become world-famous.

By the last decade of the century, though, the hopes of the evolutionists evaporated as rival hypotheses about relationships were proposed without any means of resolving the differences. Lacking fossil evidence, it was impossible to decide which morphological features were the most reliable guide to ancestry. William Bateson at Cambridge, who had worked on the origin of the vertebrates, and Thomas Hunt Morgan, originally a Brooks student, both abandoned morphology around 1900. They turned to the experimental study of heredity and went on to become founders of the new science of genetics.[20]

The technical details of the rival phylogenetic hypotheses would have been of little interest to the lay reader, and most popular science writers ignored them to concentrate on the broad outline established by Haeckel. It was certainly possible, however, for anyone with sufficient interest to gain access to information on the topic. The successive volumes of the ninth edition of the *Encyclopaedia Britannica* included a series of articles on relevant biological topics. Huxley's 'Evolution in Biology' of 1878 gave a general outline with a brief survey of the supporting evidence. It was supplemented by a series of specialized articles that addressed the problems posed by the reconstruction of past relationships. Lankester's article 'Zoology' of 1888 provided a detailed survey of the development of the whole animal kingdom, noting some of the rival phylogenies. He described a single main line of progressive evolution from which various side branches diverged, reinforced with the image of an evolutionary tree with a definite central trunk leading to the vertebrates. Lankester collected some of these articles into a single volume that could serve as a zoology textbook.[21]

[20] For details of this work, see my *Life's Splendid Drama*. On Lankester's teaching, see the biography by Joe Lester, *E. Ray Lankester and the Making of Modern British Biology*. On the American scene, see, for instance, Jane Maienschein, *Transforming Traditions in American Biology, 1880–1915*. The quotation from Wells is from his *Experiment in Autobiography*, 1: 200–1.

[21] Huxley's 'Evolution in Biology' is reprinted in his *Collected Essays*, vol. 2, *Darwiniana*, pp. 187–226. See also Lankester, 'Zoology', esp. p. 811 and Lankester, ed., *Zoological Articles Contributed to the 'Encyclopaedia Britannica'*, including 'Protozoa', pp. 1–38, esp.

These articles show that the idea of a privileged axis extending through the ascent of life had some currency within the scientific community. Lankester also tapped into the most popular assumption about the stimulus that drove evolution upward, although he chose to expose its more pessimistic underside. Kingsley's *Water Babies* had popularized the claim that progress is achieved through effort and initiative but also warned that idleness led to degeneration back to a more primitive state. The Lamarckian aspects of this were obvious to all: the inherited effects of use and disuse by the individual determined the future of the race. But the Darwinians could also explain why ceasing to use an organ would lead eventually to its elimination, since selection would favour individuals who did not waste energy growing a useless structure.

It was this insight that Lankester exploited in his little book *Degeneration: A Chapter in Darwinism*, published in 1880. He challenged the assumption that natural selection automatically led to progress by developing the point that a species adapting to a less active way of life would degenerate to a less complex state. Parasites and the barnacles that Darwin had studied were the classic examples, but Lankester also appealed to Kovalevskii's theory that the ascidians (tunicates, or sea-squirts) are a degenerate offshoot of the active and mobile ancestors of the vertebrates. Choosing a sessile lifestyle had led to disaster, and Lankester was only too willing to drive home the lesson for modern humanity:

> Possibly we are all drifting, tending to the condition of intellectual Barnacles or Ascidians. It is possible for us – just as the Ascidian throws away its tail and its eye and sinks into a quiescent state of inferiority – to reject the good gift of reason with which every child is born, and to degenerate into a contented life of material enjoyment accompanied by ignorance and superstition.[22]

Here was a warning for those who rashly assumed that the Spencerian model of progressive evolution offered guarantees for the future.

H. G. Wells, who became a close friend of Lankester, incorporated the point into an article titled 'Zoological Retrogression' written for the *Gentleman's Magazine* in 1891 and went on to use it in his story *The Time Machine* a few years later.[23] By this time confidence in the inevitability of progress was beginning to wane. But when Lankester wrote, his warning would have come as a startling reminder that the focus on effort

p. 9, and 'Vertebrata', pp. 173–83, esp. p. 183. In one of his technical surveys, Huxley too had written of the modern reptiles and bony fish as being off the 'main line' of evolution; see Bowler, *Life's Splendid Drama*, p. 425.

[22] Lankester, *Degeneration*, pp. 60–1.

[23] 'Zoological Retrogression' is reprinted in Robert M. Philmus and David Y. Hughes, eds., *H. G. Wells*, pp. 158–68.

and initiative might not be as easy to maintain as the Spencerians hoped. For the time being, however, the theme of a main line of progressive evolution ascending toward humankind remained central to the evolutionary epics produced by popular writers, whether they used it to bolster Spencer's ideology or muscular Christianity. Since the search for hard evidence had led only to debate and confusion, it is hardly surprising that those aiming at a wider readership chose to ignore the arguments raging behind the scenes.

Fossil Developments

There were some fossil discoveries that greatly helped the evolutionists' cause, although others seemed to go almost unnoticed in the popular literature until the end of the century. A few examples became iconic, their story repeated in almost every popular text, while others seem to have been hard to fit into the prevailing narrative. The discovery of *Archaeopteryx* had shown that intermediates could pose as many problems as they solved – most experts thought it came too late in the record to be the original transitional form. It was only in the late 1880s that H. G. Seeley began to report what came to be known as the mammal-like reptiles from the rocks of South Africa, thus filling an even more important gap. This time Huxley missed their significance because the fossils came to Owen at the British Museum.[24] It was in the ancestry of particular modern species that the real breakthrough occurred. Huxley had tried to reconstruct the ancestry of the modern horse using European fossils, but with little success. He then found that a far more convincing series had already been unearthed in America by O. C. Marsh.

Huxley's conversion to the American ancestry of the horse came on his much-publicized tour of the United States in 1876. This boosted his global reputation and gave him an opportunity to promote evolutionism to a different public. He was invited to give lectures across the continent (from which he made a profit of £600) and was told that even the miners in California were reading his books. The main event promoting evolutionism was a series of lectures in New York's Chickering Hall, given to capacity crowds. The *New York Tribune* issued a commemorative issue on 23 September with the full texts, while the *Daily Graphic* featured a front-page cartoon of Huxley combating Moses on 27 September. It was in these lectures that he presented the evidence that Marsh had shown him for the evolution of the modern horse from less specialized

[24] See Desmond, *Archetypes and Ancestors*, pp. 193–201.

ancestors. A diagram depicted a sequence from a small, four-toed creature through others with progressively more prominent central digits culminating in the hoof of the modern horse (see Fig. 5.3). Huxley proclaimed this 'demonstrative evidence of evolution', and shortly afterwards Marsh was able to report a five-toed ancestor that he called *Eohippus*, the dawn horse.[25]

The book version of the lectures under the title *American Addresses* sold out within a year, but it was the horse fossils that captured the attention of other writers on evolution. The case was widely cited in popular works into the twentieth century, usually with the image that Huxley had used. Horses were part of everyday life, so pinning the evidence for adaptive evolution onto this symbol had an immediate effect. The sequence also helped to make the case for Darwinism by demonstrating that evolution should not be seen as a single advance toward the human form. But the technique of arranging a few fossil species into a linear sequence ending with the modern horse could also be misleading. It presented the specialization of the horse for running on the open plains as a linear trend that seemed to be aimed at an inevitable goal. Only in the twentieth century did it become clear from further discoveries that the process of horse evolution should be seen as an irregularly branching tree.[26]

At the following year's AAAS meeting in Nashville, Marsh himself gave an address titled 'Introduction and Succession of Vertebrate Life in America', which was published both in the *American Journal of Science* and for a wider audience in *Popular Science Monthly* (it was also copied in *Nature* for British readers). This was a survey of the history of life, which, unlike many evolutionary epics, relied solely on fossil evidence. Marsh admitted that the details of most phylogenies could not yet be filled in but stressed the link between dinosaurs and birds (he had added more toothed birds to the list himself) and, of course, the horse sequence.

Marsh was a self-proclaimed Darwinian, but he was also an enthusiast for Spencer's philosophy who spoke at the famous dinner at Delmonico's restaurant in New York when Spencer visited in 1881. His address expanded the Spencerian thesis that most evolutionary developments

[25] The lectures were published as Huxley's *American Addresses*, but are most easily found as the 'Lectures on Evolution' in his *Collected Essays*, vol. 4; see pp. 130 and 132 for the diagram and the quotation. For details on the American tour, see Ronald W. Clark, *The Huxleys*, pp. 88–93, and Desmond, *Huxley: Evolution's High Priest*, chap. 4. Diarmid Finnegan shows that Huxley was one of several scientists who crossed the Atlantic to lecture; see his *The Voice of Science*, with chap. 2 on Huxley.

[26] On horse evolution, see, for instance, Bowler, *Life's Splendid Drama*, pp. 330–1 and 363. On neo-Lamarckism, see below and also Bowler, *The Eclipse of Darwinism*, chaps. 4 and 6.

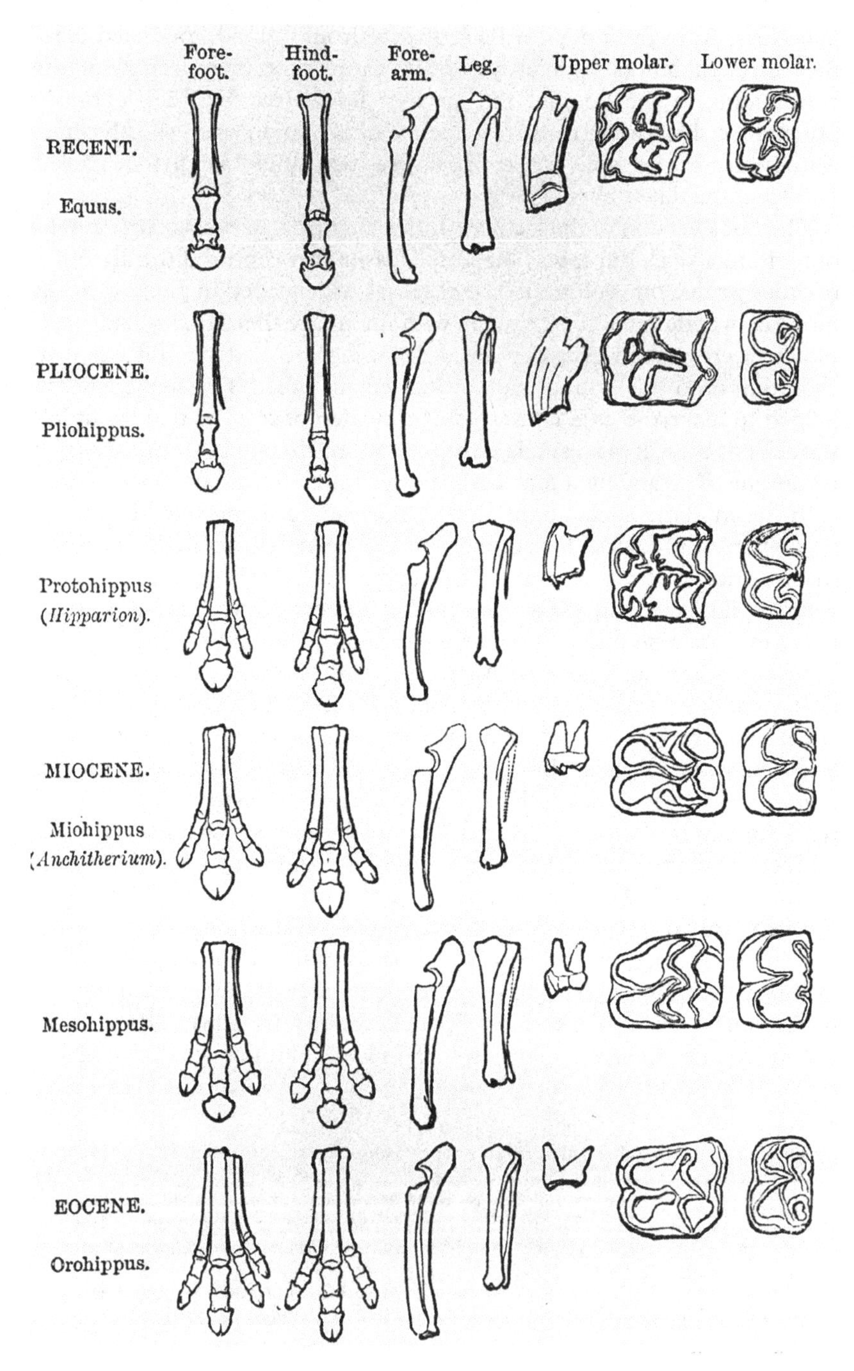

Fig. 5.3 O. C. Marsh's sequence of horse fossils showing progressive specialization, as used by T. H. Huxley and widely copied. This version from A. R. Wallace, *Darwinism* (1889), p. 388.

are progressive, using evidence from trends in the fossil record of North American mammals. Although his horse fossils could be arranged into a linear pattern, he made no effort to generalize this point – his specialized works included irregularly branching tree diagrams with many extinctions. But in every branch, he insisted, there was a tendency for the brains of successive species to become larger, indicating higher levels of intelligence. This was his 'law of brain growth', with obvious implications for society today: 'In the long struggle for existence during Tertiary times, the big brains won, then as now.'[27] Like the evolutionary epics that used living species to illustrate the main steps in the advance of life, Marsh's address was firmly tied in to the ideology of progress..

Onwards and Upwards

In the 1870s and early1880s Darwinism was still a relatively broad church and many followed Spencer in thinking that progress was inevitable because natural selection could be supplemented with the Lamarckian alternative of the inheritance of acquired characteristics. Some still saw the process as the fulfilment of a divine plan. Almost everyone assumed that if nature embodied processes driving life toward higher levels of organization, the continuation of those processes in the modern population must ensure social progress. The contested issue was whether the future state of perfection would be one of material well-being or some form of spiritual enhancement, although even Spencer thought that the end-product of evolution was not a world of ruthless individualists; the ability to get along with others was necessary for any organism living in a social group.

The term 'evolutionary epic' highlights the impact of surveys telling the story of how life has fought its way upwards to inspire confidence that further progress is inevitable. Scientific evidence was marshalled to bolster the argument, but attention inevitably focused on those elements best suited to make the case for progress. The real enthusiasts were the supporters of scientific naturalism who followed Spencer and their rivals who favoured Kingsley's version of liberal Christianity. For the Spencerians, his philosophy of cosmic progress was the true foundation for an evolutionary perspective, while Darwin had merely facilitated the incorporation of that perspective into biology. The rival position came from the still very active body of writers seeking to promote the view that

[27] Marsh, 'Introduction and Succession of Vertebrate Life in America', quotation from p. 55 of the *American Journal of Science* printing. On Marsh's links with Spencer, see my 'American Paleontology and the Reception of Darwinism'.

evolution through individual effort guaranteed not only moral but also spiritual progress, thus fulfilling God's purpose in creation. The end product in either case would be a perfected form of humanity, giving the whole process a clearly defined goal even though some creatures had to lag behind to maintain the environment.

There were underlying parallels between the rival visions that allowed Spencer's philosophy to be presented as compatible with liberal Christianity. Most popularizers paid lip-service to the theory of natural selection, seeing the elimination of the less successful new types as an unfortunate necessity. Grant Allen was one of the few who made no secret of the harsher implications of the Darwinian mechanism. But it was generally acknowledged that Spencer favoured the Lamarckian process of use inheritance as the way in which self-improvement could be transmitted to future generations, thus ensuring progress. John Tyndall's controversial address to the British Association meeting in Belfast in 1874 (mistakenly seen by many as endorsing materialism) included both Darwin's and Spencer's views in its account of evolution.[28] Others frequently blurred the distinction between Darwinism and Lamarckism by implying that struggle must inevitably lead to the emergence of higher functions. The fact that the coiner of the phrase 'survival of the fittest' was also a Lamarckian made it easier to fudge the issue.

The two most popular writers promoting Spencerianism in Britain were Allen and Edward Clodd. In America, John Fiske, although writing at a more academic level, also achieved substantial sales for his books presenting the synthetic philosophy in a form more palatable to religious believers. Of the three, Allen projected the most Darwinian viewpoint, in part because he preferred to write about familiar animals and plants. His explanations of how they were shaped by the struggle for existence inevitably focused attention on adaptation rather than progress, although he saw each story as a miniature epic in its own right. Clodd came closer to producing a broad-brush survey on the scale pioneered by Haeckel, his *Story of Creation* even echoing the title of the latter's 1876 translation. He too presented natural selection as the main mechanism of change and explained its operation in detail. Fiske painted on a broader philosophical canvas and offered an optimistic view of cosmic progress with a moral and spiritual agenda, suggesting that the transition to higher forms of life might not always be continuous.

[28] Tyndall, 'Address of the President', pp. lxxxiii–xci. The address was printed separately and again in Tyndall's *Fragments of Science*, 2: 145–214, see pp. 183–92. On the controversies sparked by the assumption that Tyndall was a materialist, see Ruth Barton, 'John Tyndall, Pantheist'.

Allen began as an enthusiastic supporter of Spencer, although they eventually fell out as Allen moved towards socialism. Nevertheless, his 1888 article 'Evolution' proclaimed Spencer's cosmic vision as a more significant achievement than Darwin's purely biological applications. In explaining how evolution worked, however, Allen was driven toward a more Darwinian approach because to avoid technicalities he concentrated on familiar species, showing how their structure and behaviour has been shaped by the constant need to adapt to the environment and to resist the assaults of rivals or predators. Two books published in 1881 collected articles from popular magazines: *Vignettes from Nature* from the *Pall Mall Gazette* and *The Evolutionist at Large* from the *St. James Gazette*. Both were published in the United States as parts of the Humboldt Library of Popular Science, a monthly magazine-format series selling for only 15 cents a copy or $1.50 for a yearly subscription.[29]

Allen's general article on evolution warned that most popular beliefs about the process were misguided, including the assumption that a great ape was our immediate ancestor. His accounts of how well-known animals and plants have been shaped by their past development stressed the constant need to survive and reproduce, often at the expense of rival species. Where Spencer saw the struggle for existence as a source of progress, Allen stressed its harsher implications – cruelty and indifference to the suffering of the weak and unsuccessful. He did occasionally address the broader history of life: 'A Big Fossil Bone', for instance, notes that successful groups tend to throw up large, specialized forms that become extinct when a rival form appears. 'Distant Relations' moves from the metamorphosis of the tadpole to the larva of the now-degenerate ascidian, which reveals the form of the first vertebrates. But most of his studies focus on adaptations and the pressures that shape them. 'The Origin of Walnuts' argues that nut shells are a consequence of the Darwinian struggle between plants seeking to protect their seeds and animals wanting to eat them. The outcome is an epic in miniature, although not necessarily one with a happy ending:

> All nature is a continuous game of cross-purposes. Animals perpetually outwit plants, and plants in return once more outwit animals. Or, to drop the metaphor, those animals alone survive which manage to get a living in spite of the protections adopted by plants; and those plants alone survive whose

[29] For details of Allen's life and work, see Lightman, *Victorian Popularizers of Science*, pp. 219–20 and 266–89, also William Greenslade and Terrence Rogers, eds., *Grant Allen*. Heather Atchison, 'Grant Allen, Spencer and Darwin', in the latter volume notes that Allen later became worried that Weismann's views threatened the Lamarckian component of Spencer's position.

peculiarities happen successfully to defy the attacks of animals. There you have the Darwinian Iliad in a nutshell.

The Evolutionist at Large includes a short poem, 'A Ballade of Evolution', in which the whole development of life is attributed to the survival of the fittest, culminating in 'our civilized hive' in which 'money's the measure of all / And the wealthy in coaches can drive / While the neediest go to the wall'.[30] These sentiments would hardly have endeared him to Spencer's wealthy American supporters.

Allen's friend Edward Clodd lost his faith through reading Huxley's science and Tylor's anthropology. He supported himself by working in a bank, but also wrote prolifically for the popular science magazines and served as assistant editor of *Knowledge*. He too was an enthusiast for Spencer's cosmic philosophy, but a Darwinian when it came to biological evolution. His first successful book, *The Childhood of the World* of 1873, reinterpreted the Bible for children in terms of the latest ideas on cultural evolution. He ruffled the feathers of religious believers even more in later books and subsequently joined with the atheists, a move that distanced him from Huxley, who feared the social disruption threatened by the radicals.[31]

Clodd's 'epic' came in the form of his *Story of Creation*, published by Longmans of London and New York in January 1888. The British edition sold for six shillings, later reduced to three shillings and sixpence, making it easily affordable. It sold 2,000 copies in the first two weeks and 5,000 within three months. A condensed version of the text with fewer illustrations was issued in 1894 as *A Primer of Evolution*, selling for only one shilling and sixpence in the United Kingdom and $1.25 in the United States. His account covered the whole field of evolution's activities from the formation of the solar system onwards. It was divided into two parts, the first describing what was known about the various structures, the second explaining how these structures had been formed by processes governed by natural law. The successive phases formed one long, continuous development: 'We began with the primitive nebula, we end with the highest form of consciousness; the story of creation is shown to be the unbroken record of the evolution of gas into genius.'[32]

[30] Allen, *The Evolutionist at Large*, quotation from 'The Origin of Walnuts', p. 171 and p. 40 in the American edition. 'A Ballade of Evolution' appears at the beginning of the English edition but at the end, p. 50, in the American. For 'A Big Fossil Bone', see pp. 17–20 in his *Vignettes from Nature*.

[31] On Clodd's career, see Lightman, *Victorian Popularizers of Science*, pp. 253–66.

[32] Clodd, *The Story of Creation*, p. 228 in the edition cited.

Clodd did not neglect the diversity of life, devoting considerable space to the plants and invertebrates. There is a survey of the rise of the vertebrates, although the Age of Reptiles gets only a brief mention with the dinosaurs referred to as 'dragons of the prime' adapted to swampy jungles. He provided images of *Archaeopteryx* and Marsh's sequence of horse fossils as examples of an intermediate type and an evolutionary specialization.[33] Yet much of the text focuses on living species to illustrate the main steps in vertebrate evolution. The invertebrate types are seen as side branches from the main line of development, with the social insects as the highest point on the articulate branch, from which 'we must descend to reach the starting point leading to the loftiest branch whose topmost twig is man'. A fold-out diagram of the tree of life separates the plants from the animals, the latter depicted with a main line leading to humanity. A chapter titled 'The Origin of Species' gives a detailed description of natural selection and stresses its role in adapting species to their environment. A brief run through the ascent of life ends with the claim that it was only when a primate species descended from the trees and stood upright that the expansion of the brain became possible – this was 'the making of man'.[34]

A chapter on the proofs of evolution includes a version of Haeckel's embryo images showing how different forms develop from a common origin and a brief reference to the recapitulation theory. *Archaeopteryx* and the horse sequence again come in to show evolutionary transformations, and there is a substantial discussion of the evidence from geographical distribution, a field now much expanded thanks to the work of A. R. Wallace. A concluding chapter on social evolution suggests that the harsher aspects of natural selection are slowly being bypassed through the rise of cooperative instincts. Between them Clodd and Allen made sure that their readers were aware of how natural selection works and how it shaped the development of life on earth. Yet throughout, the Spencerian implication that progress is inevitable in the long run was always apparent. Thanks to the efforts of these and other popularizers, belief in the evolution as a goal-directed process spread even around the British empire. In *Kim* (a story of adventure set in India, published in 1901), Rudyard Kipling introduces the Bengali spy Hurry Chunder Mukergee as a follower of Spencer who believes everything is the result of a 'process of Evolution … from Primal Necessity'.[35]

In the United States, E. L. Youmans popularized Spencer's philosophy through his *Popular Science Monthly*, but another leading supporter,

[33] Ibid., especially pp. 51 and 123. [34] Ibid., p. 113, facing pp. 132 and 182.
[35] Kipling, *Kim*, pp. 319–20.

John Fiske, adapted the synthetic philosophy to the powerful religious element in American culture. He wrote prolifically for the highbrow magazines, and his *Outlines of Cosmic Philosophy* of 1874 was reprinted nineteen times before the end of the century despite its bulk (two fat volumes). Buried within its pages was a survey of the phases of cosmic evolution, the whole depicted as a progression toward the emergence of humanity's spiritual faculties. The program was summarized in his *Destiny of Man Viewed in the Light of His Origin* of 1884, which cost only a dollar and sold 26,000 copies by the end of the century. Fiske exploited Darwin and Spencer's recognition that social life required the development of cooperative instincts to argue that progress was always toward a morally uplifting goal. The 'unknowable' power that Spencer conceded must lie behind the observable universe was transmuted into an active God who has established the laws of nature to achieve his goal of producing a spiritually mature humanity.[36]

Outlines of Cosmic Philosophy began with Spencer's law of evolution as a transition from homogeneity to heterogeneity. There followed chapters on planetary evolution, the beginnings of life and the reasons for rejecting special creation in favour of biological evolution. The second volume opens with an explanation of natural selection and then considers objections to the theory – the new fossil intermediates are introduced to provide hard evidence of continuous change. A chapter on direct and indirect adjustments to the environment argues that natural selection must be complemented by the inheritance of acquired characteristics. The rest of the volume consists of an extensive discussion of mental, moral and spiritual evolution within the human species. Fiske qualifies his endorsement of the principle of continuity, suggesting that there have been key breakthroughs at two points in the line leading to humanity: the emergence of the modern human mind (which took place only when the first civilizations were founded) and the acquisition of an immortal soul. Evolution proceeds via long periods of relatively slow change interrupted by rare episodes in which one line of development passes through to a new phase of development. Humanity is the end-point of the process: 'no race of organisms can in future be produced through the agency of natural selection and direct adaptation which shall be zoologically distinct from, and superior to, the human race'.[37]

[36] For details of Fiske's publications, see Lightman, 'Darwin and the Popularization of Evolution', pp. 12–13, and on his religious links Moore, *The Post-Darwinian Controversies*, pp. 230 and 235. More generally on Spencer's impact, see Lightman, 'Spencer's American Disciples'.

[37] Fiske, *Outlines of Cosmic Philosophy*, 2: 234–5 and 292–3, quotation from p. 321.

Design Continued

Fiske's reconciliation of Spencerianism with liberal Christianity helped it to become embedded in American culture perhaps more fully than in Britain, where Spencer was identified with indifference to formal religion. There was no shortage of efforts by other authors to modify evolutionism in ways that made it more acceptable to religious believers on both sides of the Atlantic. One of the more imaginative efforts came from two veterinarians, Albert and George Gresswell, whose *The Wonderland of Evolution* was published in Britain and the United States in 1884. The book begins with the suggestion that the 'fairy Chance' can explain the whole development of life. But then, over a series of chapters, this claim is gradually undermined. The advance of life was not without conflict – one chapter suggests that the vertebrates and invertebrates were at one time rivals for dominance – but there is also a strong Lamarckian influence. The end result was inevitable because a 'Great Intelligent Power' has been at work throughout to ensure that humanity would emerge.[38]

Perhaps the most successful contributor to the genre of Christian evolutionary epics was Arabella Buckley, who shared A. R. Wallace's enthusiasm for spiritualism. Some of her writing was aimed at juvenile readers, including her most popular work, *Winners in Life's Race*, published in London in 1882 and in New York the following year. It remained in print in both countries into the early twentieth century. She had no intention of playing down the diversity of life – an earlier book, *Life and Her Children*, was a survey of the various invertebrate types. *Winners in Life's Race* was a follow-up on the vertebrates, but from the start she made it clear that the other types were evolution's less successful efforts: in the former book 'we watched life trying different plans, each successful in its own way, but none broad enough or pliable enough to produce animals fitted to take the lead in all the world'. The vertebrates and ultimately humankind were the real winners of the race, and the personification of 'life' in this quotation suggests that it should be seen as an active, creative force trying ceaselessly to fulfil the Creator's purpose. Lightman has identified an article, probably written by Buckley, arguing that the Creator's power has been delegated to living things so their efforts to improve themselves can drive evolution forwards – in effect a version of the psycho-Lamarckism espoused by E. D. Cope and

[38] A. Gresswell and G. Gresswell, *The Wonderland of Evolution*, see chap. 3 on the vertebrate-invertebrate rivalry and p. 84 for the 'Great Intelligent Power' quotation.

Samuel Butler.[39] The harshness of the struggle for existence is marginalized to focus on the achievement of higher functions, and, like Fiske, Buckley identifies humanity as the intended goal of the Creator's plan.

Winners in Life's Race mentions the fossil record from time to time but depends mostly on living species to show the advances made by each vertebrate class. The ascidian larva gives a clue to the type's origin, and the metamorphosis of the tadpole into a frog illustrates the transition from fish to amphibian. The dinosaurs get a brief mention, as does *Archaeopteryx*, but the living reptiles and birds are the real heroes of their classes. The birds are an advance on the reptiles, although their small brains prevented them from dominating the world. Early mammalian forms, the monotremes and marsupials, are preserved in Australia, 'isolated from the great battle-fields' of the larger continents. The placental mammals have flourished because of their greater intelligence and tendency to live in family groups. Despite their many divergent types, though, the main line of development is obvious: monkeys appear early in the record and define the line leading to the apes and humanity. Our nearest relative is the gorilla, not the savage beast of caricature but a degenerate relic of our early ancestor.[40]

Buckley's purpose is summed up in her concluding chapter: 'A Bird's-Eye View of the Rise and Progress of Backboned Life'. Some groups have risen to dominance only to face decline and extinction as the conditions that first favoured them disappeared. In the end, humanity, 'the last and greatest winner of life's race, has taken possession of the earth'. The whole process has been 'the result of the long working out of nature's laws as laid down from the first by the Great Power of the Universe'. Despite the difficulties experienced by living things, there has been a growth of intelligence and of the willingness of individuals to cooperate: 'intelligence and love are often as useful weapons in fighting the battle of life as brute force and ferocity'. We may shrink from the idea that progress depends on the struggle for existence, but the growth of parental care and cooperation shows that 'amidst toil and suffering, struggle and death, the supreme law of life is the law of SELF-DEVOTION AND LOVE'.[41]

Darwinism is thus tamed to make room for moral progress. The harsh aspects of evolution are a negative factor necessary to remove the less

[39] Quotation from Buckley, *Winner's in Life's Race*, p. 9. See Lightman, *Victorian Popularizers of Science*, pp. 238–53, and on her article supporting Lamarckism p. 244.

[40] Buckley, *Winners in Life's Race*, on the birds' failure to dominate pp. 179–80; on Australia p. 186; on the monkeys and the gorilla pp. 240 and 253–5.

[41] Ibid., pp. vii–viii, 348–9 and 353.

successful attempts of life to improve itself. The efforts of individual organisms to improve themselves provides the progressive element in evolution, allowing Buckley to join Fiske in combining Lamarckism with elements of natural selection. The element of self-adaptation in this approach appealed to the American tycoons who saw free-enterprise individualism as the motor of economic progress and to liberal clergymen such as Henry Ward Beecher who argued that progressive evolution is the mechanism adopted by the Creator to achieve his ends. The traditional Christian doctrine of original sin had to be abandoned, a bold step advocated in Beecher's widely acclaimed sermons collected in his *Evolution and Religion* of 1885.[42] Other religious figures added to the chorus: *Popular Science Monthly* routinely published articles in support of theistic evolution, often written by authors with clerical rather than scientific qualifications.[43] All used the loose combination of Darwinism and Lamarckism to explain both adaptation and progress, with the human race often being seen as the predetermined goal of the process.

Darwinism Dissolving

Toward the end of the 1880s, however, the synthesis of Darwinism and Lamarckism underpinning the evolutionary epics began to dissolve. A more aggressive school of neo-Lamarckians argued for the outright rejection of the selection theory, prompting in response the emergence of a neo-Darwinism denying any role for the inheritance of acquired characteristics. There was also a revival of interest in the possibility of non-adaptive evolution driven by internally programmed variation trends. These developments pointed the way toward what would be called the 'eclipse of Darwinism' at the turn of the century (the topic of Chapter 6). As several historians have pointed out, the term 'eclipse' is misleading if it is taken to imply that Darwin's theory of natural selection had previously been the dominant element in the evolutionary world-view. Nevertheless, the 1890s saw a definite increase in the level of hostility between the Darwinians and those who doubted the efficacy of natural selection. Much of the hostility to natural selection was driven by fear of materialism, with many Lamarckians emphasizing that their theory gave

[42] For details of these religious developments, see Moore, *The Post-Darwinian Controversies*, chaps. 10 and 11.

[43] Examples of theistic evolutionism in *Popular Science Monthly* include W. Stanley Jevons, 'Evolution and the Doctrine of Design', and Lawrence Johnson's address to the Franklin Society of Mobile in June 1872 entitled 'The Chain of Species'.

living things an agency in evolution – they could choose the new habits they adopted to meet environmental challenges.[44]

In the 1860s there had been many efforts to challenge the selection theory, even by those willing to accept evolutionism. The more active opponents of materialism rejected the basic utilitarian assumption that the primary directing agent is the need for organisms to remain adapted to their environment. Theistic evolutionists including the Duke of Argyll suggested that some specific characters had no adaptive value and had been developed simply for their beauty, as though the Creator had an aesthetic motive when designing the laws of evolution. Argyll summarized his views in an article published by the *Contemporary Review* in 1871 and continued to write books and articles on the topic. W. B. Carpenter, originally sympathetic to Darwinism, argued in the *Modern Review* in 1884 that there seemed to be an element of design in the geometrical patterns of the Foraminifera shells he studied.[45]

The threat to the Darwinian perspective expanded when the possibility of non-adaptive steps was extended to the claim that a whole series of related transformations might drive evolution consistently in a direction predetermined by internal forces. In his 1878 *Encyclopaedia Britannica* article on evolution, Huxley declared that 'the importance of natural selection will not be impaired, even if further enquiries should prove that variability is definite, and is determined in certain directions rather than in others, by conditions inherent in that which varies'.[46] He still believed that natural selection would determine which trends could succeed in the outside world, but his protégé Patrick Geddes was prepared to imagine trends that proceeded independently of any adaptive value. In articles for the *Encyclopaedia Britannica* and *Chamber's Encyclopaedia* he argued that it was the balance between internal biological forces that had created structures such as flowers. He also suggested that the spines on some plant species had evolved as the result of their 'ebbing vitality' – a vision of racial senility that elicited a sharp response from Wallace, who retained the Darwinian view that they were a defence against the attacks of

[44] See, for instance, Mark A. Largent, 'The So-Called Eclipse of Darwinism'. The phrase was used by Julian Huxley to denote the rising popularity of anti-selectionist theories around 1900 and was used as the title of my own book *The Eclipse of Darwinism* as a convenient way of drawing attention to an episode that at the time had been largely ignored by historians. As my follow-up *The Non-Darwinian Revolution* made clear, this was never intended to imply that the selection theory had become dominant in Darwin's own lifetime.

[45] See Argyll, 'On Variety as an Aim in Nature', and for details of his later work Bowler, *The Eclipse of Darwinism*, pp. 49–50. Carpenter's 'The Argument from Design in the Organic World' is reprinted in his *Nature and Man*, pp. 409–63.

[46] Huxley, 'Evolution in Biology', *Collected Essays*, 2: 223.

animals. Here was the basis for a view of evolution as a process driven mainly by internal rather than external forces.[47]

The American neo-Lamarckians also modified their position to include similar anti-adaptationist views. Cope's enthusiasm for parallelism led him to conclude that once a species had acquired an adaptive behaviour pattern, the resulting trend toward specialization would become locked in and eventually lead to dangerous levels of overdevelopment. He and his colleague Alpheus Hyatt were the most literal-minded exponents of the recapitulation theory, arguing that variation was driven by an 'acceleration of growth' that advanced successive generations in a predetermined direction, thus defining the main lines of evolution (a view not unlike that adopted earlier in Chambers' *Vestiges*). This was a different application of recapitulationism to that which saw the ascent of life toward the human species repeated in the development of the modern infant. For Cope and Hyatt every family had a built-in chain of being unrelated to the main line leading to humanity. Each substantial branch of the tree of life had its own pattern leading to its own characteristic end-point. All members of the group advanced in parallel through the same scale but not necessarily at the same rate.

Hyatt worked with fossil cephalopods such as the Ammonites, where the whole life cycle is visible in the internal structures of the coiled shell. He extended the parallel between evolution and the individual's life cycle to include senility and death, claiming that the sequences he studied all ended with eventual degeneration and extinction. Species grow old and die, just like organisms, and if the environment plays a role in extinction, it is only at the very end when the animals are unable to cope any further. In the 1890s the theory that evolution is driven in predetermined directions conferring no adaptive benefit would be popularized by Theodor Eimer under the name 'orthogenesis', and palaeontologists would increasingly speculate about species being driven to extinction by trends of over-specialization or sheer degeneration. Here was apparently more convincing evidence of racial senility in evolution. Hyatt's cephalopods were frequently cited as evidence for such degenerative trends, but new examples that would allow the model to be extended to extinct vertebrates were also being unearthed by Cope and his great rival Marsh in the American West.[48]

[47] On Geddes views and his encyclopaedia articles, see Chris Renwick, 'The Practice of Spencerian Science'. Geddes went on to become a leading environmentalist and town planner. For Wallace's critique, see his *Darwinism*, pp. 428–35.

[48] Hyatt's papers were all published in the specialist literature, but were extensively quoted in Cope's *Primary Factors of Organic Evolution*, see pp. 182–92 and 405–22. For more details, see my *Eclipse of Darwinism*, chaps. 6 and 7, and Gould, *Ontogeny and Phylogeny*, chap. 4.

The American school became known as a form of 'neo-Lamarckism' because Cope and Hyatt were at least willing to allow an adaptive phase at the start of each group's development – provided natural selection was not involved. The Lamarckian element in their thinking assumed that the adaptative phase of a group's development resulted from the animals choosing a new lifestyle that modified their structure by use inheritance. But Cope had no interest in Spencer's views, and his interpretation of the process was closer to that suggested in Britain by Samuel Butler (discussed below). This was what was sometimes known as 'psycho-Lamarckism' because it gave animals the capacity to choose the habits that would shape their species' evolution. The creative activity of the organism was the expression of a non-material vital force, and in his *The Theology of Evolution* Cope suggested that the Creator had, in effect, delegated his designing power to life itself. In 1878 he gained control of the *American Naturalist*, which he used to publicize his views, and in 1887 his articles were collected in a book, *The Origin of the Fittest* – the title proclaimed the general view of the Lamarckians that it was the creation of fitter characters, not their survival, that really mattered.

The Lamarckian effect seemed plausible to many because the modern notion of heredity had not yet emerged. Today we take for granted the model of inheritance built into the popular image of genetics: a process that transmits characters unchanged through the generations. Individuals inherit genes from their parents and pass them on to their own offspring. The transmission of characteristics can be studied without worrying about how they are developed in the embryo of the next generation. This is not the way naturalists and ordinary people thought about the situation in the nineteenth century. Everyone knew that a parent's characteristics could be passed on to the offspring, but it was assumed that the parent's own body manufactured the unknown physical entity that shaped the offspring's development. On such a model it seemed reasonable to assume that any changes to the parent's body could be built into the process of inheritance. Darwin's own theory of inheritance, pangenesis, worked in this way, which is why he could allow a limited role for Lamarckism.[49]

In Britain there were some Lamarckians still willing to admit that selection plays a role in eliminating the less successful modifications induced directly by the environment. In 1888 the botanist Rev. George Henslow took this approach in his *Origin of Floral Structures*, issued in the International Scientific series. Others were less conciliatory, however,

[49] On these conceptual developments, see, for instance, Staffan Müller-Wille and Hans-Jörg Rheinberger, *A Cultural History of Heredity*, and my own *The Mendelian Revolution*.

and now began to claim that their theory made natural selection, with all its negative emphasis on struggle and death, totally redundant.

For those with literary tastes, the most visible example of this emerging hostility was provided by the novelist Samuel Butler. Originally a supporter of Darwinism, he had included a good-natured pastiche of natural selection applied to the improvement of machinery in his *Erewhon*, published anonymously in 1872. Having read Mivart he realized that one could be an evolutionist without accepting natural selection and became convinced that Lamarckism offered a rationally and morally superior alternative. He turned against Darwin when he realized that Lamarck and others had proposed this mechanism long before the selection theory had been conceived. This point was driven home in his *Evolution, Old and New* of 1879 and a series of later books on the topic. For Butler as for Cope, the theory's great advantage was that it allowed the creative activity of the organism to direct its species' evolution. He suggested that heredity was a form of memory: the developing embryo is in effect remembering the evolutionary development of its species as in the recapitulation theory. His books may have had limited readership at first; Grant Allen wrote a negative review of *Evolution, Old and New* and Butler admitted that its publication had been badly managed. But a second edition was called for in 1882 and his ideas gradually began to gain some support, with even Darwin's son, Francis, effecting a reconciliation.[50]

In 1886 Spencer published a conciliatory effort to defend the synthesis of Darwinism and Lamarckism in the magazine *Nineteenth Century*, reprinted in *Popular Science Monthly* and in his *Factors of Organic Evolution*. But this was in the context of rising tensions, aired at a session chaired by Lankester at the annual meeting of the British Association the following year.[51] Orthodox Darwinians were springing to the defence of the selection theory. From the Darwinian inner circle George John Romanes published his *Scientific Evidences for Organic Evolution* in 1882, reprinted from articles in the *Fortnightly Review*. It was a slim volume priced at only 2/6d, which also sold well in the United States. It opened with a short description of natural selection followed by accounts of how fields such as embryology and geographical distribution supported the theory. The chapter on the fossil record was only two

[50] For Butler's comments on the reception of his books, see H. Festing Jones, *Samuel Butler (1835–1902)*, 1: 155, 304, 318 and 371.

[51] See Lankester, 'Inheritance of Acquired Characters', and for details of these debates Bowler, *The Eclipse of Darwinism*, chaps. 4 and 6.

pages long, however, although Marsh's horse sequence got a mention in the survey of comparative anatomy.

Alfred Russel Wallace now became active in response to the growing tensions. He lectured across America in 1886–9 defending evolution by natural selection (although his greatest success was in San Francisco when he switched to spiritualism).[52] In 1889 he published an extensive survey under the title *Darwinism*, which opened with a detailed descriptions of natural selection and then outlined the evidence in its favour. Wallace summarized his own extensive work on geographical distribution, showing how it could be explained by the migration of species over time. He differed from Darwin on several issues, including the relevance of artificial and sexual selection, but more importantly on the application of the theory to human origins. Here he invoked the involvement of the supernatural, and attentive readers would note that his conclusion echoed Buckley's view that the whole process of evolution was designed by the Creator to ensure the emergence of humankind.[53] Nevertheless, on the basic question of natural selection versus Lamarckism, Wallace was identified as a neo-Darwinian who denied any role for use inheritance.

Darwin's closest supporters were only too well aware that their critics were driven by a fear of materialism, but unfortunately the selection theory was indeed taken up by radical thinkers opposed to organized religion. Even Huxley disapproved of atheists and secularists such as Edward B. Aveling who saw Darwinism as a valuable tool for undermining faith in a designing God. Aveling lectured widely to classes organized by the National Secularist Society, and his books were published in the International Library of Science and Freethought organized by Annie Besant and the atheist Charles Bradlaugh. They included his *Student's Darwin* of 1881 and a number of later works giving handy surveys of the theory linked to expositions of its materialist implications. Aveling also translated Haeckel's *The Pedigree of Man* for the same series.[54]

Darwinists such as Romanes and Wallace were trying to shore up an edifice simultaneously being undermined by increasingly strident

[52] See Finnegan, *Voice of Science*, chap. 4.

[53] Wallace, *Darwinism*, pp. 476–8. His later *The Ascent of Life* would expand this more theistic take on evolution.

[54] See Susan Paylor, 'Edward B. Aveling', and more generally Lightman, 'Darwin and the Popularization of Evolution', which also discusses the work of Romanes, Wallace and Drummond. It was Aveling's offer to dedicate his book to Darwin that was indirectly responsible for creating the myth that Marx had offered to dedicate a volume of *Capital* to Darwin. Aveling was in a relationship with Marx's daughter and the letter was at first mistakenly attributed to Marx; see Margaret A. Fay, 'Did Marx Offer to Dedicate *Capital* to Darwin?'

opponents of the selection theory. Chapter 6 shows how the tensions between neo-Darwinians and neo-Lamarckians increased during the 1890s to create the situation later known as the eclipse of Darwinism. Initially, the Darwinians benefitted from a new wave of interest in the study of heredity associated with the German biologist August Weismann. His theory of the 'germ plasm' envisioned a material that transmitted characters from parents to offspring with no possibility of including characteristics acquired during the parents' lifetime. Needless to say, the Lamarckians dismissed the idea as a materialist chimaera and redoubled their efforts to provide their alternative with hard evidence. Nor did the focus on heredity continue to benefit the Darwinians, since it generated renewed interest in the belief that new characters could be introduced by abrupt saltations (soon labelled 'mutations') that appeared and bred true whether or not they conferred any adaptive benefit. Into the early decades of the new century the nature of the evolutionary mechanism remained a source of controversy, giving encouragement both to the opponents of Darwinism and to religious conservatives who still retained doubts about evolution itself.

6 Challenging Darwinism

The decades around 1900 saw changes in society, in science, and in the means of communication, which interacted to complicate public understandings of evolutionism. At one level, it might have seemed that the basic idea of evolution was secure. There was widespread acceptance of a belief that the world today is the product of a sequence of developments that has taken place over a long period of time. But the question of what has driven that process of change remained open and became even more contested. There were increasing doubts about the credibility of the Darwinian/Spencerian approach that had dominated late nineteenth-century thought. Various alternatives were debated, with no sign of a resolution until the 1930s. Some religious thinkers and moralists continued to see a wider purpose in the evolutionary process. More controversially, in the United States especially, there was a renewed attempt to challenge the very basis of the evolutionists' position.

The challenge in science arose from the rise of experimental disciplines, including the study of inheritance. This field drew inspiration from the conviction that heredity rather than environment determines the individual's character. It undermined Lamarckism but initially it also generated a storm of opposition to the theory of natural selection leading to the 'eclipse of Darwinism'. Whatever the doubts raised by historians about the suitability of this phrase, it does identify a period when the theory came under increased criticism. The new science of genetics eventually rescued the selection theory from the doldrums, but its early proponents were anything but supportive. The scientific uncertainties fed into debates over how best to ensure social progress, highlighted by the concerns of those who worried that progress might not be possible in the artificial world created by modern civilization.

The confusion in the scientific community was in part the product of its increased specialization. The amateur expert had been eliminated from all but a few branches of natural history as universities and research centres created laboratories for experimental work. This was what Garland Allen called the 'revolt against morphology', which led to areas such as comparative anatomy and palaeontology being to some extent marginalized except in zoos and natural history museums, encouraging a move away from the disciplines that had underpinned the search for ancestries. The new science of genetics added yet another alternative to the list of anti-Darwinian theories, disseminated to the public via reports of its potential to improve the breeding of agricultural varieties and even human beings.[1]

There was also continued hostility to the alleged materialism of the Darwinian theory, with opponents now able to claim that the theory was no longer accepted even by scientists. Liberal religious thinkers renewed the claim that evolution must express a divine purpose and appealed to the various non-Darwinian theories. There was, however, a growing recognition that progress could not be regarded as an intrinsic law of nature. Moralists such as George Bernard Shaw used their cultural authority to promote Lamarckism but now emphasized the creativity of evolution. As Chapter 7 shows, there was a growing willingness to accept that the ascent of life has been more sporadic and opportunistic than the Victorians had imagined. For the creationists, though, the attempts to invest evolution with a moral purpose were a compromise with atheism, their hollowness revealed by the scientists' inability to agree on how the process worked. The re-emergence of Darwinism from its temporary eclipse was slow to be recognized even in science and was not widely reported in the popular media until the 1930s.

These complexities pose a problem for a survey focused on the public understanding of evolution – how does one accommodate the emergence in the 1920s of a movement rejecting not just Darwinism but the whole concept of biological evolution? Creationism invokes the supernatural rather than any scientifically testable mechanism of change, and its arguments against evolution were (and to some extent still are) just rehashes of the initial claims raised against the *Origin of Species*. But the movement cannot be ignored because it focused attention on the topic and it persuaded many of those who retained serious religious beliefs that the objections – both to Darwinism and to the basic idea of transmutation – were still valid. Those who defended evolution had to campaign in ways that minimized the effect of their opponents' arguments. The scientists of the 1920s often found it convenient to stress that there were less materialistic theories than Darwinism. It was the radical thinkers openly suspicious of organized religion who were most likely to embrace a materialistic interpretation of natural selection. Their involvement only confirmed the creationists' worst fears.

[1] The phrase 'revolt against morphology' was coined in Allen's *Life Science in the Twentieth Century*.

Media Innovations

The turn of the century saw an expansion in the scope and sophistication of the print media and the emergence of totally new means of communication, including cinema and eventually radio. By 1925 the latter was able to play a major role in publicizing the 'monkey trial' of John Thomas Scopes in Dayton, Tennessee. Books and magazines took on a more modern appearance thanks to the use of photographic illustrations. Mass-circulation newspapers and magazines gave their owners and editors immense power to shape public opinion, aided by a new journalistic style exploiting florid headlines coupled with a tendency to oversimplify issues.[2] These sources offered limited scope for communicating science, at least in a form acceptable to the scientific community. Some scientists did learn how to exploit the mass media: E. Ray Lankester wrote a column on science for the Saturday edition of the *Daily Telegraph*, the articles being subsequently collected in book form. A few scientists became celebrities, Julian Huxley being a relevant example. Their efforts were supported and sometimes offset by public figures in other areas who used their position to comment on scientific issues. H. G. Wells, who had studied briefly under the elder Huxley, was generally supportive, while popular novelists such as G. K. Chesterton and the playwright George Bernard Shaw were critical.

In addition to the press and radio, there was a substantial publishing industry devoted to self-education material aimed at readers seeking to improve themselves through study at home. The Home University Library, for instance, was launched by Williams and Norgate in 1911 and issued in the United States by Henry Holt, with copies selling for only one shilling (rising to two shillings and sixpence in the next decade) and $1 for the American editions. Scientific topics formed a substantial part of its output. There were also serial works issued in magazine format and later as books – the *Daily Mail* sponsored the *Harmsworth Popular Science* serial that ran through 1911–13. Wells' *Outline of History* (which had an introductory section on evolution) was a sensation in this format and was followed by an *Outline of Science* edited by J. Arthur Thomson, which sold even better in the United States.[3]

Popular science magazines flourished in America but were less successful in Britain. *Popular Science Monthly* changed owners in 1915 and

[2] See Gowan Dawson, 'The *Review of Reviews* and the New Journalism in Late Victorian Britain'.

[3] For details of British popular science in this period, see my *Science for All* and Peter Broks, *Media Science before the Great War*.

switched to short articles mainly on practical topics, the same format as *Scientific American*, although both had some coverage of natural history and fossil discoveries. Topics relevant to evolution also emerged in coverage of social and practical issues, including the eugenics movement's calls for controls on human reproduction and agriculturalists' demands for the breeding of improved crops. Publication of popular material on science seems to have declined in the United States until the 1920s when the scientific community actively sought to re-engage with the public. John C. Burnham has argued that a growing unwillingness of American scientists to participate in public awareness led to a decline in serious coverage of the topic, although his analysis assumed that only those sympathetic to the area provided valid analysis. Some topics were genuinely controversial both inside and outside science, so not all criticism was superficial or distorted.[4]

On the topic of evolution, the media circus was a battleground even before the creationist onslaught of the 1920s. Given the lack of agreement over the mechanism of change, this was true even when expert scientists were involved. Those from an older generation were inclined to support interpretations that still favoured Lamarckism or some form of divine plan, joining the liberal theologians who were willing to accept evolutionism provided it was presented in a non-materialistic form. Commentators from outside science joined the assault on Darwinism: Chesterton revived a traditional Christian view, while Shaw promoted the idea of 'creative evolution' driven by a mysterious life force. The fact that noted scientists found it necessary to respond confirms that they recognized the critics' ability to shape public opinion.[5] Some publishing initiatives were a direct response to the attacks. In Britain C. A. Watts issued cheap paperbacks in the Forum Series linked to the programme of the Rationalist Press Association. Several of the books provided refutations of the claim that Darwinism was no longer accepted in science. In the United States, the Little Blue Books issued by Haldeman-Julius Publications also provided cheap material linked to a radical agenda, including several texts on evolution intended to counter the creationist attacks leading up to the Scopes trial.

Fiction could itself be a powerful vehicle for exploring the implications of evolutionism for modern life. Wells' science fiction writings provide an

[4] Burnham, *How Superstition Won and Science Lost*; more generally on the United States, see Marcel LaFollette, *Making Science Our Own*; Ronald C. Tobey, *The American Ideology of National Science, 1919–1930*; and Constance Areson Clark, *God – Or Gorilla*, chap. 5.

[5] This point is stressed in Piers J. Hale, 'The Search for Purpose in a Post-Darwinian Universe'.

obvious example, in this case sympathetic to Darwinism. There were also novels that brought in topics related to prehistoric life, including the life of 'cavemen' and the popular genre of 'lost world' narratives in which humans interact with supposedly extinct creatures such as dinosaurs. The same themes emerged in the new medium of cinema and were already parodied by Charlie Chaplin as early as 1914. A movie entitled *Evolution*, released in 1923 and vetted by the American Museum of Natural History, surveyed the history of life on earth to counter the growing threat posed by fundamentalist opposition to the theory. From the scientists' perspective, though, the increasing breadth of media potentially available exposed their inability to control what was offered to the public.

The Death of Darwinism?

In 1890 the *Universal Review* published Samuel Butler's last diatribe against the theory of natural selection under the title 'The Deadlock in Darwinism'. In 1904 translations of anti-Darwinian arguments by German scientists were published in the United States to convince the public that they were, as Dennert's book is titled, 'at the deathbed of Darwinism'. As late as 1932 J. B. S. Haldane's proclamation of the theory's revival following the synthesis with genetics opened with the aphorism 'Darwinism is dead – any sermon' (in an article on the topic he attributed the phrase to Hilaire Belloc).[6] The belief that Darwin's theory had run into difficulties so serious that it might have to be abandoned altogether was widely touted in the media during the decades around 1900 and – as Haldane's reaction suggests – remained popular for some time. In an increasingly polarized situation, traditional alternatives to selectionism such as Lamarckism vied with a newer one derived from the study of heredity to replace it. Paradoxically, the hereditarian approach was even more damaging to the old Lamarckism and would eventually emerge as a new foundation for the selection theory.

Unfortunately for the Darwinians, the first generation of geneticists had linked their model of heredity to the idea of evolution by abrupt saltations (now known as 'mutations'). Biologists working in laboratories found it hard to believe that the struggle for existence was powerful enough to control the direction of change. Their opposition to

[6] Butler's article is reprinted in his *Essays on Life, Art and Science*, pp. 234–340; see also Eberhart Dennert, ed., *At the Deathbed of Darwinism*, and J. B. S. Haldane, *The Causes of Evolution*, p. 1, and 'Darwinism Today' reprinted in his *Possible Worlds*, pp. 27–44, reference to Belloc at p. 27.

Darwinism coincided with the last surge of interest in Lamarckism among biologists, so the selection theory was attacked on two fronts – although the two alternatives were also hostile to each other. It would soon be realized that small-scale mutations could provide the element of undirected variation that the Darwinians had always held to be the raw material of selection. But in the meantime the selection theory seemed to be in serious trouble, leading Julian Huxley to adopt the phrase 'eclipse of Darwinism' to denote the events at the turn of the century.[7]

The controversies in science also coincided with an intensification of hostility to the selection theory among religious thinkers and moralists who saw it as an expression of materialism and the source of social attitudes that undermined traditional values. We must be careful, though, not to identify this hostility too closely with outright opposition to evolutionism and the rise of what would eventually be called 'creationism'. All eyes tend to focus on the 'Monkey Trial' of John Thomas Scopes in 1925 as the symbol of fundamentalist hostility to the whole ideology of evolution. But this movement was largely a product of social tensions that did not emerge until after the Great War of 1914–19, and even then most religious opposition focused on the origin of humanity. Many theologians were still willing to accept evolution in the animal kingdom as long as it was seen as the unfolding of a divine plan. Darwinism remained a key target because it was so closely identified with materialism – which is why some scientists found it convenient to highlight the alternative theories to checkmate this form of attack. But the eclipse of Darwinism certainly did not mean the eclipse of evolutionism. Traditionalists were prepared to tolerate theistic evolutionism, while many liberal Christians joined the wider moral crusade to present evolution as a process driven by purposeful forces toward a spiritually significant goal.

Focusing first on the scientific debates, it was hard to conceal the extent to which Darwinism had come under fire. This was a consequence of the collapse of the more broadly based understanding of the theory that had emerged in the 1870s as Spencer's philosophy encouraged a synthesis of the selection theory with Lamarckism. The public image of 'Darwinism' changed as the term became identified with Weismann's claim that natural selection is the only viable mechanism of evolution, the position sometimes called 'neo-Darwinism'. In reaction, many of those who still endorsed Lamarckism in turn became more dogmatic, creating a rival movement known as 'neo-Lamarckism' totally hostile to the

[7] There is a section under this title in Huxley's *Evolution: The Modern Synthesis*, pp. 22–8.

selection theory. This was most active within the traditional areas of natural history and palaeontology. It was reinforced by the calculations of the physicist William Thomson, Lord Kelvin, who argued that the earth could only have supported life for a hundred million years, far too short a period for natural selection – which Darwin had seen as a very slow process – to produce the advanced species of today. Lamarckians insisted that the inheritance of acquired characteristics could work much more rapidly.[8]

Experimental disciplines including physiology and new approaches to the study of heredity moved in a different direction and became less willing to accept the traditional arguments supporting Lamarckism. They endorsed August Weismann's concept of the germ plasm as the material basis for thc transmission of characteristics between generations and his claim that modifications of the adult organism could not be incorporated into the process. The translation of Weismann's *The Germ Plasm* was published in 1893, and in the same year he engaged in a controversy with Spencer in the pages of the *Contemporary Review* that helped to define the gulf between neo-Darwinism and Lamarckism.[9] But while the neo-Darwinians followed Weismann in holding that this model left natural selection as the only viable mechanism of evolution, the experimentalists were unwilling to imagine that the struggle for existence in nature could eliminate the new characteristics they saw emerging through mutations. Genetics took over this position and thus was at first seen as a new source of opposition to Darwinism. There was now a three-way polarization of opinions between Darwinians, Lamarckians and the geneticists.

An indication of the growing tensions emerged when the Marquis of Salisbury gave his presidential address to the 1894 meeting of the British Association for the Advancement of Science. He accepted the basic idea of evolution but pointed out the new attacks on Darwinism. The vote of thanks was proposed by Lord Kelvin and seconded by the aging T. H. Huxley, who managed to politely chide the Marquis for exaggerating the level of opposition to the theory.[10] There were endless angry exchanges in the pages of *Nature*. The widely reported celebration of the centenary of Darwin's birth held in Cambridge in 1909 was obliged to include eminent speakers representing the whole range of opinions, including

[8] See Joe D. Burchfield, *Lord Kelvin and the Age of the Earth*.

[9] Spencer, 'The Inadequacy of Natural Selection', 'Professor Weismann's Theories' and 'A Rejoinder to Professor Weismann'; Weismann, 'The All-Sufficiency of Natural Selection'.

[10] Salisbury, 'Presidential Address'. The event was extensively reported in the press; see Desmond, *Huxley: Evolution's High Priest*, p. 223, and Leonard Huxley, *The Life and Letters of Thomas Henry Huxley*, 2: 375–8.

Weismann for the neo-Darwinians, Haeckel for Lamarckism and Hugo De Vries and William Bateson for the new science of heredity. In his preface to the published texts, Professor A. C. Seward conceded that 'The divergence of views among biologists as regards to the origin of species and as to the most promising directions in which to search for truth is illustrated in the different opinions of the contributors.' All accepted that Darwin had started an important movement, but they could not agree on the direction ahead.[11]

There was no shortage of literature that would fill in details of the alternatives for the reading public. The most authoritative survey was Vernon Kellogg's *Darwinism Today* of 1908, intended for both students and general readers – technical details quoted from original sources were printed as endnotes for those who wanted them. In 1911 Patrick Geddes and J. Arthur Thomson co-authored the volume *Evolution* for the Home University Library. It was sympathetic to both Darwinism and the new genetics (Thomson had written an extensive survey of the debates on heredity), but also included a chapter on Lamarckism. The collection *At the Deathbed of Darwinism* introduced American readers to the German attacks and was clearly intended to reach those with religious concerns about the theory.

The controversies in science were themselves influenced by the changing perception of Darwinism in the public arena. In the 1890s Spencer was still promoting his synthesis of natural selection and Lamarckism, but the intransigence of Weismann and the neo-Darwinians forced him to take a more militant attitude to defend the inheritance of acquired characters. Most neo-Lamarckians turned their backs on Darwinism altogether and joined Butler in dismissing it as a theory with unacceptable moral implications. Religious critics of the *Origin of Species* had pointed out the dangerous implications of seeing the world as a scene of relentless struggle, and some novelists had begun to explore the human implications of this vision, but their fears had been to some extent offset by the Spencerian compromise that allowed struggle to be seen as a force for good – any species living in social groups had of necessity to develop cooperative instincts. In the 1890s this interpretation of Spencer's ideology was replaced by a harsher interpretation of the struggle for existence that proclaimed a ruthless indifference to the fate of those less fit to survive, reviving the fears of those who had warned all along that this was the real message of Darwinism.

[11] Preface to A. C. Seward, ed., *Darwin and Modern Science*, pp. v–viii, quotation from p. vii; On the extent of the press coverage, see Marsha L. Richmond, 'The 1909 Darwin Celebration'.

One contributor to the image of nature as a scene of ruthless struggle was T. H. Huxley. Partly out of concern over the new social tensions but also as a result of personal tragedy (his daughter Marion lapsed into insanity and died), he became more conscious of the harsh implications of the Darwinian image of the world. In 1888 he attacked Spencer's ideology of unrestrained free enterprise in a widely reported speech in Manchester Town Hall, subsequently published as 'The "Struggle for Existence" in Human Society'. He was no egalitarian, but he wanted an orderly world governed by experts, not a free-for-all. In 1893 he gave the Romanes lecture in Oxford under the title 'Evolution and Ethics', stressing the need for human society to rise above the harsher aspects of the natural world. The text is sprinkled with references to the 'struggle for existence' and presents Darwinism as the 'gladiatorial theory of existence'.[12]

Huxley seems to have thought that evolution has produced a species capable of transcending the mechanism of development. For many commentators, however, the best way out of the dilemma was to recognize that the Darwinian image of 'Nature, red in tooth and claw' is misleading. This was the message of the Scottish Presbyterian minister Henry Drummond's bestselling *Ascent of Man* in 1894. He also promoted his message that evolution is driven by cooperation rather than struggle in lecture tours in North America in 1887 and 1893.[13] The anarchist Peter Kropotkin responded to Huxley's 1888 diatribe with a series of articles also in the *Nineteenth Century* later collected in his 1902 volume under the title *Mutual Aid*. He argued on the basis of his own observations that animals in the wild routinely cooperate with one another, providing a better model on which to base human society. His argument paralleled the position of liberal Christians such as Drummond who were willing to adopt the evolutionary principle but needed an alternative to the harsher model now associated with Darwinism. Many shared the view that Lamarckism offered a more satisfactory alternative to natural selection, but for those with more conservative Christian principles the best policy was increasingly to reject evolutionism altogether.

The Challenge of Creationism

Histories of evolutionism conventionally assume that once the initial religious opposition to the *Origin of Species* died down, there was little activity

[12] See Desmond, *Huxley: Evolution's High Priest*, chap. 10, for details of both speeches. 'Evolution and Ethics' is reprinted in Huxley's *Collected Essays*, vol. 9, references from pp. 51 and 80–2.

[13] See Finnegan, *Voices of Science*, chap. 5.

on this front until the emergence of creationism in the 1920s. In fact, the issues remained sensitive – this may have been an age of secularization, at least in Britain, but even here atheism struggled to gain a hearing. Many ordinary people would gain their view of evolutionism from sources that still expressed concern, if not over the basic idea, then certainly over materialistic versions including Darwinism. Even in the American South, outright rejection of evolution was relatively unusual until after the Great War, and many religious sources favoured the view that the development of life has been divinely guided. The war encouraged conservative suspicions of modern cultural developments, which began to refocus on Darwinism as an expression of dangerous materialism. Traditionalists, whether Protestant or Catholic, rejected the idea of progress and saw any threat to the unique status of the human soul as unacceptable.

The most visible component of this new anti-evolution campaign was the trial of John Thomas Scopes in 1925, accused of violating Tennessee's Butler Act, which forbade the teaching of evolution in the schools. This was the culmination of a campaign by the evangelical Protestants, sometimes called fundamentalists after the title of a series of pamphlets published just before the war. The whole episode is often seen as the opening move in the sustained campaign to promote creationism that still plagues evolutionists today. But even the term 'creationism' is hard to pin down – it covers a range of opinions and is applied to a series of initiatives on the part of different religious groups over this extended period. The most extreme version is 'young earth' creationism, which rejects orthodox geology and claims that the earth is only a few thousand years old. Its 'flood geology' argues that the strata of fossil-bearing rocks were laid down in a single catastrophic event identified with Noah's flood. This position was already being articulated in the 1920s by the Seventh-Day Adventist George McCready Price, but it played only a minor role at this point. Most fundamentalists were content to accept the earth's antiquity, invoking successive creations or even theistic evolution to explain the history of life. The one creation they did insist on was that of the human soul. Although usually associated with evangelical Protestantism, this range of positions can also be seen within the Catholic Church. Nor were the attacks confined to the United States – there were more limited expressions of the same sentiments in Britain.

Modern accounts of American creationism expose a number of myths that have accumulated around the Scopes trial, many of which were dramatized in the 1960 movie *Inherit the Wind*.[14] Opposition to evolution

[14] See, for instance, Ronald L. Numbers, *The Creationists* and *Darwinism Comes to America*; Edward Larson, *Summer for the Gods*; Clark, *God – Or Gorilla*; and my own *Monkey Trials*

had been dormant even in the southern states, but revived in response to the perceived moral decline of the 'jazz age' that emerged after the Great War. The campaign to ban the teaching of evolution in schools was active in many states, but only three – Arkansas, Mississippi and Tennessee – actually implemented the move. Although some newspapers in the big northern cities followed H. L. Mencken of the *Baltimore Sun* in seeking to discredit the trial, others were more sympathetic, while some southern newspapers actually criticized the prosecution. William Jennings Bryan, who led the prosecuting team, was in contact with George McCready Price but openly accepted the earth's antiquity and was sympathetic to theistic evolutionism – as were some of the authors who contributed to the fundamentalist pamphlets. In the aftermath of the trial, the creationists' goal of getting evolution removed from the school curriculum succeeded and their campaign subsided until the 1960s.[15]

The interchanges between Bryan and the agnostic lawyer Clarence Darrow, who led for the defence, shed more heat than light on the topic of how evolution was caused. Only the distaste that Bryan and many others felt for the concept of the 'survival of the fittest' came through. The scientists who supported the defence were led by Henry Fairfield Osborn, who rejected Darwinism in favour of a theory of parallel evolution, which allowed him to challenge the ape ancestry of humanity. Osborn had already clashed with Bryan in the Sunday editions of the *New York Times* in 1922. He did not attend the trial (nor did any other senior scientist) but did publish a book, *The Earth Speaks to Bryan*, which sold less well than some of his other works. The American Museum of Natural History where he worked was a key centre for the promotion of evolutionism throughout the decade. Numerous books and articles were written in response to the creationist challenge, although a magazine entitled *Evolution* was not a success (it ran initially from 1928 to 1932).[16]

The conviction that the spiritual element in human nature must have been the product of divine intervention was shared by the Catholic Church. When the American priest John Zahm wrote his *Evolution and Dogma* of 1896 to defend theistic evolution, his book was condemned – but the condemnation was not published and the Church hoped that the issue could be sidelined. Henri de Dorlodot's *Darwinism and Catholic Thought*, translated by Ernest Messenger in 1925, revived the argument,

and Gorilla Sermons. Numbers et al., eds., *Creationism in Twentieth-Century America*, reprints a series of creationist publications.

[15] Dorothy Nelkin, *Science Textbook Controversies and the Politics of Equal Time*.

[16] See Joe Cain, 'Publication History for *Evolution: A Journal of Natural History*'; the magazine was revived in 1937.

and Messenger's own *Evolution and Theology* appeared in 1931. Both accepted that the human soul had to be separated from the general case for biological evolution as the unfolding of a divine plan. The Church was willing to allow that case to be made, although it would not openly endorse the theory and later blocked the efforts of Pierre Teilhard de Chardin to publish on the topic.[17]

In Britain there was a tendency to dismiss the Scopes affair as a purely American phenomenon, but in fact there was significant opposition to evolution emerging from both Catholic and evangelical Protestant forces. The Catholic objections achieved high visibility because they were endorsed by two literary figures whose reputations gave them access to the media: Hilaire Belloc and G. K. Chesterton (the latter converted in 1922). Belloc attacked the evolutionism displayed in the introductory chapters of H. G. Wells' *Outline of History*, published in 1920. His critiques in the *London Mercury* and the *Dublin Review* were then extended into a book. He singled out Wells' support for Darwinism as evidence that his science was out of date. In 1927 he clashed with the anatomist Arthur Keith in the pages of *Nature* on the question 'Is Darwinism dead?' (hence Haldane's attribution of the phrase). Ten years later he was still identifying Darwinism as a key element in the materialism that was precipitating – in the title of another book, *The Crisis in Our Civilization*. Chesterton was already a much-loved novelist when he got involved, first in a 1922 attack on eugenics and then on Darwinism itself in his 1925 book *The Everlasting Man*. In 1935 he used his regular column in the *Illustrated London News* to again insist that the Darwinian view of nature was a dangerous illusion.[18]

There was also growing opposition to the efforts that had been made to liberalize the position of the other churches. When E. W. Barnes, then canon of Westminster, preached what became known as his 'gorilla sermons' in 1920, his claim that Christians needed to rethink their position on original sin in the light of evolution was vigorously criticized in the press. The head of the Salvation Army wrote to the *Times* in protest, and when an article in the popular science magazine *Conquest* referred to the sermons, it too received objections. In 1927 Barnes preached in support of Arthur Keith's Presidential Address to the British Association, which had endorsed Darwinism. Much of the

[17] See Don O'Leary, *Roman Catholicism and Modern Science*.

[18] Belloc, *A Companion to Mr. Wells'* Outline of History and *The Crisis of Our Civilization*; Chesterton, *Eugenics and Other Evils* and *The Everlasting Man*; his 1935 article 'About Darwinism' is reprinted in his *As I Was Saying*, pp. 194–9. For more details of these attacks, see Bowler, *Reconciling Science and Religion*, pp. 395–400.

criticism came from evangelical Protestants led by the physicist Sir Ambrose Fleming, who became president of the Victoria Institute, founded to show that science could be made compatible with traditional Christian beliefs. Their activities led to the foundation of the Evolution Protest Movement in 1935, which campaigned against the BBC's coverage of the topic.[19]

The effectiveness of the British campaign is hard to evaluate. Hostility to the assumed materialism of the selection mechanism was an important factor, but the focus on human origins makes it difficult to be sure whether those involved were opposed to the whole idea of evolution. Rejecting Darwinism had an obvious advantage when the scientists themselves still disagreed over the validity of the selection theory. The scientific community pretended that the objections were based on superficial misunderstandings, but the fact that prominent figures such as Keith and Haldane felt it necessary to respond suggests a deeper concern over their potential impact. The rationalist Joseph McCabe described the fulminations of Belloc and Chesterton as 'the most successful hoax that has been worked on the British public'. When McCabe debated George McCready Price, then in London, in 1925, the audience heckled the latter's efforts to defend flood geology. This suggests that either biblical literalism was a step too far for the British public or the rationalists had packed the audience with their own supporters.[20]

A Sense of Purpose

Those with a more liberal Christian perspective were keen to retain the idea of progress so long as it was conceived as a means by which the Creator achieved his ends. Evolution must be seen as a process that enhanced ethical behaviour and the emergence of spiritual qualities. As Barnes pointed out in his 'gorilla sermons' this meant abandoning the traditional belief in a fallen humanity. The same position had been articulated in Henry Drummond's *Ascent of Man* in 1894 and was central to the 'new theology' of Reginald Campbell, who preached to huge crowds at the City Temple in London during the early years of the new century. Campbell and Barnes were perhaps the most prominent

[19] For details, see Keith, *Concerning Man's Origin*, and Bowler, *Reconciling Science and Religion*, pp. 124–30. For the *Conquest* interchange, see the editorial 'Canon Barnes on the Fall of Man', II (1920–1), p. 232, in response to the article on apes by Pocock, pp. 134–7.

[20] For the McCabe quotation, see his *The Riddle of the Universe Today*, pp. vii and 13; for Wells' complaint about publishing, see his *Mr. Belloc Objects to* the *'Outline of History'*; see also Price and McCabe, *Is Evolution True?*

members of what became known as the Modernist movement in the Anglican and Nonconformist churches (not to be confused with modernism in the arts).

In the United States too there was an active Modernist movement, although it faced a more robust challenge when the wave of fundamentalism built up. Accepting evolution as the unfolding of a divine plan was also central to their liberal interpretation of the Christian message, following the lead of earlier thinkers and theologians such as Fiske and Henry Ward Beecher. In the early years of the new century the movement was led by the University of Chicago theologian Shaler Matthews and the Presbyterian preacher Harry Emerson Fosdick. Book such as Matthews' *The Church and the Changing Order* (1907) and Fosdick's *Christianity and Progress* (1922) accepted evolutionism as a foundation for the hope of spiritual progress in the future.[21]

There were scientists too who supported the liberal position. Matthews edited a series of pamphlets on science and religion that included contributions from biologist Edward Grant Conklin and geologist Kirtley Mather. The American editor Frances Mason persuaded scientists from both sides of the Atlantic to write popular articles showing how the latest developments led away from materialism. Her *Creation by Evolution* of 1928 was a Book of the Month Club recommendation and the subject of radio broadcasts. Support for the idea of evolution as a progression toward higher levels of consciousness also came from the physicist Sir Oliver Lodge, whose enthusiasm for spiritualism led him to claim that evolution was a process intended to generate immortal souls. He had a successful lecture tour in America in 1920, and books such as his *Evolution and Creation* sold well. An article on spiritualism by Lodge in the serial *The Outline of Science* attracted opposition from E. Ray Lankester.[22]

Much of this literature was vague on the actual process of evolution, although there was widespread support for the Lamarckian view that animals can direct their own evolution. Sympathetic biologists were in a difficult position because the experimental evidence for the Lamarckian effect was increasingly fragile, while Darwinism had to be presented in a way that avoided its materialistic implications. We can see this dilemma

[21] See David N. Livingstone et al., eds., *Evangelicalism and Science in Historical Perspective*; Edward B. Davis, 'Science and Religious Fundamentalism in the 1920s'; and Bowler, *Monkey Trials and Gorilla Sermons*, pp. 171–3.

[22] On Frances Mason, see Clark, *God – Or Gorilla*, pp. 99–100, and Bowler, *Reconciling Science and Religion*, pp. 45 and 138–40. On the background to Lodge's views, see Janet Oppenheim, *The Other World*, pp. 371–90. For Lankester's critique of Thomson and Lodge, see Lester, *E. Ray Lankester*, p. 213.

in the work of one of the most prolific popular science writers of the period, the Aberdeen professor of natural history J. Arthur Thomson. His books sold well on both sides of the Atlantic, and he too lectured across the United States in 1924. His whole output was coloured by a desire to show that the development of life has been something more than a purely mechanistic process. One of his popular books was called *The Gospel of Evolution*, the good news being that the process had a moral foundation in which struggle was shaped by the purposeful activity of life. He insisted that animals can respond creatively to the challenges of the environment and that this can direct the course of natural selection (a process known as organic selection). He still featured a 'tree of life' with a central trunk, but it did at least indicate that the birds and mammals branched off in separate directions (see Fig. 6.1).

Thomson was impressed with the French philosopher Henri Bergson's vision of 'creative evolution' – this might be poetry rather than science, but it gave a meaning to what the scientists were finding. Bergson had attracted wide attention in both Britain and the United States even before the translation of his *Creative Evolution* appeared in 1911. Like many others, Thomson understood Bergson's *élan vital* as an active principle emerging from the freedom that living things have to choose how to react when challenged.

There were still a few eminent figures willing to endorse the Lamarckian effect. George Bernard Shaw was one of Britain's most publicized intellectuals, his opinions and eccentricities widely commented on in the press. When he attacked Darwinism in the name of Bergson's creative evolution, scientists were well aware that he was an adversary to be reckoned with.[23] Shaw's most vitriolic critiques of Darwinism were contained in the prefaces to some of his plays, most notably *Man and Superman*, published in 1906, and *Back to Methuselah* in 1921. The earlier play was a success, but the latter was not – although the attack on Darwinism in the published version was widely noted in the press, with even the satirical *Punch* responding, 'But we are quite sure that DARWIN never intended his theory to be applied to Mr. SHAW.' In fact, Shaw's emphasis on the creativity of living things ignored the more sophisticated aspects of Bergson's thinking and was no more than a renewal of Samuel Butler's Lamarckism. He expressed his hatred of natural selection in similarly picturesque terms: 'If it could be proved that the whole universe had been produced by such Selection, only fools

[23] On Shaw's influence, see Hale, 'The Search for Purpose in a Post-Darwinian Universe'.

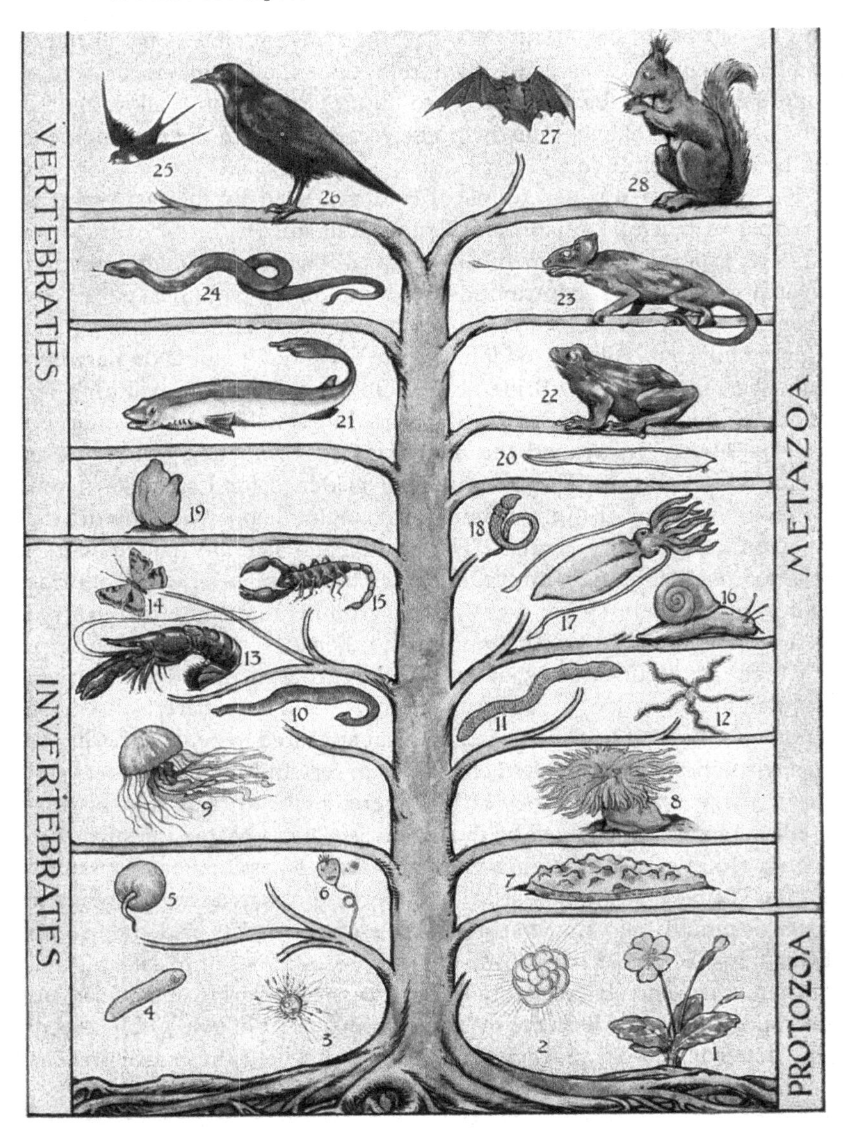

Fig. 6.1 'Tree of life' with a central trunk dividing at the top to indicate the divergence of birds from mammals. From J. Arthur Thomson, *The Outline of Science* (1922), vol. 1, facing p. 336.

and rascals could bear to live.'[24] The alternative preferred by all right-thinking people was creative evolution driven by a life force whose innovations could be transferred to future generations, allowing the whole sequence of achievements to be recapitulated in the development of the modern embryo.

Recapitulation was a key line of evidence used by the last eminent scientist to defend Lamarckism, Ernest William MacBride of Imperial College London. He aligned himself with Thomson and the reaction against materialism, contributing to both of the volumes edited by Frances Mason. Thomson used his position as science editor of the Home University Library to let him write the series' volume on heredity, published in 1924. MacBride also wrote the volume on evolution for another popular series, Benn's Sixpenny Library. Both books attacked the Darwinian theory and the new genetics, dismissing mutations as merely degenerative. The experimental evidence for Lamarckism was defended and linked to the claim that changing habits were the driving force of evolutionary change. The volume on heredity ended with a chapter on the human implications of the topic in which MacBride endorsed the hierarchy of races on the grounds that some had evolved in a more stimulating environment than others.[25]

Much of MacBride's chapter on the experimental evidence was devoted to the work of the Austrian biologist Paul Kammerer, who made a name for himself by demonstrating what appeared to be the inheritance of environmentally stimulated characters in sea-squirts, salamanders and the midwife toad. The story of Kammerer's rise to fame and his discrediting in a campaign led by the geneticists later became the subject of Arthur Koestler's *The Case of the Midwife Toad*, a passionate contribution to the claim that Lamarckism had been eliminated by a dogmatic Darwinism. In fact, the first generation of geneticists had themselves been hostile to Darwinism, while Koestler's implication that Lamarckism represents a morally superior world-view founders on his depiction of MacBride as the 'Irishman with a heart of gold', oblivious of the fact that he was an Ulster Protestant who held the native Irish in contempt.[26]

[24] Shaw, *Back to Methuselah*, p. liv, in this case quoting his own words from the preface to *Man and Superman*. The first edition cited here is a well-produced book, not a cheap publication. For the *Punch* response, see the issue of 6 July 1921, p. 2.

[25] MacBride, *An Introduction to the Study of Heredity*, chap. 9. See Bowler, 'E. W. MacBride's Lamarckian Eugenics and Its Implications for the Social Construction of Scientific Knowledge'.

[26] Koestler, *The Case of the Midwife Toad*, p. 82.

Kammerer's observations on the midwife toad were not, in fact, the most convincing element of his work, but became the focus of attention when they were subjected to an intense scrutiny that led to them being discredited. William Bateson had already questioned Kammerer's work in his *Problems of Genetics* of 1913, and his opposition intensified as Kammerer toured Britain and the United States in 1923 and 1924. By this time Kammerer had become involved with Eugen Steinach's much-publicized hormonal rejuvenation techniques, which he saw as a medical application of processes similar to those postulated by Lamarckism. Kammerer's book *Rejuvenation and the Prolongation of Human Efficiency* appeared in 1924, and Steinach was also featured in his *The Inheritance of Acquired Characters* in the same year. This hailed the prospects for the improvement of the human race if the Lamarckian process could be controlled – leading the *Daily Express* in Britain and the *New York World* to run headlines on the supposed benefits.[27]

The case against Kammerer focused on the midwife toad specimens, which could be studied to see if the supposed changes induced by a new environment were genuine. Conducted mostly in the pages of *Nature* from 1919 to 1926, with Bateson for the prosecution and MacBride for the defence, it was brought to a close when G. K. Noble of the American Museum of Natural History reported that the mating pads on the specimens' limbs had been marked with India ink. Shortly afterwards Kammerer, who was about to move to Moscow, shot himself. He had claimed that an assistant must have marked the pads when the original colour began to fade, but the details of the debate received little press attention and the whole affair soon sank from sight.

Mutations and Genes

The decline in the fortunes of Lamarckism was a result of its inability to meet the demands of the experimental study of heredity that came to the fore around 1900. With hindsight we know that genetics would eventually become the salvation of Darwinism, but in its early years the new science was equally hostile to the selection theory. Precisely because it was an experimental discipline based in laboratories and breeding centres, its practitioners were not in tune with the concerns of field

[27] For Bateson's early opposition, see his *Problems of Genetics*, pp. 199–211. Steinach appears in Kammerer's *The Inheritance of Acquired Characteristics*, chap. 50. The newspaper headlines are quoted in Koestler, *The Case of the Midwife Toad*, pp. 85–7. For a careful reappraisal of these events, see Sander Gliboff, 'The Case of Paul Kammerer'. For a brief description of the controversy over rejuvenation techniques, see Bowler, *A History of the Future*, pp. 188–92.

naturalists. They found it hard to believe that environmental pressures could generate a struggle for existence with the power attributed to it by the Darwinians. They were sucked into a wave of scepticism about the whole programme that saw adaptation as a key factor in evolution. Even Huxley had thought that new characters appeared by abrupt saltations and survived whether or not they had any adaptive value. Now this way of thinking was in the ascendence and Darwinism was just one of the casualties.

The theory of non-adaptive evolution was revived in the 1890s by the German naturalist Theodor Eimer, a vocal opponent of Weismann who championed the idea of orthogenesis, evolution driven by non-adaptive variation trends. His *Organic Evolution*, translated in 1890, detailed observations suggesting the inheritance of acquired colour changes in lizards and butterflies but also provided evidence of trends with no adaptive significance. A shorter work entitled *On Orthogenesis* was published in the Open Court Press's Religion of Science library in 1898, selling for only 25 cents (one shilling and sixpence in Britain). Eimer argued that the resemblances that the Darwinists attributed to mimicry were actually the result of similar variation trends operating within separate species, producing parallel evolution.[28]

The problem with Eimer's work was that the sequences making up his orthogenetic trends were constructed from living species that he could only assert were equivalent to a series produced in time. For this reason, the most visible switch from Lamarckism to orthogenesis came within the ranks of the palaeontologists. Cope and his fellow American neo-Lamarckians invoked use inheritance to explain the apparently linear trends they saw in the fossil record but assumed that these could sometimes continue beyond the point of utility, leading to degenerative over-development. A once-beneficial trend became embedded in the animals' constitution, acquiring a momentum that drove it to non-adaptive levels, leaving the species vulnerable to extinction. As the evidence for the Lamarckian effect became less plausible, the next generation of fossil-hunters tended to assume that even the early phases were produced by orthogenetic variation trends (this topic is explored in more detail in Chapter 7).

[28] The full title of Eimer's *Organic Evolution as the Result of the Inheritance of Acquired Characters According to the Laws of Organic Growth* suggests the relationship; see also his *On Orthogenesis and the Impotence of Natural Selection in Species Formation.* For more details, see Bowler, *The Eclipse of Darwinism*, chap. 7. Mark A. Ulett argues that the language used by Eimer and other proponents of the theory was designed to engage the public in their campaign; see his 'Making the Case for Orthogenesis'.

Eimer's more restricted claim that non-adaptive modifications could be seen in living organisms as well as in the fossil record was taken up by biologists studying heredity. They became convinced that the varieties they could observe within species were the products of discontinuous reconfigurations of the developmental process, the sudden 'jumps' known as saltations or sports of nature. Their focus on internal forces paralleled the logic of orthogenesis, except that the saltationists saw the result as an array of different forms rather than a sequence of modifications in a single direction. Unlike Darwin's minute individual differences, they were substantial transformations that produced a new, permanent variety instantaneously. Gradual modification was not involved, and new varieties seemed to establish themselves even though they conferred no adaptive benefit. Hugo De Vries called these new varieties 'mutations' – a term that would subsequently be appropriated in a somewhat different context by the science of heredity that came to be known as genetics.

The focus on the power of heredity to predetermine the organism's characteristics was inspired by Weismann's theory of the germ plasm, but many converts found his reliance on natural selection unconvincing. Darwin's cousin, Francis Galton, was also emerging as a champion of the hereditarian position, arguing that human characters are rigidly predetermined at birth, but in his *Natural Inheritance* of 1889 he called in saltations to explain the appearance of new species. His eugenics programme of controlled breeding in the human populations followed the logic of artificial, not natural selection. The theory was also endorsed by William Bateson, who abandoned his research on the origin of the vertebrates to study the discrete varieties that exist within many plant species. The Darwinians saw these as incipient species built up by the accumulation of minute individual differences over many generations. Bateson now insisted that they were produced abruptly by saltations that bred true and survived whether or not they conferred any adaptive benefit. His *Materials for the Study of Variation* of 1894 was anything but a best-seller, but it was taken seriously by many experimental biologists, and his interest in discrete character-differences prompted his recognition that Gregor Mendel's work on the hybridization of varieties might become the basis for a new theory of heredity.

Genetics would create a model of heredity based on discrete units (the genes) that predetermine the organism's structure and whose effect cannot be modified by environmental influences. By modern standards this image of the gene as a blueprint was oversimplified, but its apparent simplicity was one source of its appeal. The model was put together during the first two decades of the twentieth century, long before the role of DNA was revealed; indeed, it was made possible precisely by the fact that no one knew how the genes worked, only that they *did* work as a determinant of character. The process of embryological development could be set aside, making it easy to adopt Weismann's dogma that the effects of the environment on the organism cannot be inherited. Once the gene had been defined as a material unit on the chromosomes of the cell nucleus, the behaviour of the chromosomes could be used to explain the laws of transmission discovered by Mendel.

One biologist involved in the 'rediscovery' of Mendel's long-neglected papers was the Dutch botanist Hugo De Vries, although he soon lost interest in Mendelism and moved on to propose his 'mutation theory', a new version of the old concept of saltations. It was based on observations that appeared to confirm the sudden appearance of distinct varieties in the evening primrose, *Oenothera lamarckiana.* De Vries claimed that Darwin had been wrong to think in terms of selection gradually adding up generations of minute individual differences, although he accepted that any maladaptive mutations would soon disappear and that an occasional one better adapted to the environment would eventually replace the original form. Many of those who took up his theory preferred to reject the whole adaptationist programme in favour of Bateson's view that saltations could create stable varieties without reference to environmental pressures. Before he adopted the Mendelian approach, Thomas Hunt Morgan endorsed the mutation theory and wrote a strident attack on the Darwinian theory, his *Evolution and Adaptation* of 1903.[29]

De Vries' *Mutation Theory* was not translated until 1910, but by this time he had become well-known in Britain and even more so in the United States. Along with Bateson he spoke at the Darwin celebrations in Cambridge in 1909, both attracting press attention with their non-Darwinian views. The theory was outlined in surveys such as the Home University Library's volume on evolution and the *Harmsworth Popular Science* serial. De Vries himself lectured across the United States in 1904 and again in 1906, his lectures at Berkeley on the first tour being edited for publication by his disciple Daniel Trembly MacDougal. His theory was debated in popular science magazines and attracted press attention through its potential applications to plant breeding and agriculture. De Vries claimed that an understanding of the process that generated mutations would allow the creation of valuable new crops. MacDougal was one of many biologists who established programmes in 'experimental evolution', exploiting funds made available by the

[29] On the mutation theory, see Bowler, *The Eclipse of Darwinism*, chap. 8.

Carnegie Foundation. De Vries himself spoke at the dedication of the Cold Spring Harbor Experimental Station.[30]

The term 'mutation' was eventually taken up by the geneticists, but initially their new approach to heredity operated independently of De Vries – even though he had been one of the biologists responsible for the rediscovery of Mendel's long-ignored work on the inheritance of hybrid varieties in peas. Historians now suspect that Mendel's studies were prompted by his interest in hybridization – they were not intended as the basis for a whole science of heredity. The biologists who 'rediscovered' his long-neglected work in 1900 thus read a good deal of their own thinking into his papers. They reinterpreted his laws in terms of their own growing interest in unit characters, which in several cases was inspired by the belief that new characters are created by abrupt saltations. It was Bateson who provided the first English translations of Mendel's papers and later coined the name 'genetics' for the research programme based on the inheritance of clearly defined characteristics. He insisted that these characteristics could not be modified by environmental influences and that they usually had no adaptive significance. He was thus hostile to the Darwinians' claim (discussed below) that characteristics normally show a continuous range of variation capable of being modified by selection, dismissing this effect as merely transitory. It would eventually be recognized that most characteristics are influenced by more than one gene, so that in a large population the genes can generate a range of variation just as Darwin had supposed. But this only became apparent later, and even in 1922 Bateson gained some notoriety for a talk given in the United States in which he claimed that the mechanism of evolution was still a mystery.[31]

In the meantime Thomas Hunt Morgan had converted from De Vries' theory to found a research programme that established the modern concept of the gene as a material unit on the chromosomes of the cell nucleus. His group also observed cases where the gene underwent an internal reconfiguration that had the potential to modify the characteristic – these were the mutations we recognize today. But where De Vries'

[30] See Geddes and Thomson, *Evolution*, pp. 125–9, and Mee, ed., *Harmsworth Popular Science*, group 3, chap. 9, 4: 2237–43. On the practical side, see Jim Endersby, 'Mutant Utopias'; Helen Anne Curry, *Evolution Made to Order*; and Sharon E. Kingsland, 'The Battling Botanist'.

[31] There is a huge literature on the emergence of genetics; see, for instance, Robert Olby, *Origins of Mendelism*, and my own *The Mendelian Revolution*. For a broader perspective, see Staffan Müller-Wille and Hans-Jörg Rheinberger, *A Cultural History of Heredity*. The oversimplified nature of the early geneticists' model is explored in Greg Radick, *Disputed Inheritance*.

mutations founded discrete new varieties, these modifications fed into the existing breeding population and, in effect, extended the range of variation. Eventually, even Morgan began to accept that a harmful mutation could not get established in the population, while one conferring an adaptive benefit would flourish. This would encourage an eventual reconciliation with Darwinism, but through into the 1920s genetics was more likely to be seen as hostile to the selection theory.

There was no shortage of information available about the new theory, much of it presented via its potential for improvements in agricultural breeding and the eugenics movement's demands for control of the human population. There were accounts of Mendelism in sources such as the Home University Library's 1911 volume on evolution and in the *Harmsworth Popular Science*. Both presented it solely in terms of character transmission and introduced the topic alongside De Vries' mutation theory as another challenge to the Darwinian perspective. The Cambridge biologist Reginald Punnett published a brief survey in 1905, which expanded in later editions to include the latest American developments and the chromosome theory. MacBride's Home University Library volume on heredity in 1924 outlined the classical theory of the gene but dismissed mutations as merely degenerative. Punnett too challenged Darwinism by explaining mimicry in butterflies as the abrupt appearance of new colours by mutation, a claim featured in the pages of *Scientific American* in 1911. As late as 1924 Lancelot Hogben promoted the discontinuous model of variation in the magazine *Discovery*, insisting that Bateson might eventually be recognized as a more important figure than Darwin.[32]

Defending Darwinism

Darwinism was not without its supporters, but their work was uncoordinated in part because it came from several different ideological backgrounds. Some still wanted to defend the optimistic progressive evolutionism of the previous century, insisting that natural selection could be seen as a force for good. At the opposite extreme, an increasingly active rationalist movement wanted to use the theory as a foundation for its attack on organized religion. There were biologists too seeking

[32] See Geddes and Thomson, *Evolution*, pp. 129–41, and Mee, ed., *Harmsworth Popular Science*, group 3, chapter 18, 3: 2117–25, and chap. 21, 4: pp. 2773–9; Punnett, *Mendelism*; and MacBride, *An Introduction to the Study of Heredity*, chap. 6. For Punnett on mimicry, see his *Mendelism*, pp. 176–9, and *Scientific American*, 104 (28 January 1911), pp. 87 and 99–100. See also Hogben, 'The Present Status of the Evolutionary Hypothesis', p. 105.

to provide a new evidential foundation for the theory. Their work was highly technical and did not find its way into most popular surveys, although Karl Pearson's involvement with the eugenics movement would have alerted its supporters to the importance of selection.

The growing doubts over the experimental basis for Lamarckism provided at least some support for neo-Darwinians such as Weismann. In the 1890s H. G. Wells, then active in science journalism, at first rejected Weismann's position but then became a convert.[33] The impact of this focus on heredity forced even Spencerians such as Caleb Saleeby to abandon the Lamarckian element in his philosophy. As a major contributor to the *Harmsworth Popular Science* serial, Saleeby played down Lamarckism and presented natural selection as a mechanism capable of driving the moral progress emphasized in Spencer's thought. This was also the position taken in J. Arthur Thomson's many popular works on evolution. The same approach can even be found in another Harmsworth product, the popular *Children's Encyclopaedia.*[34]

A significant modification of the neo-Darwinist position had already emerged in A. R. Wallace's comprehensive volume *Darwinism* of 1889. Here he abandoned Darwin's original focus on individual variants in order to avoid the criticism that the effect of a single favourable new character would be swamped by interbreeding with the unchanged mass of the population. He pointed out that for many characteristics there is a continuous range of variation within the population, with most individuals clustered around the average value for the characteristic and smaller proportions tailing off on either side. Wallace provided figures to show this kind of variation in wild populations and even gave the outline of a figure similar to what we now recognize as the bell curve of normal variation. On this model there are always plenty of individuals occupying the favoured end of the range, and there is no reason to suppose that continued selection will not shift the whole population in that direction (see Fig. 6.2).

This model of continuous variation was taken up by the biometrical school of Darwinism led by the statistician Karl Pearson and the biologist W. F. R. Weldon. They analysed detailed measurements from wild populations, producing distribution curves suggesting that the range of variation was being shifted by adaptation. Both were suspicious of

[33] Wells' *Saturday Review* article 'The Biological Problems of Today' and a number of later pieces are reprinted in Philmus and Hughes, eds., *H. G. Wells*; see pp. 123–7.

[34] Geddes and Thomson, *Evolution*, in the Home University Library, and Saleeby, *Organic Evolution* and *Heredity*, both in Jack's popular series, also his *Evolution: The Master Key*; see also Mee, ed., *Harmsworth Popular Science*, group 3, sections 9 and 11, 2: 1033–41 and 1277–85, and Mee, ed., *The Children's Encyclopedia*, group 10, chap. 2, 1: 203–8.

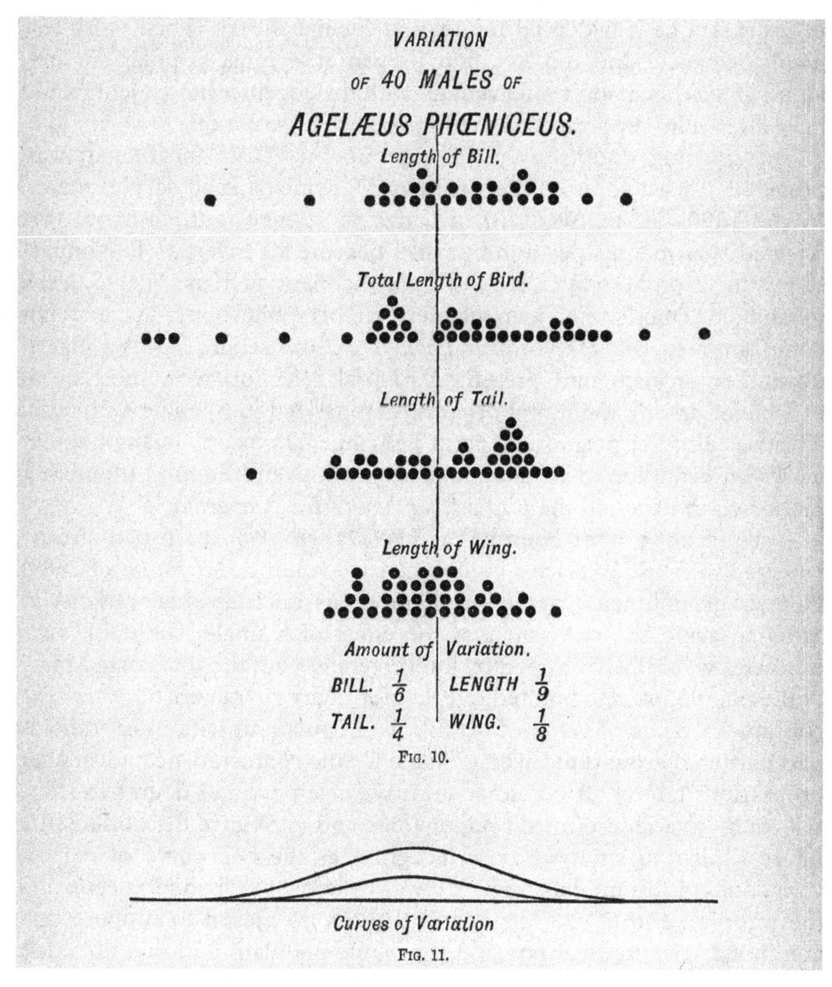

Fig. 6.2 Diagrams illustrating the range of variation within a population, from A. R. Wallace, *Darwinism* (1889), p. 64.

Mendelism – Weldon was working on a book critical of the new approach when he died in 1906 and Pearson based his analysis on Galton's alternative 'law of ancestral inheritance', although he showed that it would allow for continuous evolution without the saltations that Galton thought necessary. Pearson's techniques would eventually play a role in the reconciliation that created population genetics as the basis for a revived selection theory. Their work was highly technical and did not receive much publicity, although Weldon did describe it in *Nature* and at the 1898 meeting of the British Association. Pearson added chapters on the topic to the second edition of his well-received *Grammar of Science* in 1900. There was no equivalent work in America, though, and Kellogg's *Darwinism Today* included only a quotation from Weldon in an endnote.[35]

Geddes and Thomson's 1910 survey of evolutionism for the Home University Library mentioned Pearson's work in passing, noting that it was too technical for them to deal with. This was in the context of efforts to demonstrate the impact of predation by birds on insect populations, including those by E. B. Poulton, whose book on animal colouration in the International Scientific Series endorsed the Darwinian explanation of mimicry. However, the Home University Library volume concluded with an account of Geddes' strongly anti-Darwinian view of directed variation, fitting uncomfortably with its otherwise fairly even-handed approach. Kellogg's survey also drew attention to a more general defence of the selection theory by J. L. Taylor published in the magazine *Natural Science* in 1899.[36]

Less formal indications that evolution was driven by the survival of the fittest came from those concerned with its implications for human society. Benjamin Kidd's hugely successful *Social Evolution* of 1894 popularized the role of struggle, but transferred it to the level of competition between rival social groups (see Chapter 8 below). More direct support for Darwinism came from the rationalists, who opposed the role played by religion in society. Since the defenders of religion usually opposed Darwinism, they sprang to its defence by insisting that the theory had not, in fact, lost its scientific credibility. Their arguments needed to be presented in terms that could be understood by the general reader, which meant describing natural selection without technical details. Outlines of the Darwinian theory were usually embedded in broad-brush accounts reminiscent of the evolutionary epics of the previous generation. In Britain the campaign centred on the Rationalist Press Association (RPA), founded in 1885. This organization gained access to a wide readership via the publishing firm of Charles Albert Watts, which specialized in cheap material often in paperback. In the United States the most active source of similar material came from the firm founded by

[35] For more details, see William B. Provine, *The Origins of Theoretical Population Genetics*, chaps. 2 and 3; Jean Gayon, *Darwinism's Struggle for Survival*, chap. 8; Daniel J. Kevles, *In the Name of Eugenics*, chap. 2; and Hale, *Political Descent*, pp. 319–25. Radick's *Disputed Inheritance* discusses the book Weldon was working on at his death; see also Kellogg, *Darwinism Today*, pp. 157–62.

[36] Geddes and Thomson, *Evolution*, pp. 165 and 234–48; Poulton, *The Colours of Animals*; Kellogg, *Darwinism Today*, on Taylor pp. 153–7.

Emanuel Haldeman-Julius in Girard, Kansas, whose Little Blue Books were issued in vast quantities and included several evolutionary texts.

Watts' list included several reprints of classic texts by Darwin and Huxley, which may have helped create the impression that the theory was out of date. But the firm also issued a new edition of Haeckel's *Evolution of Man* in 1905, translated by Joseph McCabe (also published by Putnam's in the United States), with a one-volume paperback selling for only one shilling in 1907. Haeckel enjoyed renewed attention in radical circles on both sides of the Atlantic thanks to McCabe's translation of his *Riddle of the Universe* issued in 1900 by Watts and Harper's, and the follow-up *Wonders of Life* of 1905. Both linked the Darwinian theory to his wider monistic philosophy. His stock remained high until 1909, when doubts about the accuracy of his embryological images began to circulate, and sank to zero in the English-speaking world when he endorsed the German invasion of Belgium in 1914.[37]

At the start of the century the most eminent biologist associated with the rationalist position was E. Ray Lankester, a disciple of Huxley who was pressured into retiring from his position as director of London's Natural History Museum in 1907. He had already published some non-technical books on zoology and now began to write a regular article on science for the Saturday edition of the *Daily Telegraph*. In 1910 some of these were collected in his *Science from an Easy Chair*, the first of a series of similar volumes issued originally at mid-price but then reduced to two shillings and sixpence. There were frequent contributions on evolution, most notably a group in the first volume including his 'Darwin's Theory Unshaken'. Lankester was also a frequent contributor to the *Rationalist Annual*, and some of his collections were later reprinted by Watts.[38]

Presumably anxious to soften the aggressive image of rationalism presented by figures such as Haeckel and Lankester, Watts also published surveys of the ascent of life intended for juvenile readers, including Dennis Hird's *Picture Book of Evolution* in 1906, Robert Macmillan's *The Origin of the World* in 1914 and Adam Gowans Whyte's *The World's Wonder Stories* in 1916. They included brief descriptions of natural selection as the agent of progress and were reprinted through into the 1920s, Hird's text being updated by C. M. Beadnell. Whyte's book was retitled *The Wonder World* and was issued in America by Knopf as *The Wonder World We Live In*.

[37] On Haeckel's changing reputation, see Hopwood, *Haeckel's Embryos*, pp. 258–62.

[38] 'Darwin's Theory Unshaken' was reprinted in Lankester, *Science from an Easy Chair*, pp. 27–37; for more details including sales figures for his books, see Lester, *E. Ray Lankester*, chap. 13.

The rationalist defence of Darwinism also included contributions by the renegade Catholic monk Joseph McCabe, who translated Haeckel and went on to produce his own popular surveys and a plethora of anti-religious material that circulated on both sides of the Atlantic. His *Evolution: A General Sketch from Nebula to Man* appeared in 1910, and Watts published his *ABC of Evolution* in 1920. These were classic evolutionary epics leading up to the origin of humanity, but they included support for the selection theory while suggesting that there might also be the abrupt appearance of new structures by mutation. Watts published the text of McCabe's debate with the creationist George McCready Price in 1925. This was one of many responses to the claim that Darwinism is dead; others included Wells' riposte to Belloc's attack on his *Outline of History* and Arthur Keith's 1927 address to the British Association, although Keith (who was an anatomist with no understanding of genetics) suggested that hormones might somehow shape variation in favourable directions. There were several later contributions by Keith in Watts' cheap 'Forum Series', and as late as 1934 the firm issued McCabe's *The Riddle of the Universe Today*, which attacked the scientists still trying to reconcile evolutionism with religion.[39]

McCabe was also publishing actively in America, where he contributed to the debate over the Scopes trial. His *The Story of a Religious Controversy* dismissed the efforts of those biologists who sought to play down the materialism of the Darwinian theory in the hope of reducing the creationists' fears. The Little Blue Books also made the text of his debate with Price available to American readers, along with a series of more general attacks on religion. The series also contained several texts on Darwinism by the University of Iowa palaeontologist Carroll Lane Fenton.[40]

Whatever McCabe's objections, the defence of Darwinism was probably more effective when the theory was presented in a flexible form, preferably embedded in a general survey of the evidence for evolution. The rationalists thus played an important role in the public debate even though they were not yet adopting the most up-to-date version of the theory. There were, in any case, accounts of evolutionism by experts not identified with rationalism who had a more positive view of Darwinism. Museums were now increasingly willing to display evidence of evolution

[39] Price and McCabe, *Is Evolution True?*; Keith's contributions to the Forum series are his *Concerning Man's Origin*, *Darwinism and Its Critics* and *Darwinism and What It Implies*. For his idea that hormones can affect variation, see *Concerning Man's Origin*, pp. 19–20, and for Huxley on Darwinism and Lamarckism, see Huxley's *The Stream of Life*, chap. 5.

[40] See Clark, *God – Or Gorilla*, pp. 57–9 and 103–4, also Bill Cooke, *A Rebel to His Last Breath*.

and sometimes of Darwinism. In 1890 London's Natural History Museum had a display of specimens demonstrating variation, artificial selection and mimicry. Lankester's former student Edwin S. Goodrich contributed a popular survey of evolutionism to the 'People's Books' series in 1912, presenting evolution largely in terms of adaptive divergence and favouring the selection theory. The following year Robert Lloyd Praeger included a brief explanation of natural selection in a children's book on botany.[41] In Chapter 7 we shall see how H. G. Wells' hugely successful *Outline of History* of 1921 also promoted a more Darwinian vision of the progress of life, even before he recognized the need for a new interpretation of how natural selection works.

In the 1920s the first efforts to reconcile genetics with Darwinism began to emerge. Thomas Hunt Morgan's public lectures at Princeton in 1916 were published as his *Critique of the Theory of Evolution* and made it clear that only those mutations conferring some benefit in the environment would spread in a population.[42] In 1923 Julian Huxley gave a series of talks on BBC radio in which he described the latest findings in genetics and showed how they could be reconciled with the selection mechanism. The texts were then published in Watt's 'Forum Series' under the title *The Stream of Life*. He also showed how the fossil record was now revealing irregularly branching sequences of development with numerous dead twigs, much like the picture originally painted by Darwin himself. In another essay he praised Bergson as a good poet but a bad scientist, implying that natural selection could be as creative as the *élan vital*. This was a way of thinking that would lead him to join Wells in writing a follow-up to the *Outline of History* that would provide the first popular account of the emerging synthesis of Darwinism and genetics (see Chapter 9).[43]

[41] For the Museum display, see Rachel Poliquin, *The Breathless Zoo*, pp. 131–2; MacGregor, 'Exhibiting Evolutionism'; and Adelman, 'Evolution on Display'. See also Goodrich, *The Evolution of Living Organisms*, later extended in his *Living Organisms*; Praeger, *Weeds*, pp. 3–4.

[42] Morgan's *Critique of the Theory of Evolution* was reprinted in 1925 as his *Evolution and Genetics*, co-published in Britain by Oxford University Press.

[43] Huxley, 'Progress: Biological and Other', in his *Essays of a Biologist*, pp. 3–66; see p. 33. On Huxley's radio broadcasts, see Alex Hall, *Evolution on British Television and Radio*, chap. 2.

7 Reconfiguring the Ascent of Life

The evolutionary epics of the late nineteenth century promoted a vision of progress driven toward the goal of modern humanity. The tree of life had a central trunk whose side branches had diverged from the overall purpose of the ascent. This way of thinking was still being defended in some popular works as late as the 1920s, but it was also being challenged by a recognition that progress was only a by-product of the complex responses of life to environmental challenge. This was the implication of Darwin's original model of evolution, and some of the non-Darwinian mechanisms also undermined the linear image of development. Even those assuming predetermined trends had to accept that there were some that led away from the sequence leading to humanity. In the twentieth century the focus on 'creative evolution' by figures such as Henri Bergson and Bernard Shaw provided an alternative to the Spencerian model of progress as a law of nature, encouraging a more episodic and open-ended view of the ascent of life. Perhaps most obvious to the wider public, the increasing wealth of new fossils on display made it abundantly clear how diverse the tree of life had been, and how many of its branches had been pruned by global changes that had also triggered new lines of development.

The concept of evolutionary progress certainly was not abandoned, but it was being transformed by an emphasis on the inventiveness of life when faced with environmental challenge. On rare occasions, living things could respond not by narrow specialization but by creating more complex structures with sophisticated functions that opened up a range of new opportunities. There was no predetermined outcome because the steps by which new levels of organization were reached would not have been predictable on the basis of any previous trends. This was a less deterministic model of the history of life that would facilitate the re-emergence of the selection theory in the middle decades of the century.[1]

[1] This interpretation was suggested in my *Life's Splendid Drama* and is the central theme of the more recent survey, *Progress Unchained*.

The fossils shaped public awareness of this reconfiguration through the way they were displayed. The most effective representations were visual, made by artists and model-makers seeking to imagine what the creatures had once looked like. Their effect was boosted by the ever-increasing sophistication of illustrations in magazines and books. Beyond the world of print the public could encounter new interpretations in the displays of the many natural history museums. European cities had museums open to the public from the early nineteenth century – London's was rehoused in its present quarters in 1881. Thanks to the munificence of capitalists such as Andrew Carnegie, many American cities soon had museums to rival those of Europe, and newly discovered fossils were among the most popular and widely publicized specimens. Museums competed for spectacular new specimens and displayed them in ways designed to show how they fitted into the pattern of evolution. The huge exhibitions held to celebrate national anniversaries and achievements sometimes contained displays that featured evolutionary topics, including fossils and the lives of the 'primitive' peoples being incorporated into areas of European and American influence.

Changing Perceptions

The new way of looking at the fossil record began to emerge around 1870. Having largely ignored the field during the initial debate over Darwin's theory, Huxley and others began to look more closely at the evidence for clues about life's ancestry. O. C. Marsh's horse fossils became iconic, providing a clear example of a trend toward specialization – although they also encouraged the view that the trend drove inexorably toward a final end-point. Huxley's efforts to uncover the reptilian ancestry of the birds offered another way forward, and in this case had an unexpected consequence by requiring a revision of how the dinosaurs were perceived. His proposed dinosaurian ancestor, the newly discovered *Compsognathus*, had to be bipedal to be plausible since wings could not have developed from forelimbs adapted for running.

At the same time, palaeontologists began to suspect that the first-known dinosaurs, *Megalosaurus* and *Iguanodon*, were not the gigantic lizards imagined by the previous generation. They too had been bipedal, totally unlike any living reptile. This was soon confirmed by better specimens discovered in Europe, while American palaeontologists uncovered yet more bipedal forms. Through the 1870s and 1880s Marsh and his rival Edward Drinker Cope discovered a wealth of bizarre dinosaurs in the West. Here was growing evidence that the Age of Reptiles was anything but a step toward the modern world – most of its inhabitants had belonged to branches of the tree of life unlike anything still alive. Marsh also described bizarre archaic mammals from the strata immediately following the disappearance of the great reptiles. These too vanished from the record, to be replaced by the ancestors of the species we know today – including the early members of the horse family.

Curiously, though, it took some time for the implications of these discoveries to be recognised. The products of the first 'dinosaur rush' to the American West received little press attention, and it was only when specimens from the second 'rush' of the 1890s went on display in the museums that they became a talking point. The character of the evolutionary epics surveyed in the previous chapter provides a possible explanation for this delay. Many of the popular evolutionary texts used living rather than fossil species to illustrate the advance of life, presenting it as a steady advance toward the modern world. The array of extinct forms now being unearthed would be hard to fit into the pattern of steady progress taken for granted by Spencer and his contemporaries. Marsh's law of brain growth still implied that extinction was part of a more or less continuous Darwinian process, but he was now describing fossils that were difficult to fit into any scheme of constant progress. If the dinosaurs, for instance, were not just gigantic lizards but a whole range of forms totally unlike any living reptile, the tree of life must have had major branches that had been lopped off in the distant past. Extinction did not just ensure the gradual replacement of species; it could on occasion operate more like the old catastrophist model involving the mass elimination of whole faunas.

The process of rethinking the status of the dinosaurs had begun somewhat earlier but seems to have remained in the province of the specialists. Huxley and others began to suspect that *Iguanodon* had been bipedal, like the smaller species that he presumed to be ancestral to the birds. As early as 1858 Joseph Leidy had described a bipedal species, *Hadrosaurus*, from the rocks of New Jersey. In 1866 Cope described *Laelaps*, from the same area, showing it was clearly bipedal. Spectacular fossils of *Iguanodon* found at Bernissart in Belgium in 1878 confirmed the representation of the species still accepted today. It was becoming clear that the dinosaurs were not a unified group – some were indeed quadrupedal as Owen had originally described them, but others were bipeds. Harry Govier Seeley divided them into two orders, the Saurischia (lizard-hipped) and Ornithschia (bird-hipped). Early representatives of the latter may have given rise to the birds, but they had otherwise disappeared altogether along with the gigantic quadrupeds.

There were some efforts to put the new image of the dinosaurs on display to the public, not all of them successful. In 1868 Waterhouse

Hawkins (who had designed the Crystal Palace models in London) was commissioned to create models for a proposed Paleozoic Museum to be erected in New York's Central Park. A surviving image shows that some dinosaurs would have been depicted as bipedal. But when the corrupt 'Boss' W. M. Tweed took over the city administration, the plan was cancelled. Hawkins then erected a skeleton of Leidy's *Hadrosaurus* at the Philadelphia Academy of Natural Sciences. It was subsequently moved to stand outside the Natural History Museum in Washington, DC, where it gradually succumbed to the elements.[2]

What has become known as the first American dinosaur rush began in the 1870s as rival teams organized by Marsh and Cope fanned out across the 'Wild West' in search of the best locations for fossils. Marsh had the advantage of a wealthy uncle, George Peabody, who financed his expeditions and created the museum at Yale in which the specimens were displayed. Cope was independently wealthy but eventually got into financial difficulties. They were on opposite sides of the debate between Darwinians and Lamarckians, but this was not the root of their hostility – they were competing for priority in the discovery and description of spectacular extinct species. Their 'bone war' has attracted the attention of historians, alerted by the conflict that broke out in the pages of the *New York Herald* in 1890, probably at the instigation of the press baron James Gordon Bennett, Jr. There were charges of corruption, incompetence and theft, some of which were probably justified due to the haste with which they rushed to gain priority. Nevertheless, their efforts revealed some of the large and bizarre dinosaurs that have become iconic representatives of the Age of Reptiles. *Stegosaurus* was named in 1877, *Brontosaurus* in 1879 (although this was an error on Marsh's part), *Triceratops* in 1887 and *Tyrannosaurus* in the following decade.[3]

Given the fascination with these creatures that grips the modern imagination, it would be easy to assume that they must have had a strong impact when discovered. But scholars who have studied the events suggest that the press response was rather muted until the 1890s. There seems to have been a general decline in the level of interest in evolutionism after the excitements of the mid-1870s. What interest there

[2] For more details on these displays, see Edwin H. Colbert, *Men and Dinosaurs*, pp. 63–9, and Adrian Desmond, *The Hot-Blooded Dinosaurs*, pp. 43–7. For the image of Hawkins' designs, see W. J. T. Mitchell, *The Last Dinosaur Book*, pp. 130–1, or Lukas Rieppel, *Assembling the Dinosaur*, p. 52.

[3] Of the many accounts of the 'bone war' I have found, David Rains Wallace's *The Bonehunters' Revenge* is the most useful. Most accounts finger the reporter William Hosea Ballou as the instigator of the 1890 press feud, but Wallace thinks Bennett played the more active role.

was in the fossil hunters focused on the adventurous nature of the expeditions and, of course, the feud. Paul D. Brinkman's study of the second dinosaur rush in the 1890s argues that the dinosaurs uncovered in the earlier episode attracted attention only in the scientific community. He finds few references to dinosaurs in the *Washington Post* for the period 1877–95, while from 1895 onwards interest expanded rapidly. David Rains Wallace's account of the first rush also argues that it did not capture the public imagination and notes that a survey of the *New York Times* and *Tribune* for the period yields few references. Museum displays of complete skeletons became common only around 1900. For some time the Peabody Museum at Yale was one of the few places where they could be seen. It was mentioned by Arabella Buckley in 1882, one of a limited number of references to these discoveries in the evolutionary epics of the period.[4]

Fossils Hit the Headlines

The second 'dinosaur rush' began in the 1890s, stimulating a wave of public interest on both sides of the Atlantic. Surveys of the fossil record again became popular. Richard Lydekker, based at the Natural History Museum in London, included articles on palaeontology in his regular contributions to *Knowledge* magazine. The article titled 'Giant Land Reptiles or Dinosaurs' was reprinted in his *Phases of Animal Life* of 1892 (also issued in the United States). It had images of the reconstructed skeletons of *Iguanodon*, *Stegosaurus*, and *Triceratops* (the latter two based on Marsh's drawings). In 1890 *Punch* carried a cartoon of Marsh as a circus ringmaster directing his archaic mammals while standing on a *Triceratops* skull. By then European museums had begun to get copies of the Bernissart *Iguanodon* specimens. A well-illustrated survey by the Rev. H. N. Hutchinson entitled *Extinct Monsters and Creatures of Other Days* was published in 1892. Hutchinson complained in his preface that despite the general popularity of natural history, little up-to-date information on the latest discoveries had been available to the public. His was the first of many surveys based on the fossil record to appear from this point onwards.[5]

[4] See Paul D. Brinkman, *The Second Jurassic Dinosaur Rush*, pp. 2–3, and Wallace, *The Bonehunters' Revenge*, pp. 163–4 and 174–5. R. V. Bruce notes a declining interest in evolutionism during the 1880s; see his *The Launching of Modern American Science*, p. 355. Buckley's reference to the Peabody Museum is in her *Winners in Life's Race*, p. 339 (her book was also available in the United States).

[5] Lydekker, *Phases of Animal Life*, chap. 8. For the *Punch* cartoon, see Wallace, *The Bonehunters' Revenge*, p. 259. For Hutchinson on the lack of information, see his *Extinct Monsters and Creatures of Other Days*, preface, p. x, in the edition cited.

In the United States the public was attracted to the fossil displays in the natural history museums opening around the country, each anxious for publicity. The money for the museums and their expeditions to the West came from the captains of industry who saw this as a way of demonstrating their public spirit. Most visible was Andrew Carnegie, who wanted giant dinosaurs to attract crowds and reporters to his museum in Pittsburgh, founded in 1895 and extended in 1905. He was rewarded in 1899 with the discovery of a giant species of *Diplodocus*, the skeleton of which was displayed to great acclaim – the museum had to be extended to accommodate it. The remains of other bizarre dinosaurs such as the horned *Triceratops* were also on view. Carnegie's biggest rival was Osborn, who headed the department of vertebrate palaeontology at the American Museum of Natural History in New York where a new Dinosaur Hall opened in 1903 to show off a huge *Brontosaurus*. He too sought press attention; he and his staff authored numerous articles in popular magazines such as *Scientific American*. These were illustrated with reconstructions of what the animals might have looked like in life by Charles R. Knight, who also painted murals for the museum walls. There were life-sized models on display and smaller copies for sale to the public. In 1911 *Scientific American* featured life-sized representations of dinosaurs including *Iguanodon* (see Fig. 7.1)[6]

Carnegie was anxious to gain credit for his efforts around the world and distributed casts of his *Diplodocus* to foreign museums. London's Natural History Museum officially received its copy in May 1907 at an elaborate ceremony widely reported in the press. Here too it joined an ever-growing display of other bizarre forms (see Fig. 7.2). The museum's director, E. Ray Lankester, had already published a well-illustrated book, *Extinct Animals*, based on lectures at the Royal Institution, and in 1910 an expanded edition of Hutchinson's *Extinct Monsters* was issued. There were articles on the fossil discoveries for popular magazines such as *The Field* and *Illustrated London News*. The satirical *Punch* had a series of cartoons entitled 'Prehistoric Peeps' by Edward Tennyson Reed, which often featured dinosaurs.[7]

[6] See Brinkman, *The Second Jurassic Dinosaur Rush*, also Rieppel, *Assembling the Dinosaur*, and Tom Rea, *Bone Wars*. More generally, see Mitchell, *The Last Dinosaur Book*, especially chap. 25, and on archaic mammals Wallace, *Beasts of Eden*. Ronald Rainger lists the articles in *Scientific American* and other magazines; see his *An Agenda for Antiquity*, p. 282, note 112. On the artwork, see Sylvia Massey Czerkas and Donald E. Grant, *Dinosaurs, Mammoths and Cavemen*. On changing methods of display, see Karen A. Rader and Victoria E. M. Cain, *Life on Display*.

[7] See Reed, *Mr. Punch's Prehistoric Peeps*; on the museum displays, see Susan Snell and Polly Parry, eds., *Museum through the Lens*, and on the Carnegie presentation Ilja Nieuwland, *American Dinosaur Abroad*.

SIXTY-SEVENTH YEAR

SCIENTIFIC AMERICAN

THE WEEKLY JOURNAL OF PRACTICAL INFORMATION

NEW YORK, APRIL 8, 1911

Iguanodon – This specimen towers twenty-five feet in the air, making the trees around look small.

A DINOSAUR THAT ROAMED THE EARTH MILLIONS OF YEARS AGO. – [See page 352.]

Fig. 7.1 Cover of *Scientific American*, 8 April 1911, showing the reconstruction of the dinosaur *Iguanodon* at Carl Hagenbeck's animal park near Hamburg.

Fig. 7.2 *Diplodocus* skeleton on display at London's Natural History Museum. Images like this were widely reproduced, reproduced here from H. N. Hutchinson's *Extinct Monsters and Creatures of Other Days* (1910 ed.), p. 155.

Reed's cartoons sometimes depicted dinosaurs in the company of cavemen, and there were numerous fictional accounts of ancient creatures surviving into the present. The genre had been pioneered by Jules Verne, whose *Journey to the Centre of the Earth* was translated in 1871. At the turn of the century stories in which modern humans encounter remnant populations of supposedly extinct species became especially popular. In Britain the *Strand Magazine* and *Pearson's* featured examples by a number of authors. The classic is Sir Arthur Conan Doyle's *The Lost World* of 1912, which locates species from the whole sequence of geological periods surviving on an isolated South American plateau. In America, Edgar Rice Burroughs returned to Verne's theme in *At the Earth's Core* (serialized in 1914) but relocated the creatures to the South Pole in *The Land That Time Forgot* (serialized in 1918).

In 1922 Doyle was involved with the making of a movie version of his story using models to represent the dinosaurs. This attracted considerable publicity, with the press anxious to make it clear that the filming involved trick photography – apparently fearing the public would think the scenes were real. There were in fact occasional reports of dinosaur sightings in the African jungles, and the marsupials of Australia were often seen as survivals of an early mammalian fauna. But for dinosaurs identical to the fossil species to be alive today they must have lingered on in some remote location for millions of years, so perhaps the public may not have fully grasped the extent of geological time (the age of the earth as estimated at this time was much less than we accept today, but the end of the Age of Reptiles was put at around four million years ago.). Those who realized that living dinosaurs were implausible could visualize the vast array of living forms that has disappeared in the course of evolution's ascent.[8]

Trends and Transitions

The fossils encouraged new ways of thinking about the ascent of life just when the debates over the various evolutionary mechanisms were most active. In one respect the study of trends in the fossil record of individual groups encouraged the most extreme non-Darwinian position, that of nonadaptive orthogenesis. Palaeontologists were among the most active supporters of this theory, coupled with the ideas of evolutionary parallelism and racial senility. The apparently bizarre structures developed by many dinosaurs and archaic mammals seemed to imply an impulse driving the groups toward overspecialized goals that exposed them to the risk of extinction. Yet the very fact that each of the groups was heading off toward its own unique goal added to the diversity of the tree of life. When whole branches of the tree seemed to have disappeared altogether, it became difficult to argue that the line leading toward humanity is the main line of progress. Eventually, even the evidence for predetermined trends began to wilt under pressure from new discoveries that could not be fitted into such neat arrangements.

It was also becoming evident that the rate of change was not continuous – the palaeontologists were being forced by growing evidence to see evolution as an episodic process with outbursts of change apparently triggered by environmental disruptions. The final extinction of groups such as the dinosaurs was quite abrupt, while the various branches of a new type emerged in an outburst of radiation. The disappearance of the dinosaurs seemed to have opened the way for the rise of the mammals, which were now known to have emerged from an ancient branch of the reptiles

[8] Burrough's *At the Earth's Core* was also published in Britain in 1923. For magazine stories by Frank T. Bullen, Wardon Allan Curtis, George Griffith and Cutliff Hyne, see A. Kingsley Russell, ed., *Science Fiction by Rivals of H. G. Wells*. On the *Lost World* movie and press reports, see Rieppel, *Assembling the Dinosaur*, pp. 191–6.

unrelated to that leading to the dinosaurs. Geologists were finding evidence that the earth's physical development had itself been discontinuous in a way reminiscent of old-style catastrophism. There were long eras of relative stability interrupted by episodes of change quite rapid in geological terms. Evolutionary theories, including Darwinism, would have to be modified to take this element of discontinuity into account.

The new ideas on the history of life were transmitted to the public through magazine articles and books, but perhaps most effectively through the ways in which specimens were displayed and interpreted in museums. The experts were anxious to make sure the public understood the significance of the new evidence, but like the tycoons who bankrolled the museums, their efforts were directed by broader agendas. Osborn's position at the museum in New York placed him in a good position to promote his vision of evolution to the public, a vision significantly at variance with Darwinism. In the early years of the century, he abandoned Lamarckism and became convinced that the linear sequences revealed by the fossils must have been produced by inbuilt forces pushing variation in a fixed direction, driving all the species within a group in parallel toward the same goal. In some cases, as with the horse family, the goal seemed beneficial, but Osborn's studies of fossil elephants showed parallel lines evolving strange tusk-shapes while the archaic mammals known as Titanotheres ended up as gigantic forms with strange horns almost as impressive as the larger dinosaurs. At the New York museum there were life-sized models and displays showing how the ornaments had expanded over time. The end-products were clearly over-specialized and may have contributed to the group's extinction, implying law-like forces that had been able to evade adaptive constraints but only for a limited time. In the horns 'a tendency or predetermination to evolve in breadth or length orthogenetically appears to be established, flowing in one direction like a tide'.[9] Some of Osborn's evolutionary trees looked like candelabras, with many branches spreading out at the start only to continue on in parallel straight lines.

The displays made it all too evident that the dinosaurs were unlike the reptiles that had survived through to today. Carnegie's *Diplodocus* was mounted in a pose that had the animal standing firmly on all four legs,

[9] Henry Fairfield Osborn, *The Titanotheres of Ancient Wyoming, Dakota and Nebraska*, 2: 844; for the Museum displays, see Rainger, *An Agenda for Antiquity*, pp. 163–9, and Wallace, *Beasts of Eden*, pp. 95–100. More generally on his ideas, see Rainger, *An Agenda for Antiquity*; Brian Regal, *Henry Fairfield Osborn*; Clark, *God – Or Gorilla*, pp. 88–95, and Bowler, *Life's Splendid Drama*, pp. 339–52. Osborn's general surveys of evolution include his *The Age of Mammals* and *The Origin and Evolution of Life*; for the candelabra-style evolutionary tree, see the latter, p. 236.

quite unlike a lizard or crocodile. Some authorities claimed that its huge weight would have required it to spend most of its time partially immersed in water. There was a suggestion, published in *The Field* and echoed in the *Harmsworth Popular Science*, that it should have been mounted with the legs splayed out like a modern reptile. This gained little credence (although it was taken more seriously in Germany) and the specimen remained mounted on all fours as if to confirm that the dinosaurs were not ancestral to any living reptile.[10]

Knight's paintings brought the dinosaurs to life, and some museum mountings depicted them in life-like poses. In 1915 the American Museum of Natural History mounted *Tyrannosaurus rex* as though in combat to give an impression of ferocity. The great herbivores including *Iguanodon* and the giant *Diplodocus* appeared as lumbering, docile brutes hampered by their minute brains and cold-bloodedness. The contrast was nicely captured in Conan Doyle's *The Lost World*, where the *Iguanodon* is herded for meat by cavemen while the carnivores remain a fearsome threat. Their sheer size, coupled with the strange horns and other excrescences of some species, could be seen as the end-products of evolution taken over by non-adaptive trends leading to racial senility and extinction. Marsh's suggestion that the horns of *Triceratops* had become so large they had contributed to its extinction was still cited in *Scientific American* in 1911.[11]

This was a very non-Darwinian view of evolution, and we have seen how Osborn used it to head off some of the criticism coming from the religious opponents of evolutionism. Yet he was far from being the only palaeontologist to endorse the idea of predetermined development. At the Natural History Museum in London, Sir Arthur Smith Woodward promoted it in the literature accompanying the displays and made it the topic of an address to the British Association in 1909. He even smuggled it into the Museum's guide to fossil humans (including the Piltdown remains), making it clear that 'As a rule [the trend] passes the limit of utility, becomes a hindrance, and even contributes to the extermination of the races of animals in which it occurs.'[12] Of course

[10] On the arguments over how *Diplodocus* should be mounted, see Desmond, *The Hot-Blooded Dinosaurs*, p. 136; Rieppel, *Assembling the Dinosaur*, pp. 104–5, and Mee, ed., *Harmsworth Popular Science*, vol. 5, group 5, chap. 26, pp. 3093–105.

[11] For Conan Doyle's depictions, see *The Lost World*, pp. 182–5; On *Triceratops*, see Harold Shepstone's articles on 'Monsters of Bygone Ages', *Scientific American*, 104 (8 April 1911), p. 364, also Hutchinson, *Extinct Monsters*, p. 185.

[12] Woodward, *A Guide to the Fossil Remains of Man in the Department of Geology and Palaeontology at the British Museum (Natural History)*, p. 3. See also his 'President's Address' and Bowler, *Life's Splendid Drama*, pp. 355–61.

this was not the case for the expansion of the brain in the human family, allowing Woodward to join Osborn in presenting this element of the mechanism as progressive. One of Osborn's terms for directed evolution was 'aristogenesis', implying the development of the better qualities of each group.

It was certainly true that there had been a replacement of small-brained types with those displaying higher levels of intelligence and cooperation. But the overall picture emerging from the museum displays was one of multiple trends driving off in various directions away from the path toward humanity. In the case of the archaic mammals, like the dinosaurs, a whole collection of families had emerged, developed and gone extinct entirely separated from the ancestors of the groups that later expanded to generate the modern fauna. Some palaeontologists favoured a theory of racial senility in which groups such as the dinosaurs had risen to dominance but had eventually run out of evolutionary energy and had degenerated toward extinction. Perhaps each group had a built-in life cycle analogous to the rise and fall of human empires, a view still expressed in a survey by W. E. Swinton as late as 1934.[13]

This was hardly a picture that would encourage anyone to see humanity as evolution's predetermined goal. Even Osborn had to admit that progress was a more irregular process that often got led astray by short-term goals. He conceded that the archaic mammals were 'the first grand attempts of nature' to exploit the class's potential – in effect a failed experiment.[14] This point was even more apparent when it came to the dinosaurs. There were not enough fossils here to create evolutionary sequences, but the huge and bizarre forms now on display could easily be depicted as the end-products of nonadaptive orthogenetic trends, apparently successful at first but ultimately leading nowhere. Most authorities now conceded that the dinosaurs were not the ancestors of the reptilian species alive today (although they might be linked to the birds). They were certainly not the ancestors of the mammals, which were now being traced back to an older and unconnected reptilian group, the Theromorpha, as explained in Hutchinson's surveys.[15] It had been known since early in the nineteenth century that primitive mammalian insectivores had lived through the Age of Reptiles, totally overshadowed

[13] Swinton, *The Dinosaurs*, pp. 177–81, and more generally Bowler, *Life's Splendid Drama*, pp. 355–7.

[14] Osborn, *The Age of Mammals*, 96.

[15] Hutchinson, *Creatures of Other Days*, pp. 80–7, and *Extinct Monsters*, pp. 111–17. He notes that the fossils are already on display in the Natural History Museum.

by the dinosaurs. Only when the giant reptiles had disappeared did the class get a chance to expand.

Attributing the decline of a whole family to a loss of evolutionary energy seemed reminiscent of the vitalist form of Lamarckism, but as the evidence from both the geological and fossil records improved it became clear that the major transformations in the history of life were driven by changes in the environment that were often quite abrupt by geological standards. This was not quite a revival of the catastrophism popular in the early nineteenth century, but the new model did imply a level of discontinuity not recognized by the uniformitarian followers of Lyell, including Darwin. There were long periods of stable conditions interrupted by dramatic episodes of mountain-building and continental uplift. These were not catastrophic in the sense that they wiped out whole populations overnight – the theory of mass extinction by asteroid impact did not appear until the 1970s – but over a period of a few thousand years (trivial in geological terms) the climate changed and species specialized for the previously stable conditions had been driven to extinction. Only those with more generalized lifestyles would survive and expand into the new world created by the upheaval. This was a view of the history of life that was compatible with Darwinism's reliance on undirected variation, provided the theory was modified to incorporate episodes of abrupt environmental change.

In the 1890s Joseph LeConte had suggested that there were cycles in the history of life and had noted certain 'critical points' where transitions had been rapid. Richard Swan Lull concluded his 1917 textbook on evolution with an essay titled 'The Pulse of Life', which attributed each surge forward to a change in the climate that drove many species to extinction while stimulating the diversity of the survivors. Osborn's subordinate at the American Museum, William Diller Matthew, published a much-quoted essay, 'Climate and Evolution', in which he suggested that the Age of Reptiles was ended by an episode of continental elevation producing a cooler, drier environment that the dinosaurs could not adapt to. Matthew's guide to the dinosaur display at the Museum also stressed the role of the environment.[16] Even Osborn conceded that the great reptiles 'in the climax of their specialization and grandeur' had vanished quite suddenly, although he was less sure about the cause: 'One of the most dramatic moments in the life history of the world is the extinction of the reptilian dynasties, which occurred with apparent suddenness at the close of the Cretaceous, the very last chapter in the "Age

[16] Matthew, *Climate and Evolution*, p. 110; also his *Dinosaurs*.

of Reptiles".' The archaic mammals were the first 'pulse' of diversity by the class that was no longer overshadowed by the dinosaurs, soon to be swept away by a second wave that laid the foundations of the families that form the modern fauna.[17]

In this new interpretation, evolution was tightly constrained by environmental factors that could sometimes precipitate bursts of mass extinction followed by adaptive divergence from the survivors. The results were unpredictable on the basis of any trends observed in the preceding period of stability. After millions of years during which the triumphant dinosaurs had overshadowed the primitive mammals, their sudden elimination had only then given the apparently superior class its chance to expand. There was still overall progress, but it operated in an episodic and irregular manner that would have been unthinkable to the earlier Darwinians – or the Spencerians – for all their emphasis on the role of the struggle for existence.

Redrawing the Tree of Life

Even where evolution had proceeded gradually during periods of stability, the expanding fossil evidence was beginning to undermine the plausibility of the orthogenetic theory. Matthew's comments in Museum publications were increasingly critical of Osborn's position. Even in the case of the horse family it became clear that the classic sequence endorsed by Huxley was not a simple process of specialization: there was a transition from browsing to grazing linked to environmental change, and the process was by no means as direct as the early representations implied. The Museum still displayed the horses in a linear trend, but its guide (authored by Matthew) noted that there were side branches leading off to extinction, a phenomenon even more apparent in other families such as the camels. Matthew thought the mechanism was probably natural selection working within an environment subject to unpredictable changes. There were no inbuilt forces pushing evolution in predetermined directions.

The foundations were being laid for a new version of Darwinism with evolution represented as a process of unpredictable branching in which there were no predetermined goals, progressive or otherwise. Images representing horse evolution as a branching tree with many dead branches now began to proliferate in popular texts (see Fig. 7.3). This interpretation of horse evolution also figured prominently in

[17] LeConte, *Evolution*, pp. 258–9; Lull, *Organic Evolution*, pp. 687–9; Osborn, *The Age of Mammals*, p. 97.

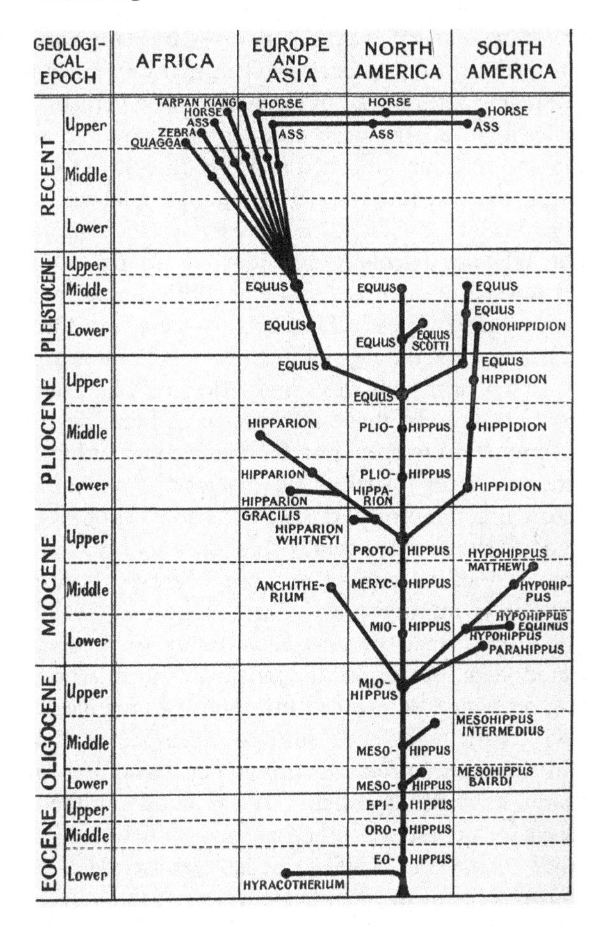

Fig. 7.3 Horse evolution represented as a branching tree. From C. M. Beadnell, *A Picture Book of Evolution* (1934 ed.), p. 10, adapted from R. S. Lull, *Organic Evolution* (1917), p. 611.

the introductory material of the Home University Library's survey of evolution.[18]

The old model of the tree of life with a central trunk did not disappear from popular representations overnight. Eminent scientists shared the liberal theologians' vision of evolution as a process under divine control, some quite explicitly, others via a sense that the process must have an underlying moral direction. The claim that humanity is the predetermined

[18] Geddes and Thomson, *Evolution*, pp. 23–32.

goal of the history of life on earth was still endorsed by some palaeontologists and was embedded in the images used in many popular books.[19] There were side branches, of course, but these adaptive divergences always deflected the species involved into dead ends. In the revised edition of his study of evolution and religion published in 1898, Joseph LeConte made this point by insisting that the path toward the ultimate goal was 'a straight and narrow way'. E Grant Conklin's *The Direction of Human Evolution* of 1921 argued that a 'wider teleology' was obvious in the sequence of advances that led toward humanity. As late as 1933 the palaeontologist Robert Broom, whose discoveries in South Africa were transforming the picture of human origins, asked if the coming of humanity was accident or design and came down firmly on the side of design. The main line of development was constituted by those species that had remained generalists, avoiding the temptation to specialize for a single way of life.[20]

In one of his many articles J. Arthur Thomson wrote of 'the age-long man-ward adventure that has crowned the evolutionary process upon the earth'. This language shows how even a biologist trying to develop a more sophisticated way of inserting a purpose into evolutionism could still invoke the image of humanity as its intended goal. Yet, as in the Home University Library text, he was also willing to promote a less rigidly goal-directed view. He began to transform the interpretation of progress by taking on board Bergson's philosophy of creative evolution. The emphasis on creativity allowed him to visualize evolution as a process that could sometimes produce entirely new structures and functions moving life into new realms of being. It was an open-ended striving toward higher levels of mental and moral activity, which had eventually allowed one branch of the tree of life to achieve consciousness. In this sense, visualizing humanity as the goal of evolution was merely a product of a very human tendency to see our species as having special value; it did not imply that humanity was the predetermined goal. This was a view echoed in the psychologist Conwy Lloyd Morgan's idea of emergent evolution, in which entirely new functions including consciousness appeared at key points in the advance of life. The idea of creative

[19] The frontispiece to Adams Gowans Whyte, *The Wonder World We Live In* is an example from a children's book. Clarke, *God – Or Gorilla*, reproduces this and other good examples from Benjamin C. Gruenberg, *The Story of Evolution* (1919) and Willian King Gregory, *Our Face from Fish to Man* (1929), see pp. 140–51.

[20] LeConte, *Evolution*, p. 90; Conklin, *The Direction of Human Evolution*, pp. 18–21; and Broom, *The Coming of Man*. See Lester D. Stephens, *Joseph LeConte*; Alexander Pavuk, 'Biologist Edwin Grant Conklin and the Idea of a Religious Direction of Human Evolution in the Early 1920s'; and on Broom, Bowler, *Reconciling Science and Religion*, pp. 134–5.

evolution was pushing even conservative thinkers toward a less structured, open-ended model of progress.[21]

Despite his open support for Lamarckism, even Bernard Shaw caught the spirit of this transition. After the Great War – which, along with unrestrained capitalism, he blamed on the influence of Darwinism – Shaw became more pessimistic about the implications of his vision of nature's creativity, at least for the future of the human race. He doubted that modern humans had enough energy to continue the race's progress toward the superman predicted in his earlier play and worried that the universal life force might push us aside in favour of a rival: 'The power that produced Man when the monkey was not up to the mark, can produce a higher creature than Man if Man does not come up to the mark.... Nature holds no brief for the human experiment: it must stand or fall by its results.'[22] He was even prepared to condone the eugenic movement's efforts to cull the weaker components of the race, which for Shaw meant those who stood in the way of the creative few.

Those more sympathetic to Darwinism were already willing to adopt a more open-ended model of progress. In the 1890s David Starr Jordan gave lectures to university extension courses in California in which he insisted that evolution was not necessarily progressive and that humanity was not its intended goal. In Britain the *Harmsworth Popular Science* openly proclaimed the contingency of evolution, suggesting that if the great reptiles been a little more active, the primitive mammals might have been wiped out. There was an implication that something on the human level would eventually emerge, but it might not have been humanity as we know it: 'Had the vital link been snapped, had the ascending line of which man is the summit and glory been severed, from what other type would Nature have fashioned her greatest son?' J. B. S. Haldane would later joke that if humanity were wiped out, the rat would be a good candidate to develop a replacement.

There were now suggestions that life elsewhere in the universe could progress toward intelligence in very different ways. This was the point made in H. G. Wells' *The War of the Worlds*, where the invading Martians are described as totally alien. In 1921 Matthew published an article speculating about life on other worlds and suggesting that it might be

[21] The quotation is from Thomson's 'The Influence of Darwinism on Thought and Life', p. 217. See also Conwy Lloyd Morgan's *Emergent Evolution*. On Thomson and Lloyd Morgan's views, see Bowler, *Reconciling Science and Religion*, pp. 136–40, and on Thomson's popular science writing *Science for All*, pp. 233–40.

[22] Shaw, *Back to Methuselah*, p. xvi.

so different that we would hardly recognize it.[23] Here the contingency of evolution guided only by natural selection acting within changing environments is explicitly acknowledged.

The most widely publicized expression of this new perspective came in Wells' *Outline of History* of 1920 – the work that Belloc objected to so strongly. Guided by his friend Lankester, Wells used the introductory section to survey the development of life on earth in Darwinian terms, but modified to take the new element of discontinuity into account. The advances are described as episodic, with the abrupt appearance of new types following long periods of adaptive stagnation. The original version has a chapter title, 'The Invasion of Dry Land by Life', to drive home the impression that here was an innovation that transformed the whole situation.[24] Environmental stress is a stimulus to evolution, but there can be no clear-cut line of progress: a visitor from another world would have been unable to predict the next progressive step on the basis of what had gone before. Such a visitor would have found the Devonian mudfish, for instance, 'a very unimportant side-fact in that ancient world of great sharks and plated fishes ... a poor refugee from the too-crowded and aggressive life of the sea'. But they nevertheless 'opened the narrow way by which the land vertebrates rose to predominance'.[25] The rise of the mammals to dominance is presented as something that could not have happened until the great reptiles had been eliminated, probably by climatic change. A later chapter uses this event to illustrate the unpredictable nature of revolutions in human history. Looking at the world in the Middle Ages, no one could have predicted that Europe, long overshadowed by the great empires of the East, would ever rise to dominate the world – just as no one looking at the earth at the end of the Mesozoic could have realized that the great reptiles were about to be replaced by the mammals.[26]

Wells would soon link up with Julian Huxley to produce a follow-up serial that would substantiate this open-ended vision of evolution,

[23] See Jordan, 'Evolution – What It Is and What It Is Not' in his *Foot-Notes to Evolution*, pp. 54–74, esp. p. 69. Also Mee, ed., *The Harmsworth Popular Science*, group 5, chap. 1, vol. 1, p. 47; Haldane, 'Man's Destiny' in his *The Inequality of Man and Other Essays*, pp. 142–7, see p. 146; Matthew, 'Life on Other Worlds'.

[24] I have used both the first book version of the *Outline of Life* (1920), which follows the original monthly parts, and the definitive edition of 1923, which keeps the text description of the invasion of the land but drops the chapter title. See chap. 4, pp. 15–18, of the 1920 edition and pp. 11–14 for the equivalent text in the 1923 version. All references below are to the latter.

[25] Wells, *Outline of History*, p. 26; on the palaeontologists endorsing this view, see Bowler, *Life's Splendid Drama*, p. 432.

[26] Wells, *Outline of History*, p. 492.

providing the first mass-market popularization of the emerging synthesis of natural selection and genetics: *The Science of Life*. Darwinism was now expanding its influence, no longer the preserve of the rationalists who used it to attack organized religion, although Huxley himself could not bring himself to abandon the hope that evolutionary progress had a moral purpose. Natural selection was seen as a process that could create new structures and functions, not just modify those already in existence. Whatever Huxley's doubts, for many biologists it was clear that evolution was an open-ended, opportunistic process that had no predetermined goal. The 1930s would see the 'Modern Synthesis' take its first steps taken toward the dominant position that it would occupy in the life sciences of the late twentieth century.

8 Social Evolutionism

The question of human origins had underpinned much of the debate following the publication of Darwin's ideas. Conservative religious believers rejected anything that would deprive us of our supposed intellectual and moral superiority over brute creation. Liberal Christians were more willing to accommodate aspects of evolutionism, either by insisting that the soul was divinely implanted in a body formed by natural processes or by seeing the process of evolution as an ascent toward a spiritual goal. The former policy would make it difficult to see a link between biological and social evolution, while the latter required an element of the old argument from design to be incorporated into the theory. Materialists and rationalists saw no point in an evolutionary theory that could not apply to the most important step in the ascent of life and rejected any role for natural theology. For Herbert Spencer and his followers, evolution was a seamless progression that could not be divided into distinct biological and social phases. In practice, though, the division between conservative and radical positions was not clear-cut, because most commentators sought compromises that avoided the harsh implications of Darwinism.

The late nineteenth and early twentieth centuries were dominated by what eventually became known as 'social Darwinism' – even though some of the links were with non-Darwinian mechanisms. Social *evolutionism* was a collection of different ways of thinking about society identified by a focus on competition as the driving force of change and/or the assumed progressive nature of the process. Spencer's philosophy became an inspiration for many, and because he saw individual competition as the motor of change, it was naturally associated with Darwinism – Spencer himself having coined the phrase 'survival of the fittest'. But Spencer was as much a Lamarckian as a Darwinian and saw evolution as a force that would produce a human race perfectly adapted to social life. His was a theory of moral development, which is why liberal Christians such as Fiske could promote it. Lamarckism could, in fact, be used to underpin either a model based on self-improvement or one in which formal education was used to instil the social virtues. In either case, learned habits could become instincts biologically embedded in the race.[1]

Commentators who constructed schemes of social and cultural evolution became household names. The opinions of Spencer, Fiske and Kidd were circulated in articles and reviews for those who found their ponderous tomes rather daunting. Spencer's frequent illnesses were reported in the newspapers, along with his latest ideas, and his likeness appeared everywhere, including on a card issued by a brand of cigarettes. The commentaries were, however, often distorted in ways that the thinkers themselves did not approve of. When Spencer visited the United States, he was uncomfortable with the enthusiastic reception he received from the robber barons of the Gilded Age. The emphasis of the popular literature also shifted with the times, as confidence in the inevitability of progress was shaken by a fear of degeneration and the Great War of 1914–19 revived a version of social Darwinism based on national rather than individual competition.

From the start there were some who welcomed a secondary aspect of Darwin's theory, the possibility that selection could operate between groups as well as between individuals. Spencer's liberalism was eventually challenged by a tide of imperialism in which the struggle for existence between nations and even races was seen as the real driving force of progress. This drew on anthropologies wedded to a hierarchical model of human diversity, creating an ideology with disastrous consequences for those races deemed 'inferior' by the triumphant Europeans. The assumption that character is predetermined by racial origin also encouraged the more general claim that heredity determines an individual's position in society. The eugenic movement's efforts to improve the human race were based on this ideology, applied more by analogy with artificial than with natural selection.

Accepting that race theories are aspects of social evolutionism forces us to recognize that the topic is not defined solely by the mechanism of change. Appealing to the struggle for existence as the motor of progress is certainly relevant, but so is the whole apparatus by which anthropologists sought to explain how the various forms of humanity had evolved from an ape ancestry. Debates over whether natural selection or Lamarckism offered the best model of social evolution interacted with equally divisive arguments over how and why we transcended our animal origins.

[1] The classic notion of social Darwinism was defined in Richard Hofstadter's *Social Darwinism in American Thought*. For a critique, see Robert C. Bannister, *Social Darwinism*, and for more general surveys Greta Jones, *Social Darwinism and English Thought*; Mike Hawkins, *Social Darwinism in European and American Thought*; and Piers Hale, *Political Descent*.

Indeed, the relevance of the mechanism depended on the belief that it had generated upward progress. The public's understanding of how evolution works may have been influenced as much by the well-publicized controversies over fossil hominids as by the efforts of those who tried to impose the structures of Darwinism and its rivals onto the pattern of human history.

The popular belief that humans have descended from a ferocious ape ancestor highlighted the worrying implications of the struggle for existence. The link was suggested by the physical similarities, and these played a role in defining stages in the transition from bestiality to humanity. The presumption that a hierarchy of forms led directly from ape to human shaped attitudes both to hominid fossils and to non-European races. The Neanderthals came to be seen as the 'missing link', the classic model for the 'ape-man' halfway between the old and the new. The supposedly 'lower' races of today, especially those identified as 'savages', were used to illustrate the later phases of the ascent from the Neanderthal type toward modern, presumed to be European, humanity. Academic anthropologists eventually turned their backs on the scheme that defined 'primitive' modern cultures and races as relics of past stages of development, but in popular debates over how we became human that model persisted well into the new century.

Discoveries of fossil hominids and the ensuing debates about their significance for evolution probably received more coverage in the popular media than any direct applications of the Darwinian theory to society. In the late nineteenth century, the fulminations of Spencer and his critics filled the better-quality magazines as the captains of industry sought to justify their economic power. But the transition to a form of social Darwinism based on national and racial struggle was presented to the wider public more in commentaries on the fate of earlier stages in the ascent toward the modern world. Extinct forms of humanity – even those deemed not to lie directly on our ancestral line – were headlined along with the most spectacular dinosaurs. Artists provided imaginative representations of what the 'ape-men' might have looked like. and experts tried to explain their demise. Accounts of 'savage' races were often penned by explorers and missionaries, while novelists such as Rider Haggard celebrated the Europeans' conquests of other races. Unfortunate individuals from 'primitive' tribes were displayed in exhibitions well into the twentieth century.

Up from the Ape

Evolutionists were haunted by the image of humanity struggling to rise above the savagery of an ape ancestry. Social evolutionism had to explain how the supposedly violent lifestyle of the gorilla could be transformed into that of modern humanity. The process was presumed to be correlated with the expansion of the brain, the loss of the huge brow-ridges and teeth, and the adoption of a fully upright posture. Preconceived ideas about these transformations shaped both the popular and the scientific interpretations of human origins. There must have been a sequence of developments leading from ape to modern human, but opinions differed on the order of events and the chief driving forces.

The popular obsession with imagined intermediates between apes and humans continued to express itself in cartoons and caricatures. Until his death in 1882, Darwin himself was often lampooned by giving him ape-like features.[2] More seriously, cartoonists and anthropologists continued to depict the 'lower' races as retaining some aspects of ape physiognomy. The cartoons were not simply the product of popular misunderstanding. Darwinists might protest that their theory derived ape and human from a common ancestor that has not survived, but the scientific consensus moved toward the view that our ancestry did indeed lie among the great apes, the gorilla or the orangutan being the most common candidates. The much reproduced frontispiece to Huxley's *Man's Place in Nature* may not have helped – although intended to show the physical similarities between apes and humans, it was all too easily read as a historical sequence whose intermediate phases needed to be identified.[3] Far from disappearing in the early twentieth century, the idea of an ape ancestry continued to be promoted by scientists in popular books such as W. K. Gregory's *Our Face from Fish to Man* and E. A. Hooton's *Up from the Ape*.[4]

At a detailed level things were not quite so straightforward. Misia Landau has suggested that the rival theories often took on a narrative structure reminiscent of folk-tales and adventure stories in order to emphasize humanity's triumph over the challenges it faced.[5] The transition from ape to human involved several different physical transformations, each with implications for how and why the process might have occurred. A variety of intermediate forms were imagined by popular artists, reflecting genuine disagreements among the experts. How were the enlargement of the brain and the consequential changes in skull structure related to the adoption of the bipedal posture that freed the hands from the requirements of locomotion? Did the two processes work

[2] Chapter 10 of Janet Brown's *Darwin: The Power of Place* describes many of these caricatures.

[3] On the influence of Huxley's image, see Gowan Dawson, 'A Monkey into a Man'. A good example is a *Scientific American* sequence from 1876 reproduced in Schwartz, ed., *Streitfall Evolution*, p. 29.

[4] Clark, *God – Or Gorilla*, chap. 10, deals with the popularity of the image in the 1920s.

[5] Landau, *Narratives of Human Evolution.*

independently, and was one more important than the other? Most authorities shared the common assumption that an increased level of intelligence is our most significant distinguishing characteristic. This encouraged the belief that the enlargement of the brain must have been the main driving force, with the adoption of an upright posture being a later addition. Such a 'brain-first' theory would imply that the earliest missing link, when found, would have an enlarged brain while still not standing fully upright.

The alternative view that the transition to bipedalism came first had been suggested by both Darwin and Haeckel. On this model the human family became separated from the apes before the brain began to expand, so the increased level of intelligence might be a consequence of the freeing of the hands for tool-making. This suggestion was largely ignored until the 1930s. The ape-men imagined by both anatomists and cartoonists were often slouching creatures, even when depicted with stone tools. Fossils that did not fit this preconception were ignored or misinterpreted.

A prominent exponent of the 'brain-first' theory was the anatomist Grafton Elliot Smith. He presented it in an address to the British Association in 1912 and in a semi-popular collection on *The Evolution of Man* in 1924. Like his rival Arthur Keith, he was regularly consulted by newspapers and magazines seeking copy on the latest discoveries. He acknowledged the environment as a stimulus to progress but postulated a trend toward brain development running through primate evolution and culminating in the emergence of modern humanity. Our ancestors did not stand upright before getting their bigger brains; it was the ape-like forms whose intelligence developed most rapidly that recognized the advantages of moving out of the trees and adopting bipedalism. Smith's model was not unilinear; there were multiple lines of primate evolution, but all moved in the same direction, toward bigger brains. This was a theory of evolutionary parallelism based on an orthogenetic trend, similar to that endorsed by palaeontologists. There had been multiple efforts to achieve the goal: Asia and Africa were 'the laboratory in which, for untold ages, Nature was making her great experiments to achieve the transmutation of the brute substance of some brutal Ape into the divine form of Man'. Smith added that he did not really intend to imply Teleology (note the capital T) but employed this terminology purely for convenience.[6] Few of his readers could have escaped

[6] Smith, *The Evolution of Man*, p. 77. He was by no means the only authority at the time to imply that nature was experimenting to achieve its goal; see, for instance, Hooton, *Up from the Ape*, p. 390, and for other examples Bowler, *Theories of Human Evolution*, pp. 169–73 and 213–18.

the impression that evolution has a predetermined goal with a moral significance.

Prevailing assumptions also influenced ideas about where the main developments took place. Darwin had suggested Africa on the grounds that it was the home of the best-known apes, the gorilla and the chimpanzee, but Haeckel had argued that the orangutan's location in Asia suggested that this might have been the original home of humanity. Africa soon fell out of favour, in part because Europeans saw its modern inhabitants as less evolved and suspected that its tropical environment did not encourage progress. By the early twentieth century Henry Fairfield Osborn and others became convinced that central Asia's colder and drier climate provided the challenging conditions most likely to have allowed the 'dawn man' to develop intelligence and cooperative behaviour. Applying his own model of parallel evolution, Osborn could present the apes as Africa's less successful efforts to progress in a similar direction. Expeditions to central Asia were sent out in the hope of discovering the missing link there, while discoveries pointing to Africa's role were ignored.[7]

Problematic Fossils

The lack of fossil evidence for the origin of humanity had generated the original concerns over the 'missing link'. The Neanderthal remains were a possibility but were regarded as anomalous by many authorities. Those who did take them seriously, including Huxley, noted the ape-like features of the skull but realized that its large cranial capacity ruled it out as a half-way house between ape and modern human. At best it was a primitive member of our own species, less advanced than the 'lowest' living races, a position endorsed in Armand de Quatrefages' *The Human Species* in the International Science Series. This impression was sustained by archaeological evidence associating Neanderthals with one of the early stone tool-making cultures, the Mousterian. An image published in *Harper's Weekly* in 1873 showed a Neanderthal with relatively modern features and holding a well-made stone axe. There were suggestions that the lowest surviving races were little more advanced than the Neanderthals – in 1911 W. J. Sollas' *Ancient Hunters* described the Australian aborigines as 'the Mousterians of the antipodes', while Madison Grant saw Neanderthals in Ireland![8]

[7] For details, see Rainger, *An Agenda for Antiquity*, and Regal, *Henry Fairfield Osborn*.

[8] De Quatrefages, *The Human Species*, chap. 26 (he uses the term 'Canstadt race' referring to another fossil discovery). The *Harper's Weekly* image is reproduced in Schwartz, ed., *Streitfall Evolution*, p. 144; for the Sollas reference, see Sollas, *Ancient Hunters and Their*

Sollas soon abandoned this position, which had already been rejected by most authorities. Discoveries at Spy in Belgium in 1886 solidified the notion of a distinct Neanderthal race or species and focused attention on the ape-like features of the skull. The cranium might be large, but it was shaped differently to that of a modern human, allowing the Neanderthals to be identified as a much earlier stage of development, perhaps as the long-missing link. Cope suggested this interpretation in his *American Naturalist* in 1893, and Haeckel provided a diagram in his *The Last Link* of 1898 showing the outline of the Neanderthal cranium as significantly less expansive than a modern form. Gustav Schwalbe became the leading defender of the Neanderthals as a distinct species forming the penultimate phase in the advance from the apes. He defended this view at the 1909 Darwin centenary celebrations and in an article in the Little Blue Books series entitled 'From Monkey to Man'. In 1911 it was endorsed in Arthur Keith's first popular account of human origins.[9]

The real purpose of Haeckel's *The Last Link* was to draw attention to a discovery that vindicated his earlier predictions. He had even given his hypothetical link a name, *Pithecanthropus*, and it was his suggestion that Asia might be the scene of the transition that led the Dutch medical anatomist Eugene Dubois to begin searching in Java (then in the Dutch East Indies). In 1891–2 he unearthed the skull and thigh-bone of a creature he called *Pithecanthropus erectus*. The cranial capacity was halfway between ape and human, and the skull had prominent brow-ridges, similar to those of the Neanderthals. This seemed to fit the bill as far as the skull was concerned, but the femur indicated that the creature had stood fully upright, and this was not what most authorities – except Haeckel – were looking for. While Haeckel saw the discovery as support for his belief that the earliest ancestors of the human family had already achieved bipedalism, other anatomists suspected that the upright posture was not consistent with so undeveloped a mental capacity. Perhaps the bones had not even belonged to a single individual. Haeckel conceded that at the 1895 International Zoological Congress fully half of the experts who gave an opinion did not support his position.[10]

The controversy over how to interpret Dubois' finds flourished mainly in the technical and popular science journals. Several contributions

Modern Representatives, p. 144, and for Madison Grant, see his *The Passing of the Great Race*, p. 108. For details of these discoveries, see, for instance, John Reader, *Missing Links*; Marianne Sommer, 'The Neanderthals'; and my own *Theories of Human Evolution*.

[9] Cope. 'The Genealogy of Man'; Haeckel, *The Last Link*, p. 25; Schwalbe, 'The Descent of Man'; and Keith, *Ancient Types of Man*, pp. 78–9, 93 and 118–19.

[10] Haeckel, *The Last Link*, pp. 22–6; see also the 1907 edition of *The Evolution of Man*, 2: 263–4. For details, see Bert Theunissen, *Eugene Dubois and the Ape-Man from Java*.

appeared in *Nature*, while Keith and Marsh published favourable reactions in *Science Progress* and *Science*, respectively (the latter now emerging as an American equivalent of *Nature*). Keith's *Ancient Types of Man* had a short chapter on the 'fossil man from Java', presenting it as a primitive version of the Neanderthal type. Because of the upright posture, even those who endorsed Dubois' interpretation saw the creature as too late in the sequence to be a real intermediate between ape and human. Dubois produced a model for display at the World Exhibition in Paris in 1900, but he eventually became discouraged and began to refuse access to the original specimens.

By the time Keith's book appeared in 1911, both the scientific and popular views of the Neanderthals had begun to change. Keith made a brief reference to a new discovery made in 1908 at La Chapelle aux Saints in France, but along with most authorities he soon began to see this specimen as evidence that the image of the Neanderthals as a late intermediate between ape and human was incorrect. Marcellin Boule described the specimen as an old man but failed to realize that it had been crippled with arthritis. He argued that its stooped posture was typical of the Neanderthal type and proved that this was a much more primitive form than had hitherto been recognized. Far from allowing the type to be seen as a link back to the apes, Boule insisted that it was comparatively recent, so there was no time for evolution to make the transition to modern humanity. The Neanderthals were thus expelled from human ancestry – they were a separate line of development that had advanced in a similar direction but more slowly. Boule insisted that his model was an improvement on the simple linear model, but he invoked the concept of parallelism to give the impression of multiple lines advancing in the same direction. The Neanderthal line had independently achieved at least some aspects of humanity, including the ability to make tools, but lagged behind the line leading to the modern species.[11]

Boule's discovery was headline news, with experts such as Keith being called upon to provide their interpretation of its significance. The new image of the Neanderthals was promoted by well-known writers such as H. G. Wells and soon caught the popular imagination (see Fig. 8.1). The *Illustrated London News* hailed the new specimen as 'the most important anthropological discovery for fifty years' and provided an image depicting the type as little more than an ape with a club. Other images were less brutal but still made the Neanderthals appear significantly less human

[11] Boule's semi-popular study was translated as *Fossil Men: Elements of Human Palaeontology*; see Michael Hammond, 'The Expulsion of the Neanderthals from Human Ancestry'.

Fig. 8.1 Restoration of 'Neanderthal Man' with brutal features. From H. G. Wells, *The Outline of History* (1920), vol. 1, p. 49.

and more ape-like in appearance. The frontispiece of Osborn's *Men of the Old Stone Age* of 1916 showed them in this way, and Charles Knight's mural for the Hall of the Age of Man at the American Museum of Natural History followed suit three years later.[12]

The sudden popularity of the thesis that the Neanderthals do not lie on the main line of human ancestry helps to explain the wave of enthusiasm that greeted the discovery of human remains at Piltdown in the south of England in 1912. Decades later these were shown to be artificially modified bones that had been 'planted' at the site, but at the time no

[12] *Illustrated London News*, 27 February 1909, pp. 300–1 and 312. For Osborn's frontispiece and the Knight mural, see Clark, *God – Or Gorilla*, pp. 202–7.

one seems to have suspected the hoax (or fraud), and 'Piltdown man' was accepted as a discovery of major significance. Subsequent to the exposure there has been much speculation trying to identify the culprit, coupled with expressions of surprise that so many experts could have been taken in. Sir Arthur Smith Woodward of the Natural History Museum in London provided the first descriptions of the finds, while authorities such as Keith and Smith threw their hats into the ring with reconstructions of the imaginary creature's features.

Nationalism almost certainly played a role – the British were anxious to have fossils from their own country to rival those found in France. But most of those who took the finds seriously were committed to the expulsion of the Neanderthals from our direct ancestry and were looking for evidence of a 'pre-sapiens' type showing a higher level of mental development. Opinions differed on the cranial capacity of the skull (which was in fragments), but all agreed that it had a more modern appearance than that of the Neanderthals. The jaw seemed ape-like (it was in fact an ape jaw with the teeth filed down), but it was widely accepted that this was a form that had advanced further toward full humanity in the crucial area of brain development. The British press especially was wildly enthusiastic, the *Illustrated London News* featuring images of the excavation site and an imagined reconstruction of this 'discovery of supreme importance to all interested in the history of the human race'. The remains featured prominently in Woodward's guide to the human fossils in the Natural History Museum.[13]

The most extreme application of parallelism came from Osborn, who applied it to the apes themselves, thereby pushing them too off the line leading to modern humans. Like Keith, Osborn at first endorsed the expulsion of the Neanderthals and used the extended timescale of the line leading to *Homo sapiens* to widen the alleged differences between the modern races. In 1922 he hailed the discovery of a fossil tooth in Nebraska as evidence that pre-modern humans had already moved into North America, only to see the tooth dismissed as that of a pig. He moved on to promote a more radical theory in which modern humans had emerged not from the apes but from a 'dawn man' that had evolved in the harsh and hence stimulating environment of central Asia. The great apes were a distinct primate family evolving in Africa, their similarities to humans being the product of parallel development. Osborn

[13] *Illustrated London News*, 28 December 1912, pp. iv–v and 958, images reproduced in Reader, *Missing Links*, pp. 62–7. See also Woodward, *Guide to the Fossil Remains of Man*, pp. 8–23. For more details in the discovery and the later literature generated by the exposure of the fraud, see, for instance, my *Theories of Human Evolution*, pp. 35–8.

promoted this revision of the human family tree in the Museum literature and the popular press.[14]

The new theory allowed Osborn to counter the creationists' insistence that evolutionism necessarily demeaned human dignity by linking us with the apes. It also fitted his belief that modern human races all had distinct and long-established characters. In his *Man Rises to Parnassus* of 1927 he claimed that the races were equivalent to distinct species, perhaps even distinct genera.[15] The theory was used to justify much-publicized expeditions to central Asia led by Roy Chapman Andrews. Important fossil discoveries were indeed made, but not of ancient humans, although Davidson Black's discovery of 'Pekin man' in 1927–9 did something to maintain the theory's credibility. This was at first described as a new genus, *Sinanthropus*, but was soon recognized as a variant of the *Pithecanthropus* type already known from Java.

Despite Osborn's efforts to publicize his exclusion of the apes from human ancestry, his theory was greeted with scepticism by most experts. Many did, however, accept a significant level of parallelism, and it was widely agreed that Africa was unlikely to be the original home of humanity. Most assumed that the expansion of the brain had preceded the adoption of bipedalism. It was hardly surprising, then, that news of a new fossil with the potential to undermine this consensus would be greeted with suspicion. In 1924 Raymond Dart announced the discovery in South Africa of the skull of a hominid child to which he assigned the name *Australopithecus africanus*, claiming that this was ancestral to modern humanity. The skull had no brow-ridges and had human-like teeth. The position of the foramen magnum (through which the spinal nerves run) suggested that the creature had walked upright, yet the cranial capacity seemed scarcely larger than that of an ape. This was a combination of characters that would vindicate Darwin's predictions that humans had emerged in Africa and had first been defined by their upright posture rather than the expansion of the brain. The find attracted considerable press attention (see Fig. 8.2), but few experts took it seriously – it was an immature specimen so claims about the adult form were just speculation.

The situation would only begin to change in the 1930s when Robert Broom began to discover more specimens of Australopithecines, including adult skeletons. The type now had to be taken seriously as the

[14] See Regal, *Henry Fairfield Osborn*; Rainger, *An Agenda for Antiquity*, pp. 145–51; Clark, *God – Or Gorilla*, esp. chap. 6; and Bowler, *Theories of Human Evolution*, chap. 5 and pp. 176–81.

[15] Osborn, *Man Rises to Parnassus*, p. 199.

Fig. 8.2 'An Early Pliocene Entertainment', cartoon depicting early hominids learning to walk upright. From *Punch*, 26 June 1929.

foundation of the human family, making *Pithecanthropus* (later renamed *Homo erectus*) a plausible candidate for the intermediate stage leading to modern humans. The assumption that humanity was the climax of a built-in trend toward brain development began to unravel, to be replaced with the far more Darwinian view that the crucial turning point had been an adaptive modification driven by our ancestors' move from the trees to the open plains. The openly teleological viewpoint embedded in the brain-first theory that had prevailed since the 1860s now faded from view.

Races and Relics

The palaeoanthropologists who invoked multiple lines of human evolution were merely extending the position developed in the 1860s by the polygenists and cultural evolutionists. They had already created the model in which parallel lines ascended the same scale of complexity to explain the survival of 'lower' races and cultures in the modern world. Admittedly, E. B. Tylor's cultural hierarchy had not been meant to condemn those peoples with a lower level of development to permanent inferiority – they could advance if properly led, but only over many generations. Eventually, though, even Tylor succumbed to the prevailing view that some races were trapped by their inferior mental capacity. In his semi-popular *Anthropology* of 1881 he suggested that to account for low levels of development 'one should partly look for an explanation of this in differences of intellectual and moral powers between such tribes as the native Americans and Africans, and the Old World nations who overmatch and subdue them'.[16] Spencer doubted that modern savages were the equivalent of ancient cultures and recognized the role of the environment in shaping characters. But his *Study of Sociology* of 1873 (whose chapters had also been published in *Popular Science Monthly*) proposed a scale of complexity against which each race's achievements could be measured, a position echoed in Fiske's *Outlines of Cosmic Philosophy*.[17]

By this time the United States had its own version of cultural evolutionism in the form of Lewis Henry Morgan's *Ancient Society*, published in 1877. He defined three levels of development: savagery, barbarism and civilization, and insisted that they formed a 'natural as well as a necessary

[16] Tylor, *Anthropology*, p. 74; for details of his and Lubbock's earlier positions, see Chapter 4 above, and on the later developments George W. Stocking, Jr.'s *Victorian Anthropology* and Bowler, *The Invention of Progress*, chap. 4.

[17] Spencer, *The Study of Sociology*, chaps. 14 and 15; Fiske, *Outlines of Cosmic Philosophy*, vol. 2, chaps. 21 and 22.

sequence of progress', all stages of which could still be seen around the world today. He too was convinced that those who retained a lower level of culture were constrained by the limitations of their mental and moral powers. The Aryan races of Europe, now dominating the world, were at the highest level. He did, however, recognize that each society was shaped by how its technology adapted to the local conditions, a perspective that eventually led to his book being promoted by socialist groups.[18]

Cultural evolutionism now blended seamlessly into the racial ideology that underpinned the thinking of the experts who focused on the physical features differentiating the races. Polygenists imagined separate origins for the races, but the expansion of the timescale of human antiquity by the archaeologists made it possible to retain the notion of distinct races even if they had diverged from a common ancestor in the distant past. Some would lag behind the others, retaining ape-like characters and a lower intelligence resembling the ancestral forms of higher types. Their antiquity ensured the stability of the types in the modern population, even when mixed together. The level of development would depend on the degree of stimulus provided by the environment in which each race had originated, which might also generate physical differences that were not derived from an ape ancestry. Lamarckians who thought the environment could directly influence character nevertheless assumed that the effect would be so slow that no improvements could be made in the modern world. This racist ideology would persist in popular culture long after it had been repudiated by professional anthropologists.[19]

Cultural anthropologists argued over the nature of primitive society, but popular literature tended to focus on the racial composition of particular regions. British readers were fascinated by efforts to reconstruct the racial prehistory of the country as a series of invasions by successively higher cultures. The idea of a single Aryan race invading from the east evaporated as archaeologists unearthed evidence of different racial and cultural types appearing over a long period of time. The British were proud of their Anglo-Saxon or Teutonic heritage, which they contrasted with that of the Celts found in Ireland and France. Works such as W. Boyd Dawkins' *Cave Hunting* of 1874 and Thomas Beddoes' *The Races of Britain* of 1884 popularized the identification of these

[18] For details, see Adam Kuper, 'The Development of Lewis Henry Morgan's Evolutionism', and Thomas R. Trautmann, *Lewis Henry Morgan and the Invention of Kinship*. The later edition of *Ancient Society* cited in this book's Bibliography was issued by a publisher specializing in socialist books.

[19] See George W. Stocking, Jr., *Race, Culture and Evolution* and the same author's *Victorian Anthropology*. There is a huge literature on the race question; see, for instance, Nancy Stepan, *The Idea of Race in Science*, and John S. Haller, *Outcasts from Evolution*.

ancient races by their head shape, the round-headed dolichocephalics partially replacing the long-headed brachycephalics.

The identification of races depended on the assumption that they had fixed mental and physical constitutions. These were not necessarily arranged into a hierarchy; the difference between Anglo-Saxon and Celt, for instance, was seen as one of temperament rather than overall ability. Even so, the native Irish were still caricatured as ape-like in cartoons whenever the political situation in Ireland became more contentious (see Fig. 8.3). As late as 1924 E. W. MacBride's contribution to the Home University Library on heredity ended with a diatribe against the Irish as relics of a Mediterranean race adapted by Lamarckism to a less-stimulating environment. Anti-Irish prejudice also spread to the United States, where Madison Grant insisted that 'ferocious gorilla-like specimens' equivalent to Neanderthals could still be seen in the west of Ireland.[20]

Under the influence of Morgan and Spencer, American anthropologists evaluated the native peoples of the continent and the newly emancipated black African population, usually condemning them to positions of permanent inferiority and occasionally announcing their imminent extinction. Leading figures were J. W. Powell and W. J. McGee of the Bureau of American Ethnology and Daniel Brinton of the University of Pennsylvania. From a broader evolutionary perspective, the works of Fiske, Joseph LeConte and Alexander Winchell were influential, all promoting a hierarchy of races even while seeking to reconcile their views with religion. Many of these experts were committed to a Lamarckian rather than a Darwinian perspective, but all invoked a linear hierarchy of development. E. D. Cope used the recapitulation theory to show how the 'lower' races were retarded in their individual ontogeny and thus did not advance beyond the level of primitive humanity.[21]

These views did not disappear in the new century. W. Z. Ripley's *The Races of Europe* of 1899 became a classic on both sides of the Atlantic, establishing the Teutonic, Alpine and Mediterranean as the three main divisions. Madison Grant's *The Passing of the Great Race* of 1916 became hugely popular in later editions (endorsed by H. F. Osborn), warning of the threats posed by immigration of inferior but quick-breeding races. In Britain the philologist Augustus Henry Keane promoted the fixity of

[20] MacBride, *An Introduction to the Study of Heredity*, chap. 9; Grant, *The Passing of the Great Race*, p. 108; for a collection of cartoons, see Curtis, *Apes and Angels*.

[21] For details, see Haller, *Outcasts from Evolution*; Livingstone, *Adam's Ancestors*; Hawkins, *Social Darwinism in European and American Thought*, chap. 8; and Gould, *Ontogeny and Phylogeny*, chap. 5.

TWO FORCES.

Fig. 8.3 Cartoon depicting a native Irish figure labelled 'Anarchy' with ape-like features threatening 'Hibernia', who is protected by a female figure, the conventional image of Britannia. From *Punch*, 29 October 1881.

racial types and the inferiority of the non-whites in books and articles, including several for the *Encyclopaedia Britannica*. Keane's insistence on the inferiority of the black races became influential among the colonists in South Africa. His *The World's People* of 1908 included an evolutionary tree with a central scale of brain capacity to define the racial hierarchy and explained diversity in terms of adaptations to different environments.[22]

All this continued despite the growing hostility of cultural anthropologists who now repudiated the evolutionism of the previous century. W. H. R. Rivers in Britain and Franz Boas in America argued that it was impossible to rank cultures into the linear hierarchy established by Tylor and Morgan. This did not prevent the physical anthropologists, aided by a horde of non-academic writers, from continuing their efforts to defend the old-style ranking. Keane wrote for a serial edited by H. N. Hutchinson, *The Living Races of Mankind*, published in 1900 in twenty-four fortnightly parts at a price of sevenpence each, and reissued in an expanded version five years later. Most of the authors were zoologists, colonial administrators and military men, and the lavish photographs were designed to highlight the alien character of the non-white races. As late as the 1920s scientists such as Julian Huxley, who would later challenge overt racism, still made no secret of their belief in the inequality of the black and white races.[23]

The depiction of non-European peoples as relics of the past also persisted in novels of adventure and exploration, in museum displays and exhibitions, and in the cartoons and caricatures that appeared frequently in the popular press. In 1887 a hirsute individual was shown as Krao, 'the missing link', at the Westminster Aquarium in London, and international exhibitions sometimes put whole families of 'primitive' people on display to promote an evolutionary narrative of progress. Novels of exploration and adventure by famous writers such as Henry Rider Haggard and Rudyard Kipling (they had colonial experience in Africa and India, respectively) routinely appealed to these stereotypes, although they sometimes singled out other races for praise. The theme of tension between the races became increasingly popular at the turn of the

[22] Keane, *The World's People*, chap. 1, diagram on p. 3. On Keane and other British anthropologists of the period, see Stepan, *The Idea of Race in Science*, chap. 4.

[23] On Hutchinson's project, see Bowler, *Science for All*, pp. 150–1, and on Huxley's early racism Elazar Barkan, 'The Dynamics of Huxley's Views on Race and Eugenics', and the same author's *The Retreat of Scientific Racism*, esp. p. 186. On the revolution in cultural anthropology, see Adam Kuper, *Anthropologists and Anthropology*, and Hamilton Cravens, *The Triumph of Evolution*.

century, just one example of the more Darwinian element that has come to characterize social evolutionism in most historical accounts.[24]

Struggle Transformed

The assumption that the races have come into conflict whenever one or another has expanded its territory became central to the ideology of imperialism and was justified by the claim that the struggle for existence has been at work throughout human history. Extending the assumption into the modern world created one form of what became known as 'social Darwinism': if struggle has been the motor of progress up from the ape, its application in the modern world must represent the best way forward. Darwin had certainly acknowledged the competition between varieties and species (assumed to be equivalent to human races), but initially most political thinkers focused on his primary mechanism of natural selection acting between the individuals within the same population. Natural selection was seen as a justification for free-enterprise capitalism – critics such as Marx said it was no more than an application of that ideology to nature itself. The main source of this policy came from Spencer, though, and his interpretation of struggle was as much Lamarckian as Darwinian. In the later decades of the nineteenth century, his philosophy remained influential in the United States even when the British were turning more toward imperialism.

The claim that a free-enterprise, individualistic society generates economic progress became popular among the industrial magnates of America's 'Gilded Age'. They saw their success as an outcome of the 'survival of the fittest', but since it was Spencer who coined that term they saw themselves more as his followers than Darwin's (the term 'social Darwinism' did not emerge until later).[25] Thanks to the influence of Richard Hofstadter's classic *Social Darwinism in American Thought*, historians used to assume this was the dominant ideology of the period, at least in that country. More recent studies have shown that the situation was much more complex.

The parallel between natural selection and a ruthless individualism was noted in the early debates over Darwinism, but was sidelined because Darwin and Spencer both accepted that if humans evolved in

[24] See Sadiah Qureshi, 'Dramas of Development'; an advertisement for the Krao display is reproduced on p. 268 and in Schwarz, *Streitfall Evolution*, pp. 116 and 381. More generally, see Qureshi, *Peoples on Parade*, part 3. On imperialism in popular novels, see Wendy R. Katz, *Rider Haggard and the Fiction of Empire.*

[25] On the origins of the term, see Donald C. Bellomy, 'Social Darwinism Revisited'.

social groups, they must have been endowed with cooperative instincts that became the basis of morality. Spencer saw this as driven by a Lamarckian effect, although he also noted that individuals who cannot engage with the process needed to be eliminated. Spencer became far more aggressive on this latter point toward the end of his career, which is why his system was taken up by his American followers to justify their competitive behaviour. But he had always insisted that struggle encouraged individual effort and initiative – eliminating the unfit was only a secondary function. Robert Bannister notes that small businessmen were less enthusiastic about the unrestrained application of his philosophy because they were the ones being gobbled up by the tycoons. Applying the model to industrial competition was in any case self-contradictory, since once a monopoly was achieved, further competition was impossible. In Britain, the growing enthusiasm for imperialism diminished Spencer's influence and switched attention to the struggle between nations and races. The two versions of social Darwinism were actually incompatible, since a focus on national competition led to a desire to minimize conflict within the population.[26]

It could be argued that the ordinary reader would hardly trust the casual remarks of a business tycoon as a guide to a scientific theory. But celebrity culture was already coming into existence at the time, allowing those with a high profile to influence public opinion on topics in which they had little expertise. Comments made by industrial magnates were reproduced in the popular press and had wide influence on ideas about evolution, especially for those with no particular interest in the details. The critique of the harsh image of Darwinism in Huxley's *Evolution and Ethics* suggests that this interpretation of the theory had become popular in the last decade of the century. As Spencer was campaigning to defend the Lamarckian mechanism against Weismann's attacks, his works were being cited to support the image of Darwinism as a guide to ruthless business tactics.

In Britain the state was expanding its role in alleviating the sufferings of the poor alongside a rising tide of enthusiasm for colonial expansion. Both of these trends were anathema to Spencer, who fought back by taking an increasingly harsh position, culminating in a series of essays in the *Contemporary Review* for 1884 collected in book form under the title *The Man versus the State*. He argued that the legislators' efforts were

[26] Critiques of Hofstadter's *Social Darwinism in American Thought* include Bannister's *Social Darwinism* and Cynthia Eagle Russett, *Darwin in America*. In the different but even more complex situation in Britain, see Jones, *Social Darwinism and English Thought*, and Pier Hale, *Political Descent*.

counterproductive because they encouraged the idleness of the undeserving poor. He had always insisted that such ne'er-do-wells should suffer the consequences of their unwillingness to support themselves, and now he argued that those who would be eliminated in a state of nature were being protected and allowed to breed. By checking the operations of the iron law of the 'survival of the fittest' (his own term, after all), the state was expanding the ranks of the unfit. Only if the freedom of the individual was restored would humanity progress to a state in which all would be perfectly adjusted to the social environment.

In the United States the captains of industry were already taking Spencer's call for unrestrained free enterprise as a license to expand their enterprises without mercy for rivals. The firms that allowed themselves to be gobbled up by the more powerful were the 'unfit', which should be eliminated to ensure economic progress. Although hailed by magnates such as Andrew Carnegie, Spencer was never comfortable with a system justifying a drive toward monopolies that had swept aside all their competitors. Nevertheless, Carnegie, John D. Rockefeller and James J. Hill all defended their rapacity with appeals to the survival of the fittest. William Graham Sumner used Spencer's books as course texts at Yale and derided state intervention as contrary to the natural laws of competition. When opponents tried to get him dismissed in 1880, the case hit the headlines. Two years later even more press attention focused on Spencer's visit to the United States where he was feted by Carnegie and met other tycoons along with Fiske, the preacher Henry Ward Beecher and the palaeontologist O. C. Marsh. The trip concluded with a banquet at Delmonico's restaurant in New York where all the main luminaries praised Spencer's gospel of progress through struggle. Exhausted by his travels, Spencer was only too anxious to board the ship for home.[27]

In *The Man versus the State* Spencer noted that many who were anxious to protect the poor were only too willing to endorse the enslavement or extermination of 'inferior races' in other parts of the world.[28] Late Victorian imperialism switched the focus of attention onto what we would now call group selection in which the struggle for existence took place between tribes, nations and races. Colonial expansion implied the subjection of non-white races and encouraged rivalry between European nations. In North America the white population was expanding westward

[27] See Hofstadter, *Social Darwinism in American Thought*, chaps. 2 and 3, and Barry Werth, *Banquet at Delmonico's*.

[28] Spencer, *The Man versus the State*, p. 143. The edition cited in the bibliography has a useful introduction by Donald Macrae.

at the expense of the indigenous inhabitants, and in the twentieth century the United States too began to acquire foreign territories. The claim that warfare is a natural state for humanity predated Darwinism, but the analogy with evolution added a new dimension by implying that the results were progressive. Darwin argued that tribal competition had played a role in shaping human nature, while biogeographers inspired by his theory assumed that species tended to expand their range and would exterminate the original inhabitants of the territory they invaded. Archaeologists and palaeoanthropologists began to cite race conquest as a driving force of human evolution, although those who endorsed it were seldom enthusiastic about the primary mechanism of selection acting within a population.[29]

Walter Bagehot's *Physics and Politics* of 1872 had insisted that the national cohesion maintained by religion was vital for the struggle against rival states, a theme repeated in Benjamin Kidd's *Social Evolution*, although he at least thought miliary conflict would diminish. Others took it for granted that preparation for war was essential. Karl Pearson's *National Life from the Standpoint of Science* of 1901 called for a eugenic programme to raise Britain's level of fitness to face external threats. It was widely believed that the newly united Germany was determined to demonstrate its superiority by a war of aggression. The American biologist Vernon Kellogg, author of the survey *Darwinism Today*, visited the German army while his country was still neutral in the early phase of the Great War and warned that its leaders were convinced that they were engaged in a national struggle for existence. His *Headquarters Nights* recording his experiences was one of the sources used by the religious opponents of Darwinism to back their claim that the theory had a malign influence on morality.[30]

The suggestion that the process of European expansion was an application of nature's struggle for existence was a convenient way of justifying these activities. The writings of African explorer and conservationist Frederick Courtney Selous – who inspired Rider Haggard – provides a case study of how the Darwinian notion of struggle could be extended to the conflict between the races. Eminent psychologists including William MacDougal and G. Stanley Hall also insisted that an element of innate pugnacity in the human character made war inevitable. Arthur Keith

[29] On biogeography and imperialism, see Bowler, *Life's Splendid Drama*, chap. 8 and pp. 435–40, and on human evolution *The Invention of Progress*, chap. 4, and *Theories of Human Evolution*, pp. 223–37. Sollas and Keith are two examples of experts who appealed to race conflict but had little time for individualistic natural selection.

[30] For details, see Paul Crook, *Darwinism, War and History*, especially chap. 5, and Hawkins, *Social Darwinism in European and American Thought*, chap. 8.

built a whole theory of human evolution on the claim that race conflict had always been the driving force. In the United States the same idea was used to justify the treatment of the First Nation tribes and to argue that the now-emancipated black population would soon decline to extinction now that it was subject to open competition with the whites.[31]

Darwinian language was a powerful rhetorical tool used by the advocates of both capitalism and imperialism. Yet these ideologies were in play before Darwin published, suggesting that the public perception of his theory was shaped by changes in cultural values. 'Survival of the fittest' became a catch-phrase even for those who had only a limited appreciation of the scientific theories. Far from promoting the harsher implications, Darwin, Spencer and their immediate followers had tried to downplay them by arguing that their theories could actually explain the evolution of *moral* behaviour. Such expectations did not disappear, but they were increasingly overtaken by a vociferous focus on the element of struggle in the evolutionary mechanism.

A Lamarckian Alternative

There was, however, another tactic provided by the Lamarckian mechanism, provided one assumed that the process was not as slow as some race-theorists imagined. In Spencer's system, competition stimulated individuals to improve themselves, and the Lamarckian effect allowed these improvements to accumulate over the generations – although he insisted that everyone should be taught in the school of real life, not in the halls of academe. For anyone prepared to allow the state or at least some relevant authority to control the way children are brought up, Lamarckism offered the possibility that the effects of formal education could become cumulative and transform human nature for the good. This offered a way out of the stalemate that Huxley had run into when he separated morals from humanity's evolutionary origins.

This strategy was exploited by Lamarckians from a variety of backgrounds, and the diversity of their ideological positions may have limited the impact of their arguments. In Britain, Bagehot had allowed a role for the Lamarckian effect in the shaping of human nature even though he focused on the struggle for existence between nations. The whole notion of a natural world based on conflict was attacked in Kropotkin's *Mutual Aid*, and although as an anarchist he wanted no state control of society,

[31] Hawkins, *Social Darwinism*, chap. 8, on Selous pp. 204–5; see also Hofstadter, *Social Darwinism in American Thought*, chap. 9. The role of instinct is stressed in Crook, *Darwinism, War and History*, chap. 5.

he did see moral improvements in successive generations accumulating to affect the race. In later writings he openly endorsed the Lamarckian mechanism against Weismann's assault.[32]

Kropotkin's world-view held little room for the vitalistic form of Lamarckism promoted by Samuel Butler and Bernard Shaw in their equally vigorous assault on Darwinism. The inheritance of acquired characteristics was a process that could be exploited by rival philosophies; it could also be explained in materialistic terms, as Paul Kammerer claimed in his much-publicised visits to Britain and the United States in the 1920s. He generated newspaper headlines by promising to create a race of supermen through the application of the Lamarckian effect, but linked his appeal to that of the rejuvenation process of Sergei Voronoff. Hormones, not some mysterious life force, drove the process of evolution. Kammerer's approach preserved the optimistic tone of the Lamarckian theory, but the materialist version could also lead to more pessimistic conclusions. The *Harmsworth Popular Science* of 1911–13 endorsed genetics, but argued that alcohol was a poison that debilitated the drinkers' germ plasm. Genes could be permanently damaged, leading to the inheritance of acquired negative characters.[33]

In the United States, sociologist Lester Frank Ward was the leading opponent of social Darwinism. In a series of articles and books beginning with his *Dynamic Sociology* of 1883 he attacked Sumner's campaign for unrestricted free enterprise and called for more structured efforts to shape society and the future of the race. He appealed to the Lamarckian effect to justify his hopes that the moral benefits of education could become implanted in the race as moral instincts. The same principle was a key component of Joseph LeConte's efforts to synthesize evolutionism and Christianity: 'All our schemes of education, intellectual and moral, though certainly intended mainly for the improvement of the individual, are glorified by the hope that the race is thereby elevated.' Neither applied this message to the 'lower' races, and Ward endorsed the claim that race-conflict is inevitable. These ideological confusions limited the impact of Ward's challenge to Sumner's social Darwinism, and LeConte's teachings also fell out of fashion at the turn of the century. As in Britain, Kammerer's visit would have helped to remind the public

[32] For details of the British opposition, see Jones, *Social Darwinism and English Thought*, chap. 5, and Hale, *Political Descent*, chap. 5.

[33] Mee, ed., *Harmsworth Popular Science*, group 12, chap. 32; IV, pp. 3891–6. On Kammerer and Voronoff, see Bowler, *A History of the Future*, pp. 189–92.

that Lamarckism was still in play in the 1920s, but this was a different version of the theory and very much a flash in the pan.[34]

The Lamarckian model depended on a pre-genetic view of heredity which encouraged the belief that individual development recapitulates the evolutionary history of the species. Because new variations were seen as *additions* to growth, past phases of evolution could be preserved in the development of the embryo, each pushed back further as new additions are bolted on. Applying this model to the human situation implied that the growing child must pass through the stages of our evolutionary past. This vision of development was broad enough to attract even Darwinians such as G. J. Romanes, who constructed a pattern of mental evolution based on a parallel between the mental powers of animals and the stages of maturity in the human child. James Sully's *Studies of Childhood* of 1895 used similar parallels, in his case explicitly linked to Lamarckism. In the United States, Alexander Chamberlain's *The Child: A Study in Evolution* of 1900 put an optimistic slant on the idea, while G. Stanley Hall's influential *Adolescence* explained the problems of this phase as a recapitulation of our savage past. The recapitulation model was also applied to the races, as in Kipling's poem 'The White Man's Burden' with its depiction of 'Your new-caught sullen peoples / Half savage and half child'. Edgar Rice Burroughs depicted Tarzan's development as 'following the evolution of his ancestors, for had he not started at the very bottom' (having been raised as an ape).[35]

The Spectre of Degeneration

The parallel scales of mental development postulated by the recapitulation theory had another unfortunate application: they could be seen as a map of what might happen if our species began to lose its higher functions. Lankester's *Degeneration* showed that the progressive trend of evolution could be reversed if the external stimulus to action was removed. The last decade of the century saw a growing fear that future progress was threatened by a variety of sources, each linked to a different

[34] On Ward's Lamarckism, see Hofstadter, *Social Darwinism in American Thought*, chap. 4, and on the comparison with Sumner, p. 70. The quotation is from LeConte, *Evolution*, pp. 97–8; see Stephens, *Joseph LeConte.*

[35] *Collected Poems of Rudyard Kipling*, pp. 334–5, and for the Burroughs quotation, *The Return of Tarzan*, p. 164. See Sally Shuttleworth, *The Mind of the Child*; J. R. Morss, *The Biologizing of Childhood*; Gould, *Ontogeny and Phylogeny*, chap. 5. Roisin Lang, 'Victorian Autobiography, Child Study, and the Origins of Child Psychology', also stresses the non-Darwinian character of this approach. On Romanes, see Richards, *Darwin and the Development of Evolutionary Theories of Mind and Behavior*, chap. 8, and on Hall, David Muschinske, 'The Nonwhite as Child.'

biological mechanism. The recapitulation theory suggested the actual course that the degeneration might take: loss of the higher mental functions would give the earlier child-like or savage elements control of an individual's behaviour. Both the Darwinian and the Lamarckian theories could explain the loss, by either the relaxation of selection or the lack of direct stimulus to development. Some worried that modern civilization was shielding the population from the pressures that would be encountered in a state of nature, allowing the less-developed to survive and breed. Alternatively, Max Nordau's influential *Degeneration* (translated in 1895) suggested that the psychological pressures imposed by modern life were undermining mental stability.

Lankester's point was that a species taking up an inactive lifestyle would lose the higher faculties and structures it no longer needed. Spencer suggested that modern savages may have degenerated from a higher past state due to their poor environment. His system would allow for either a Darwinian or a Lamarckian cause, but as the evolutionary debates became more polarized, each side began to stress its own explanation of how the downward slide would operate. The Darwinians became more aware of the problem in the light of Weismann's theory of heredity, which led him to postulate a process of 'panmixia' in which the loss of a once-useful character could occur because of the relaxation of selection pressure. If modern society cushioned everyone from the pressures once faced by our distant ancestors, the least able would survive and even breed faster than their betters, bringing down the overall standard. This was the position taken up by Pearson, who highlighted the poor quality of the recruits to the British army for the war against the Boers in South Africa. The *Harmsworth Popular Science* – while preferring Mendelism to Pearson's Darwinism – also ended its section on eugenics with a warning that every previous civilization had degenerated as the pressure of selection was relaxed.[36]

As a neo-Darwinian, Pearson had no time for Lamarckism, but that theory had always allowed for the loss of characters no longer being used. Its supporters could endorse Lankester's point by arguing that modern social conditions did not encourage the development of the higher aspects of human nature. The link between Lamarckism and the recapitulation theory also gave them a model for the direction that would be taken on the downward descent. If individual development follows the same scale as past evolutionary progress, loss of the highest stages would

[36] Mee, ed., *Harmsworth Popular Science*, group 12, chap. 36, 6: 4369–76, and on the panmixia debate Hale, *Political Descent*, chap. 7. See J. Edward Chamberlin and Sander L. Gilman, eds., *Degeneration*, and Daniel Pick, *Faces of Degeneration.*

leave the adult at an immature level corresponding to ancestral forms. Chamberlain had included a chapter on 'The Child and the Criminal' in his book, arguing that criminals were adults who retained an earlier, child-like stage of development. The leading exponent of 'criminal anthropology', the Italian psychologist Cesare Lombroso, saw the immature form more as an ape-like savage than an innocent child. Criminals could be recognized by the physical marks of savagery reminiscent of the apes. If the conditions of modern life are somehow encouraging such a failure to mature, the degeneration of the race would be marked by an ever-increasing proportion of atavistic delinquents.[37]

Perhaps the most widely read illustration of this fear was H. G. Wells' 1895 story *The Time Machine* with its prediction of a future world in which the human race has indeed degenerated. In the first phase of the story the effete Eloi have been so long exposed to comfort and luxury that they have lost the higher mental functions. Wells then introduces a twist that depends on the confusion built into the recapitulationist notion of the child as savage. The Eloi are indeed child-like but are prey to the Morlocks, descendants of the working classes now adapted to an underground environment in which they have degenerated into something far more savage and ape-like. By imagining a divergence in humanity's future, Wells simultaneously exploits fears of racial degeneration while exposing the inconsistency in the claim that the child is only one step removed from the ancestral ape.

A belated application of the recapitulation model to human affairs deserves a brief mention, even though it was seldom recognized as such at the time. Sigmund Freud's psychoanalytic theories had a huge impact on early twentieth-century thought and popular culture. They were based in part on the belief that the unconscious mind is a relic of early stages in human evolution buried beneath the newly acquired conscious functions. Freud wrote an essay titled 'Phylogenetic Fantasy' in 1915 acknowledging the contribution of recapitulation theory, but this was not published at the time, and the works that made him famous do not explore the link. It was left for modern scholars to recognize the role played by this aspect of evolutionism. Even so, the claim that subconscious desires are linked to the sex-functions suggested that they might be relics of the past. By implying that the psychological pressures of modern culture allow these ancient desires to overwhelm the

[37] On Lombroso and criminal anthropology, see Gould, *Ontogeny and Phylogeny*, pp. 120–5, and Hawkins, *Social Darwinism in European and American Thought*, pp. 74–80.

conscious mind, Freud sustained the fear of racial degeneration into the new century.[38]

Eugenics and Evolution

By the end of the nineteenth century there was a growing fear that the 'unfit' were outbreeding their superiors and undermining the character of the race. The eugenics movement sought to redress the balance in the hope of staving off degeneration and perhaps even improving the human stock. It drew on traditional views on the power of heredity that had been refined by Darwin and turned into an ideology, almost a substitute for religion, by his cousin Francis Galton. The movement did not take off at first, perhaps because it was inconsistent with Spencer's emphasis on the improving effects of hard work and individual effort. While Lamarckism remained a viable option and suspicion of state intervention was rife, Galton's proposals for applying the principle of selection to the human population struggled to gain a hearing. Only with the emergence of the new science of heredity and a growing willingness to allow state intervention did it seem reasonable to prevent the unfit from breeding in order to save the race.

Was this an extension of social Darwinism? Eugenics certainly depended on the claim that there were inherited differences between individuals, and even Huxley was no egalitarian. The theory of natural selection played a role for some eugenists, including Galton's lieutenant Karl Pearson. As a neo-Darwinian he could see the application of artificial selection to humanity as a replacement for the natural process now checked by civilization – while transferring the struggle for existence to the national level. But the first geneticists were not Darwinians: they saw new characters as produced by mutation without reference to environmental constraints. Especially in the United States, eugenics flourished in conjunction with the application of genetics to animal breeding and horticulture. It was artificial selection applied to the human race, and even anti-Darwinians conceded that it should be effective. American eugenists were also more concerned with the threat posed by the immigration of 'inferior' races, using a racial hierarchy that was not derived solely from Darwinism.[39]

[38] See Frank J. Sulloway, *Freud, Biologist of the Mind*, and Lucile B. Ritvo, *Darwin's Influence on Freud*, chap. 5.

[39] There is a huge literature on eugenics; for good surveys, see Daniel Kevles, *In the Name of Eugenics*, and Alison Bashford and Philippa Levine, eds., *The Oxford Handbook of the History of Eugenics*.

Inspired by Darwin's views, his cousin Francis Galton used his own experiences – including explorations in Africa – to focus on the range of abilities in the human population. His *Hereditary Genius* of 1869 sought to establish that individual levels of intellectual capacities are inherited. He argued that if those with the lowest levels were having more children than their superiors, the results would downgrade the population as a whole. In 1883 he coined the term 'eugenics' to denote his proposed social programme to reverse the trend by discouraging the least able from breeding (negative eugenics) and encouraging the best to have bigger families (positive eugenics). His approach depended on a rejection of the belief that education and experience can develop a person's character. For Galton, it was nature (i.e., heredity), not nurture (education and environment) that determines character, the ideology later known as hereditarianism. He instituted a programme of anthropometry to measure the variation of characters in the population and used the data to support his arguments for state action to influence rates of reproduction.

Darwin's theory may have inspired Galton and hence the movement he founded, but there was no one-to-one connection between the theory of natural selection and calls for the control of human breeding. Galton tested Darwin's own theory of heredity, pangenesis, and rejected it (significantly, it allowed for an element of Lamarckism). His own 'law of ancestral inheritance' was based on the belief that all previous generations can play a role in determining an individual's inheritance. There was no equivalent to the geneticist's model of single characters transmitted as units from parent to offspring. Galton did not believe that selection acting on a continuous range of variation could produce new characters, favouring the theory of evolution by saltations. Pearson used his statistical skills to show that natural selection *could* work with continuous variation, paving the way for later developments in population genetics. At the time, however, no link with genetics was possible: Pearson conducted a vitriolic debate with Bateson, arguing that the geneticists' use of small-scale breeding experiments to trace individual characters was irrelevant for the study of whole populations. The geneticists shared Galton's belief that new characters could arise only from sudden rearrangements of the germ plasm. On their model, there could be no equivalence between the processes endorsed by the Darwinians and the artificial selection of unit characters.

Galton attracted some attention through his Anthropometric Laboratory, set up at the 1884 International Health Exhibition held in London's Science Museum. Some 9,000 visitors had their physical measurements taken to provide him with evidence for his survey. Yet his campaign to promote eugenics achieved little success until circumstances in both science and social attitudes changed at the turn of the century. In biology, Weismann's theory and later genetics focused attention onto heredity and entrenched the belief that it predetermines the nature of the individual. There was a growing willingness to tolerate state intervention to maintain the health of the population. Pearson led the campaign to promote Galton's programme, focusing especially on the need to limit the reproduction of the least able members of society – usually identified as the 'feeble-minded' – if necessary by confining them to institutions, segregated by sex (sterilization was never favoured in Britain). In 1904 Galton obtained a position for Pearson at University College London, from which he defended Darwinism and called for eugenic reform. The Eugenics Education Society was founded to coordinate the campaign, which achieved limited success with the passing of a Mental Deficiency Act in 1913.

Some support came from the professional classes and others seeking to limit government spending on alleviating the conditions of the poor. Those advocating more stringent measures to protect public health were supportive, but many medical doctors were opposed. Socialists were often hostile, although some conceded the value of a eugenic programme only if conditions were improved to establish a level playing field. These were all members of the more literate classes – most working-class people were hostile to efforts to control their behaviour, and there is evidence that many had never even heard about eugenics.[40]

The scientific basis for the policies remained in dispute. Caleb Saleeby, who contributed the section on eugenics in the *Harmsworth Popular Science*, attacked Pearson's version of Darwinism in the name of genetics. He insisted that selection applied to the normal range of variation in the population was fruitless since this was produced by environmental factors – he even called for better health care and education to help the disadvantaged. But mental deficiency was a clearly defined character controlled by a recessive gene, and those carrying this gene should be prevented from reproducing. Many geneticists would eventually jump on this bandwagon. In 1928 Darwin's son Leonard published his *What Is Eugenics?* in Watts' cheap Forum Series of books. Not surprisingly, it mentioned natural selection and ignored genetics, yet even Darwinism was sidelined by a focus on the artificial selection used by animal breeders. MacBride's contribution on heredity to the Home University Library called for restrictions on the breeding of the least fit, whom he identified with the Irish, an inferior type shaped by a poor environment.

[40] For a useful survey of evidence for the wider dissemination, see John Macnicol, 'Eugenics and the Campaign for Voluntary Sterilization in Britain between the Wars'.

This emphasis on race was unusual in Britain, as was MacBride's open support for compulsory sterilization. In the 1920s younger scientists associated with the re-emergence of Darwinism still endorsed eugenics; Julian Huxley, for instance, wrote frequently in support of the underlying principle of hereditarianism.[41]

Eugenics was, if anything, more influential in the United States than in Britain. Galton's work was acknowledged here, but Pearson's biometrical Darwinism was largely ignored in favour of Mendelism. The geneticist Charles Davenport used Carnegie funding to establish a centre for the study of experimental evolution at Cold Spring Harbor in 1904, and the Eugenics Records Office was founded there in the following year. The link between genetics and animal breeding was emphasized, with the American Breeders Association setting up a eugenics committee in 1906. Promotion of the programme emphasized family pedigrees, with the example of the Jukes family (studied in the late nineteenth century) being revived to suggest that the descendants of a feeble-minded individual would inherit the harmful gene. The American Eugenics Society provided displays at events such as the Sesquicentennial Exposition held in Philadelphia in 1926. At local fairs around the country there were booths displaying genetical charts proclaiming that characters such as feeble-mindedness, criminality and pauperism could be inherited, along with claims that they could be eradicated in three generations if strict eugenic measures were introduced. The fear of degeneration sparked increasing calls for the compulsory sterilization of those thought to be carrying harmful genes, with twenty-four states eventually passing legislation to allow this.

The threat of contamination by inferior racial types entering the population characterized much American eugenics. The original white colonials were increasingly disturbed by the influx of immigrants from Eastern Europe and further afield. The prospect of 'inferior' types breeding prolifically in North America now generated a concern that the genetic quality of the population was being undermined. Henry Fairfield Osborn was the most visible scientific proponent of these ideas, openly supporting the demagogue Madison Grant's calls for immigration to be controlled. In 1924 the Immigration Restriction Act led to the establishment of facilities such as the one on Ellis Island in New York to

[41] Mee, ed., *Harmsworth Popular Science*, group 12, chaps. 17, 18, 20, 29; 3: 2091–6, 221–6, and 4: 2449–56 and 3531–6; Leonard Darwin, *What Is Eugenics?*, see pp. 7–8 on natural selection; MacBride, *An Introduction to the Study of Heredity*, chap. 9. For Huxley's popular contributions, see Barkan, *The Retreat of Scientific Racism*, p. 186. R. A. Fisher was an ardent eugenist but his support was published mostly in the movement's own literature.

vet the racial background of prospective immigrants and deny entry to those deemed unfit.

Cultural anthropologists led by Franz Boas now rejected the belief that races could be ranked into a hierarchy with whites at the top. But the eugenics movement reached deep into the ranks of the white population through promotion in books, pamphlets and exhibitions. To what extent this activity shaped popular ideas about evolutionary biology is hard to judge – the Darwinian elements were interspersed with various alternative perspectives. The prominence of genetics encouraged the analogy with artificial selection, and most geneticists were at first hostile to Darwinism. Osborn's defence of evolutionism against the creationists' attacks was based on his own non-Darwinian model of evolution. In the Scopes trial debates, Bryan had cited the imperialist version of social Darwinism to highlight the theory's moral dangers, but Osborn and his followers stressed that their theory did not have these harmful implications. Yet those theories promoted the racist element in the eugenics movement, and there were many conservative Christians who shared the belief that the races were distinct creations. The link to race science would encourage a belief in the struggle for existence *between* populations, not the individualistic view of selection acting *within* a population that was only just being revived from its 'eclipse' in the 1920s.

9 The Evolutionary Synthesis

During the 1920s the geneticists remained hostile to Lamarckism but gradually softened their position on natural selection. It was conceded that wild populations contain many genetic variants with overlapping effects, creating a range of variation for each character that might be changed if some variants bred more prolifically than others. The type of statistical analysis used by Pearson and the biometricians could now be modified to create a science of population genetics that would show how a gene conferring a reproductive advantage would expand in the population (and vice versa). Far from undermining the plausibility of Darwin's theory, genetics turned out to be its salvation. In 1930 R. A. Fisher's *The Genetical Theory of Natural Selection* provided a technical basis for the rapprochement, supported also by the work of J. B. S. Haldane and (using somewhat different techniques) Sewall Wright in the United States. Julian Huxley's 1942 survey *Evolution: The Modern Synthesis* gave the revised Darwinism its name: it was an 'evolutionary synthesis' of the once hostile forces.

In its simplest form the synthesis was a combination of genetics and natural selection, but as the date of Huxley's book suggests, the full application of the new approach took effect only in the course of the 1930s. Another of the founders, Ernst Mayr, later insisted that there was also a coming together of different branches of biology that had not been able to communicate up to that point. Mayr was a field naturalist with an interest in geographical distribution and pointed to the role of Theodosius Dobzhansky's *Genetics and the Origin of Species* of 1937 in showing how other disciplines could make use of the new insights provided by population genetics to replace what for some had been an inclination to support Lamarckism. The recapitulation theory once favoured by Lamarckians had already been largely discredited by the work of Gavin de Beer and others, while George Gaylord Simpson's *Tempo and Mode in Evolution* of 1944 brought palaeontology into line with the new Darwinism. Stephen Jay Gould argued that in its original form the new approach still included some non-Darwinian elements – Simpson, for instance, rejected orthogenesis but retained the idea of small-scale non-adaptive changes produced by what Wright called genetic drift. Then the consensus hardened to focus more or less exclusively on natural selection.

More recent studies have thrown doubt on the suitability of the term 'synthesis' to describe what was going on. There was still a debate over the level of genetic variation in wild populations, eventually settled in favour of the view that it was substantial. It was necessary to head off an attempt to retain a role for macromutations, which Mayr dismissed as the theory of the 'hopeful monster'. In practice the various disciplines involved still did pretty much their own thing, and the claim for an element of unity centred on Darwinism was to some extent a rhetorical device used to counter the threat that the old natural history sciences might be marginalized by the increasingly influential biomedical areas. The air of solidarity was also employed to promote the new Darwinism as the basis for a humanistic philosophy retaining a residue of the belief that we are the product of a meaningful progression with cosmic implications. Whatever the technicalities, the synthesis became the basis for a coherent world-view presenting a rather optimistic take on the materialism of the Darwinian theory. As the next chapter shows, conservative social and religious movements would soon wake up to the renewed threat.[1]

Our purpose here focuses on the efforts made to put the new evolutionism before the general public. Expositions of the synthetic theory itself are only part of the story; equally important are its applications to popular topics such as natural history and the fossil record, and its impact on wider social and religious issues. The process took place as the opportunities for communicating with the public continued to expand into new media, triggering further tensions between the scientific community and the authors, journalists and broadcasters who controlled how the material could be presented. Popular science was strongly focused on practical applications and technology, with only the most spectacular theoretical innovations receiving significant coverage. Natural history and big fossil discoveries could still attract some attention, but this did not necessarily focus on new interpretations. The new Darwinism was hard to present as a major initiative since it combined existing work in a complex way and seemed to raise no wider concerns that had not already been articulated.

[1] On the first steps in the synthesis, see William B. Provine, *The Origins of Theoretical Population Genetics*. For the later debates over the nature of the process, see Ernst Mayr and Provine, eds., *The Evolutionary Synthesis*; Vassiliki Betty Smocovitis, *Unifying Biology*; and for a short overview, Bowler, *Evolution*, chap. 9.

The scientific community was divided on how to tackle the problem of dealing with the popular media. Some called for more responsible coverage, but many were reluctant to get involved themselves. There were calls for a more coordinated approach, the most successful of which was the Science Service founded in the United States in 1921. This provided material for both journalists and broadcasters on the new radio stations, but although the latter shared the newspapers' fascination with the Scopes trial, material on evolution was limited. Britain's centralized BBC had both education and entertainment in its remit and did broadcast some material on evolution (thereby attracting complaints from the local creationist lobby). Gerald Heard became a popular commentator and, despite having no training in science, fronted programmes discussing its broader implications. The British scientific community was lucky in that it produced two high-profile figures who were both good communicators and supporters of the synthesis, Julian Huxley and J. B. S. Haldane. But both had many other fish to fry and both had viewpoints that were not shared by the whole community – Huxley retained elements of an older teleological worldview, while Haldane became a Marxist.[2]

The process of communication was not necessarily by direct exposition of the new version of natural selection. With a few prominent exceptions, the initial re-emergence of Darwinism was not widely noted in the popular science literature, and older versions of evolutionism were still on display into the 1940s. On the other hand, the museum-going public had already been exposed to a more open-ended and less directional model of evolution thanks to the fossil discoveries outlined in the previous two chapters. Simpson's rejection of orthogenesis built on the diversity and irregularities already recognized in the fossil record in previous decades. Julian Huxley could bring the topic before readers and radio listeners interested in natural history – he was for a time the director of the London Zoo – although much popular literature in the field still tended to bypass the topic of evolution. The efforts of Huxley and others to promote the human implications of Darwinism, aided by high-profile discoveries of hominid fossils, did most to expose the new version of the theory to public scrutiny and, ultimately, to renewed attacks by conservative religious forces.

[2] On developments in science communication, see Ronald C. Tobey, *The American Ideal of National Science*; Dorothy Nelkin, *Selling Science*; Marcel C. Lafollette, *Making Science Our Own* and *Science on the Air*; and (more controversially) John C. Burnham, *How Superstition Won and Science Lost*. For British developments, see Bowler, *Science for All*; Hall, *Evolution on British Radio and Television*; and Tim Boon, *Films of Fact*. On Heard, see Bowler, *Science for All*, pp. 212 and 257.

Coming Together

Even at the end of the century, Richard Dawkins lamented the stranglehold the image of the chain of being still held on popular evolutionism. Representations of the tree of life with a central trunk running purposefully up toward humanity as the goal of creation still abounded even as the Darwinian synthesis was being assembled and promoted. Julian Huxley himself endorsed one book with such an image as its frontispiece in 1944, and David Attenborough's classic TV series *Life on Earth* followed a similar model.[3] But it was becoming harder to escape the fact that the history of life did not fit easily into a scheme of predetermined progress. We have seen not only the growing recognition of this point by the palaeontologists, but also how it was used in the introductory section of H. G. Wells' hugely successful *Outline of History*. In 1938 the well-known science writer Ellison Hawks picked up the same point in explaining the struggle for existence:

> This struggle – the modification, advancement, and adaptation – from the earliest forms of life to Man, has not been the ladder of progress that some people consider evolution to be. We can picture it better by thinking rather of a tree with many branches, some of which have not yet ceased growing, whilst others are now flourishing as specialized groups of animals such as the mammals, the birds, the fishes and the reptiles. Other branches appear to have ended blindly and abruptly when Nature chanced to produce a form of life that seemed to be almost a failure although adapted to certain conditions. For example it seemed to be impossible to find a place for the giant reptiles in the conditions that followed their creation – conditions not brought about by any sudden change, but the results of constant and persistent changes going on in the world.[4]

Yet Hawks conveyed no sense that a new form of Darwinism was emerging. He still saw mutations as the only real source of new developments, a view that had been endorsed in T. H. Morgan's *The Scientific Basis of Evolution* some years earlier.

Morgan still held to a very simplified version of selectionism, although he actively repudiated efforts to see any mystical progressive trend in evolution. Nor, with two major exceptions, was there significant recognition that a new version of Darwinism was emerging in the popular science literature of the interwar decades. As early as 1920 there was a

[3] For Dawkins' complaint, see his 1992 article 'Progress'; Huxley wrote the foreword to Eileen Mayo's *The Story of Living Things* in 1944, a book written for older children with the tree as its frontispiece. Attenborough's series is described below.

[4] Ellison Hawks, *The Marvels and Mysteries of Science*, p. 365; on mutations, see pp. 374–7; see also Morgan, *The Scientific Basis of Evolution*, chap. 5 (and note the attack on Lamarckism in chap. 9).

discussion of industrial melanism in the peppered moth, later an icon of the new Darwinism, in the British magazine *Conquest*. This attributed the darkened form to a mutation and attributed its spread in industrial areas to an ability to resist pollutants (not to camouflage as in the modern explanation). In 1924 another magazine, *Discovery*, ran an article by respected biologist Lancelot Hogben that gave no support to the selection theory. In the United States, *Popular Science Monthly* published Dr E. E. Free's series on evolution in the 1920s, which focused mostly on the history of life but with a brief reference to mutations as the source of new characters. The topic of harmful mutations came up in the frequent calls for a eugenic policy. School textbooks still included some material on the development of life, but avoided the term 'evolution' to escape criticism from the creationist lobby.[5]

Fisher's *Genetical Theory of Natural Selection* concluded with open support for eugenics, but this was a technical work far beyond the comprehension even of many biologists. The two major figures seeking to alert the public to the emergence of a new relationship between genetics and Darwinism were Huxley and Haldane. The former's 1926 radio talks published as *The Stream of Life* had included a brief discussion of the better understanding of variation being offered by genetics. Huxley was also responsible for the biological material in the fourteenth edition of the *Encyclopaedia Britannica* (now based in the United States) and used Haldane as one of his contributors, thus ensuring a more sympathetic approach to Darwinism. Issued in 1929, this edition was vigorously promoted in home sales as an educational tool.[6]

Haldane's semi-popular book *The Causes of Evolution* of 1932 was a direct response to Belloc's claim that Darwinism was now outdated. On the contrary, he showed, it had acquired a new and more secure foundation through the recognition that the normal variation shown by any large population is built up from a series of mutations accumulated over previous generations and preserved at low frequencies. Selection acts to boost the level of those genes that turn out to confer reproductive advantage and drive down that of those that are harmful. Haldane also made it clear that this mechanism fitted in well with the latest discoveries in areas such as the fossil record and biogeography that were increasingly demonstrating the open-endedness of the process.

[5] H. Onslow, 'Black Moths', *Conquest*, 2 (1921): 491–6 and 534–40; Hogben, 'The Present Status of the Evolutionary Hypothesis'. See also *Popular Science Monthly*'s online Archive Gallery, 'In Defence of Evolutionary Theory'. On textbooks, see Nelkin, *Science Textbook Controversies and the Politics of Equal Time*, and Adam R. Shapiro, *Trying Biology*.

[6] On the publication of this edition of the encyclopaedia, see Bowler, *Science for All*, pp. 144–6.

Wells' *Outline of History* had sold like hot cakes, and he joined with Huxley to follow it with another blockbuster, *The Science of Life*, which appeared in 1931 on both sides of the Atlantic. It was initially available as a fortnightly serial. This was a general survey of the life sciences with a generous helping of evolutionism that highlighted the new version of the selection theory. The fifth part was an outline of the progress of life on earth, continuing the theme pioneered in the *Outline of History*: progress was to be seen as a series of unpredictable innovations, not as a built-in trend. The element of successive achievements was displayed in the section's title, 'The History and Adventures of Life', and each major step was presented as an expansion into new environments, such as the 'conquest of the dry land'. Some of the evolutionary trees had no central trunk, and there was a clear recognition that once-successful branches had disappeared completely, the extinction of the dinosaurs being attributed to a fairly rapid change in the climate to which they could not adjust. The description of the reconciliation between genetics and the selection theory came before the outline of the fossil record, so the reader was aware that the latter should not be seen as a preordained ascent. After explaining how mutations are the ultimate source of the variation shown by all populations, the effects of selection were described and the section ends with a chapter dismissing the *élan vital* and the notion of a 'mystical evolutionary urge'.[7]

Haldane and Huxley's writings would have given anyone with an interest in evolutionism fair warning that something was stirring in the field with the potential to establish Darwinism as the major theoretical foundation. The credibility of the old non-Darwinian mechanisms was on the decline, if not yet eliminated altogether. As yet, though, the full implications of the emerging relationship between genetics and the selection mechanism had not been explored across the sciences. Many working palaeontologists and field naturalists were still sympathetic toward Lamarckism and orthogenesis, while well-known public figures such as Bernard Shaw were still actively promoting 'creative evolution'. In the course of the 1930s, though, the word spread and naturalists such as Ernst Mayr, who had taken it for granted that Lamarckism was the most obvious explanation of adaptive change, began to see that natural selection might do a better job. A key step was the publication of Theodosius Dobzhansky's *Genetics and the Origin of Species* in 1937, a book designed to show field workers how the genetical selection theory

[7] Wells et al., *The Science of Life*, books 4 and 5; for the tree of life, see p. 449, and on the extinction of the dinosaurs p. 453. On the publication of the book, see Bowler, *Science for All*, pp. 104–7.

(whose technicalities many found obscure) could be applied in practice. Until its effects began to take hold, there were few efforts to keep up the momentum of popularization initiated by Haldane and Huxley. The real emergence of what would be greeted as the new evolutionary synthesis came only in the 1940s.

Consolidation

The key works establishing the revival of Darwinism appeared when the world was convulsed by war. Huxley's *Evolution: The Modern Synthesis* appeared in 1942 when Britain still faced a Europe occupied by Nazi Germany (first editions contain the usual stamp indicating that it complied with wartime economy standards). Mayr's book on speciation appeared in the same year, and Simpson's on the fossil record two years later. Huxley's was the more accessible survey, although he admitted that it was written mainly to convince other specialists that they should take the results of population genetics more seriously. It received positive reviews and continued to sell, but it would have been unrealistic to expect serious expansion into the world of popularization until societies and economies had begun to recover.

Historians disagree over exactly what Huxley meant by a 'synthesis' and over what he hoped to achieve. It is now widely accepted that there was only a limited coming together of the various theoretical mechanisms and their applications to fieldwork and the history of life. Fisher, Haldane and Wright had developed significantly different models of how selection could affect the genetical make-up of a population (Fisher insisted it could work only very slowly) and the role of small isolated sub-groups (a key element for Wright and Mayr). There was a fairly general agreement that acquired characteristics were not inherited, driven home by the controversy sparked by T. D. Lysenko's efforts to promote non-Darwinian mechanisms in the Soviet Union. But there was still a lingering suspicion that adaptation to the environment was not the whole story. For all its Darwinian emphasis, Huxley's book still conceded that there might be some non-adaptive characters and trends, and even Simpson admitted a role for slightly discontinuous 'quantum evolution'. These non-Darwinian elements were eliminated only as the Darwinian element 'hardened' in the course of the following decade.[8]

[8] On nonadaptive trends, see Huxley, *Evolution*, pp. 504–16. The disagreements over what the synthesis meant to him and others, and over the breadth of his outreach, are visible in various contributions to C. Kenneth Waters and Albert Van Helden, eds., *Julian Huxley*.

In the end, Huxley's effort was a triumph of rhetoric, an effort to convince his fellow biologists that they should play down their differences and realize the advantages of showing a united front to a scientific community and public increasingly obsessed by the potential benefits of the biomedical sciences. In principle Huxley should have been able to spread the initiative to a wide public. His position at the London Zoo (1935–42) allowed him to promote a vision of 'general biology' that took evolution for granted.[9] He was now a frequent broadcaster on the BBC, including its popular 'Brains Trust' programme. After leaving the Zoo (unwillingly) he was eventually appointed to UNESCO and became a figure on the world stage promoting both the practical and the ideological benefits of science. But most of this activity was devoted to issues other than evolution theory (although we shall see that the concept of evolutionary progress was central to his world-view).

Huxley did get an article on the new Darwinism into the magazine *Discovery*, but the non-specialist science literature of the mid-1940s reveals few contributions on the Darwinian theory. The publishers of the successful Penguin paperbacks began a regular *New Biology* magazine in 1945 and a *Science News* the following year, but apart from a few items on the history of life and human origins, there was little on evolution and nothing on Darwinism through the rest of the decade. In 1948 a revised edition of the *Encyclopaedia Britannica* recognized the new approach, with articles on evolution and population genetics by Sewall Wright.

Perhaps the best pointer to future developments was the publicity generated by David Lack's *Darwin's Finches* in 1947. The Galápagos Islands had slipped from biologists' attention since Darwin's time – an expedition from the California Academy of Sciences in 1905 had shown no interest in the birds as evidence of adaptive evolution. The term 'Darwin's finches' was coined at a British Association meeting in 1935 that had commemorated Darwin's visit, but when Lack did his own work there, he did not at first realize the adaptive significance of the finches' beaks. By the time he wrote his book, though, he had adopted Darwin's own interpretation, thus introducing what would become an icon of the islands as an illustration of adaptive divergence. Lack himself subsequently published an article on the topic in *Scientific American*, and the paperback reprint of his book in 1961 noted that in the meantime it had become widely accepted that most characteristics are adaptive.[10]

[9] See Joe Cain, 'Julian Huxley, General Biology, and the London Zoo, 1935–42'.

[10] Huxley, 'Darwinism Today', *Discovery*, n.s. 4 (1943): 6–12 and 38–41. For facsimiles of Wright's articles, see his *Evolution: Selected Papers*, pp. 524–38. See Lack, *Darwin's Finches*, and 'Darwin's Finches', *Scientific American* 188 (1953): 67–70; for his claim

Through the 1950s and 1960s the scientific community presented the new Darwinism as a consensus adopted by almost all experts, whatever their internal differences. The BBC produced several programmes featuring evolution in this period, often linked to a humanist viewpoint (and drawing further complaints from the creationists). For ITV, Granada broadcast effective programmes on primate ethology filmed at London Zoo. In the United States, CBS broadcast a series from the Princeton Bicentennial meeting in 1946 with material from Dobzhansky and Haldane.[11]

There were also many non-specialist books such as A. J. Cain's *Animal Species and Their Evolution* and John Maynard Smith's *The Theory of Evolution*. Cain's appeared in Hutchinson's University Library in 1954 and was reprinted as a Harper paperback in 1960; Smith's was the first in a Pelican series on biology and was revised in 1966 to include new material on the role of DNA – which was seen as the death knell of Lamarckism. Loren Eiseley's *Darwin's Century* of 1958 was a much-reprinted survey of the history of evolutionism vindicating Darwin's original insights. Gavin De Beer's 1963 biography of Darwin included photographs illustrating H. P. D. Kettlewell's new interpretation of industrial melanism as the result of selective predation on moths lacking appropriate camouflage. This became something of an icon for the synthesis because it could be simplified to provide a clear statement of how natural selection works.[12] De Beer had also produced a new edition of the Natural History Museum's *Handbook on Evolution* to accompany a display commemorating the 1958 centenary of the publication of the Darwin-Wallace papers. We have seen in Chapter 7 that even before the synthesis emerged, museum literature on both sides of the Atlantic had already begun to reflect a more open-ended view of the history of life compatible with a loosely Darwinian viewpoint.

The most flamboyant celebration of the centenary of the *Origin of Species* was the one held at the University of Chicago in 1959. This paraded a host of experts, attracted a press-corps of twenty-seven and

that views had changed, see the preface to the Harper edition of his book, p. v. On the earlier expedition, see Matthew J. James, *Collecting Evolution*, esp. pp. 68–9, and on the British Association meeting p. 242 and more generally Edward J. Larson, *Evolution's Workshop*. Lack was eventually confirmed in the Anglican Church and adopted a less Darwinian position.

[11] On the BBC's productions, see Hall, *Evolution on British Radio and Television*, chap. 3, and on Granada's, see Miles Kempton, 'Commercial Television and Primate Ethology'. On the CBC broadcast, see LaFollette, *Science on the Air*, pp. 207–9.

[12] See De Beer, *A Handbook on Evolution*, preface, and for the peppered moth his *Charles Darwin: Evolution by Natural Selection*, pp. 192–3, with photographs between pages 196–7.

inspired a half-hour-long movie made by *Encyclopaedia Britannica*. Huxley spoke on the origins of Darwinism, but it was his Convocation Address in the University Chapel that generated the most vivid headlines. He used it to promote his humanistic 'Evolutionary Vision', and in this case the public response was anything but positive. Despite his often-repeated claims that humanity could still be regarded as the high point of the development of life, he noted that 'orthodox Middle-Westerners' were shocked, along with Sol Tax, the conference organizer.[13] His use of the synthesis to underpin this wider vision depended, of course, on the assumption that it could be extended to include the origins of the human race.

Human Ancestry

This assumption was highlighted in November 1953 when the Natural History Museum in London announced that the celebrated Piltdown remains had been shown to be fraudulent – an artificially modified ape jaw combined with a fragment of a relatively modern human skull. Newspapers on both sides of the Atlantic headlined the story, provoking shock among a public that had been led to believe the find to be iconic evidence of how we evolved from an ape ancestry. *Punch* and other magazines featured cartoons lampooning the event, and it was even debated in Parliament. Science Service in Washington mistakenly claimed it had provoked a fistfight at one London meeting. In an effort to head off criticism, the Museum organized a display to explain what had happened.

The affair seemed to reflect badly on a scientific community that had been taken in for so long, but in fact the Piltdown remains had already begun to seem an anomaly that did not fit in with an emerging reinterpretation of how the human species had evolved. Piltdown had seemed plausible because everyone at the time it was discovered had assumed that the expansion of the brain had led the way in lifting the human line above its ape origins. By the 1940s it had become clear that Darwin had been right all along: the fossil evidence now suggested that the human family had been defined by the adoption of an upright posture adapted to walking in an open environment, with the expansion of the brain coming only much later. As J. S. Weiner noted at the start of his book on the exposure, by focusing attention onto an adaptive rather than a

[13] See Vassiliki Betty Smocovitis, 'The 1959 Darwin Centennial Celebration in America'; for Huxley's views, see his *Memories II*, pp. 191–2.

progressionist vision of the process, this transformation helped to bring the field more into line with a Darwinian mode of explanation.[14]

The first Australopithecine fossil, discovered in 1924 by Raymond Dart, had pointed the way toward the reinstatement of Darwin's hypothesis but had been widely dismissed at the time as an aberrant ape. The situation began to change in the following decade when Robert Broom, already an authority on the origin of the mammals, began to find more Australopithecine remains, also in South Africa. These were highlighted in magazines such as the *Illustrated London News* as evidence for a new way of understanding the first hominids. They had stood upright, yet had brains no bigger than an ape's. As early as 1931 the popular broadcaster Gerald Heard had insisted that the crucial step in the origin of humanity was the descent from the trees and the adoption of an upright posture. For Broom and an increasing number of experts it now seemed plausible to treat the Australopithecines as the ancestors first of the *Pithecanthropus* types (soon renamed *Homo erectus*) and then of the earliest modern humans. Museums began to rework their displays – the Natural History Museum in London published W. E. Le Gros Clark's *Antecedents of Man* in 1949, later reprinted as a Harper paperback in the United States. Broom himself wrote a short account of his work in 1950, issued by the rationalists' favourite publisher Watts, even though Broom made clear his own belief that a supernatural plan was involved. Louis Leakey, whose discoveries were also treated as newsworthy, revised his *Adam's Ancestors* to accept that the Australopithecines had changed the picture, although he still thought them a sideline to the true ancestry of humans. In 1964 his discovery of *Homo habilis* provided evidence of the first stage in the expansion of the brain.[15]

Broom may have seen a supernatural element in human evolution, but Huxley was one of the few promoters of the Darwinian synthesis to suggest that there might be something unique in our position at the head of nature. The revised understanding of human origins fitted perfectly into a Darwinian vision in which adaptation to new ways of life displaced progress toward a predetermined goal as the driving force. At one level,

[14] There is a huge literature on the Piltdown affair, mostly inspired by efforts to identify the culprit; for my own take on the subject, see my *Theories of Human Evolution*, pp. 35–8. For samples of the press coverage of the exposure, see Charles Blinderman, *The Piltdown Inquest*, pp. 77–81 (with the *Punch* cartoon on p. 80), and Ronald Millar, *The Piltdown Men*, pp. 208–17. On the Science Service report, see J. S. Weiner, *The Piltdown Forgery*, p. 69, and on the event as a fulfilment of Darwin's prediction, chap. 1.

[15] Broom, *Finding the Missing Link*; see pp. 97–101 for his views on the supernatural element (a topic on which he had already published his *The Coming of Man*). For some of the public reception of his discoveries, see Reader, *Missing Links*, chap. 6. Like Clark's *Antecedents of Man*, the 1953 revision of Leakey's *Adam's Ancestors* was also reissued as a Harper paperback. For Heard, see his *The Emergence of Man*, pp. 28–9.

the family tree of humanity was simplified – treating each new fossil as a distinct species was no longer in fashion, as when the Java and Peking remains were unified as *Homo erectus*. The Neanderthals were now seen as a primitive form of modern humanity, not a completely different branch. But at another level, the main branches of primate evolution were treated as an irregularly branching tree with no central trunk leading toward ourselves. The separation of the human from the ape lines of development was no longer a triumphant surge toward higher intelligence, but an adaptation to life on the open plains. Precisely why the human line had eventually begun to gain a bigger brain became a key question, with one much-discussed but highly controversial theory being Sherwood Washburn's claim that big-game hunting had spurred the need for tools and the intelligence to make them. Challenging this gendered vision of 'man the hunter' would eventually spur a public fascination with the work of female primatologists such as Jane Goodall.[16]

Wider Implications

Indeed, from the1960s onwards the application of evolution theory to the study of human nature generated endless controversies leading to claims that a new social Darwinism was emerging. As will be seen in Chapter 10, not all of these applications were derived from the mechanism of natural selection – the suggestion by Robert Ardrey and others that we have preserved the aggressive instincts of our ape ancestors seemed little removed from the nightmares of the Victorian period. But there were certainly efforts to argue that if natural selection is the driving force of evolution, we cannot have escaped the element of selfishness that ensures survival and reproduction in a competitive environment. In the 1940s and 1950s, however, the revival of Darwinism was presented in a more positive light. It appeared as part of a progressive social movement, helping to undermine the scientific foundations that had been used to sustain the ideologies of racial division and eugenics. It was also promoted by at least some of its supporters as the basis for a view of humanity's place in the world that confirmed our unique status. All the backers of the modern synthesis were convinced that evolution is progressive and that humanity is the highest form to have emerged in the history of life. This was a significantly different version of progress from that popular in earlier decades: humanity could not be seen as the

[16] On Washburn, see Donna Haraway, *Primate Visions*, pp. 211–17, and on Goodall and Dianne Fossey, pp. 179–85 and 263–8. Goodall had been inspired by Leakey, who distrusted the claim that the move onto the open plains was a key step.

intended goal of creation, but our species is still the product of the last and most significant innovation in the diversification of life. For Huxley and Dobzhansky especially, this sense of uniqueness allowed us to retain something of the old sense that life has a spiritual purpose.[17]

Scientific racism did not die out altogether. In the last years of his life (he died in 1940) the Lamarckian E. W. MacBride wrote to the press openly in support of Nazi policies. In 1949 Arthur Keith produced his most substantial account of his belief that competition between rival racial types was the motor of progress. Entitled *A New Theory of Human Evolution*, it was anything but an expression of the new Darwinism since Keith had no interest in genetics or selection acting on variation within a population. By the 1930s the application of population genetics to the diversity existing within the human population had begun to undermine the old model in which the distinct character of each racial type was guaranteed by its antiquity. Huxley – who had expressed overtly racist views in the 1920s – now joined with anthropologist A. C. Haddon to produce *We Europeans*, published in 1935 and reprinted as a Pelican paperback four years later. It was an outspoken critique of the traditional race sciences, arguing that there were no significant genetic barriers between different populations and that the ethnic groups in Europe were highly intermingled. It openly ridiculed the Nazi's image of an ideal Teuton – would he be as blond as Hitler, as tall as Goebbels and as slender as Goering? At the same time psychologists such as Otto Kleinberg in America were joining the cultural anthropologists in challenging the claims that the intelligence levels of blacks and whites were intrinsically different, pointing instead to the effects of upbringing and education. As the horrors of the Nazi death camps became public knowledge, scientists came under increasing pressure to repudiate the support offered to the idea of distinct racial types. Leslie Dunn and Dobzhansky produced their *Heredity, Race and Society* in 1946. In 1950 UNESCO issued its famous 'Statement on Race', drafted by Huxley, Kleinberg and geneticist Hermann Muller, firmly nailing the colours of the scientific community to the anti-racist mast.[18]

The situation with respect to eugenics was more complex. In the 1930s many biologists had become uncomfortable with simple-minded claims

[17] Michael Ruse's *Monad to Man* explores the role of progressionism in the synthesis; see especially chaps. 9–11. My own *Progress Unchained* argues that this vision (anticipated as we have seen by H. G. Wells and others) was very different from that popular in the previous century.

[18] For the anti-Nazi comment, see Huxley et al., *We Europeans*, p. 26 (Haddon later expressed regrets at having become involved in the project). For further details, see Elazar Barkan, *The Retreat of Scientific Racism*, and Kevles, *In the Name of Eugenics*, pp. 132–8.

that 'feeble-mindedness' and other negative characters could easily be eliminated by sterilizing those carrying the relevant genes. It was not that scientists such as Huxley and Haldane doubted a role for genetics in determining human character, or that they thought everyone had the same level of mental competence. But they did accept that upbringing and education could mask the genetic differences, so that without a level playing field for all it was hard to identify those who might be 'inferior'. Population genetics also played a role by showing that simply preventing those allegedly disadvantaged from breeding would have little immediate effect if the genes involved were recessive. R. A. Fisher was an ardent eugenist, yet his model of selection implied that it would take many generations for a eugenic policy to bring about significant improvement in the population. The new Darwinism was thus at least consistent with opposition to old-style eugenics, although public attitudes were again influenced by the knowledge that the Nazis had actually killed those who they thought might transmit harmful genes. Open support for eugenics largely disappeared in the 1950s, although similar policies often remained in place under other names, and the focus on heredity as a determinant of character would revive as understanding of the genes expanded following the recognition of DNA as the basis of their activity in 1953.

For Huxley, at least, the expectation of future human progress had a wider dimension than mere biological improvement. In the interwar years he had gained notoriety with his book *Religion without Revelation* in which he had rejected the belief in a personal God but sought to retain something of the old vision of a spiritual purpose in life. This expanded into the philosophy of humanism, which he promoted relentlessly in articles, radio broadcasts and books such as *The Humanist Frame* of 1961. His controversial address to the Chicago conference honouring Darwin in 1959 was an expression of this faith explicitly linked to its evolutionary foundations. The progressive element in biological evolution offered a new way of defining humanity's unique role in the cosmos. We may not be the preconceived goal of progress in the sense envisioned by theistic evolutionists, but we are the last and the only conceivable pinnacle of the process. Huxley drew on Broom's position to argue that although the tree of life has many branches, they have been increasingly sidelined into dead-end specializations. At each crucial evolutionary innovation, the number of species with the potential for further progress has been whittled down, leaving humanity as the only form of life now capable of moving on to a higher level of mental and moral development.[19]

[19] On Huxley's humanism, see, for instance, Bowler, *Reconciling Science and Religion*, pp. 150–3, and on his progressionism, Ruse, *Monad to Man*, pp. 331–8.

There were elements of the traditional viewpoint here, and to the consternation of his peers Huxley openly endorsed the revival of the theistic vision of evolutionary progress by the Catholic priest Pierre Teilhard de Chardin. He wrote an enthusiastic introduction to the English translation of the latter's *Phenomenon of Man*, posthumously published in 1959, with a cheap paperback in 1965. Despite a scathing review by the biologist Peter Medawar, the book was a sensation. For many non-scientists, Teilhard's mysticism offered an alternative to the materialism usually associated with Darwinism. Temporarily at least, Huxley's action linked the modern synthesis back to an older tradition in evolutionism.[20]

Thanks to his Russian Orthodox upbringing before coming to America, Theodosius Dobzhansky also retained elements of the religious way of thought and wrote books for a wider audience expounding his vision of progressive evolution. He too wrote enthusiastically about Teilhard's efforts to present humanity as the goal, but unlike Huxley he appreciated the difficulty of reconciling this with the new Darwinism. There might be a general (if irregular) tendency for progressive innovations to occur, but this did not mean that humanity could be seen as the inevitable outcome. At the level of local adaptation, each evolutionary modification depends on the local conditions, and if those conditions were slightly different the outcome would not be the same. Extended to the whole tree of life, this meant that each of the major branches would be a unique product of the circumstances in which it emerged. Some branches will represent higher levels of organization, but there will be many ways of advancing, and the line leading to humanity is but one of the routes by which our level of mental and moral capacity could be reached. In his *Biology of Ultimate Concern* of 1962, Dobzhansky presented evolution as a process groping its way upward, an inefficient way of proceeding whose outcome could not have been predicted. If intelligent life had evolved on planets other than the earth, it would be unlikely to be in anything like the human form.[21]

The same point was made by the palaeontologist George Gaylord Simpson, who shared the view that evolution is progressive in the long run without attributing any spiritual significance to the outcome. His *The Meaning of Evolution* of 1949 depicted the process as opportunistic and open-ended, using the example of the kangaroos in Australia to argue

[20] For Huxley's introduction, see Teilhard de Chardin, *The Phenomenon of Man*, pp. 11–30; Medawar's review is reprinted in his *The Art of the Soluble*, pp. 81–92.

[21] Dobzhansky, *The Biology of Ultimate Concern*, pp. 120–8; see Mark Adams, ed., *The Evolution of Theodosius Dobzhansky*.

that there were multiple innovations capable of advancing to higher levels. In *This View of Life* he challenged Teilhard's image of humanity as the inevitable outcome of a progressive trend and in a chapter titled 'The Nonprevalence of Hominids' insisted that evolution did not necessarily advance toward the human form. There had been many crucial turning points in the chain of events that had led to humanity, and 'If that causal chain had been different, *Homo sapiens* would not exist.' He too insisted that if life was found on other worlds, it might have progressed toward intelligence but not necessarily toward the human form.[22]

The possibility of evolution proceeding differently on other worlds had already been mentioned in previous decades. It is implicit in the Martians of Wells' *War of the Worlds* and was discussed explicitly by Haldane and W. D. Matthews. Now – Huxley excepted – it was becoming the dominant view among biologists that evolution might not have generated humanity even if one assumed that it must eventually progress to our level of intelligence. The ability of later science fiction writers to envisage aliens who do not resemble humans physically or psychologically, and might even be totally incomprehensible to us, suggests that this way of thinking was gaining some hold over the public imagination. Although it did not require an understanding of the latest thinking on natural selection, it was a view of humanity's place in the world conveying an image of the history of life in tune with the broader Darwinian viewpoint.

The founders of the modern synthesis succeeded in gaining a degree of public acceptance that went largely unchallenged until after the centenary of 1959. By promoting it as the basis for an optimistic world-view based on progress, they were able to evade the traditional fears about materialism and social Darwinism, at least for the time being. Whether one wanted to follow Huxley's humanistic vision of humanity as the last best hope of life on earth or the less structured image of nature groping its way upward, the new version of Darwinism could still be used to give us a significant role in the cosmos. In 1959 Hermann Muller expressed the confidence of many in the scientific community that Darwinism was now sufficiently secure that it should be reincorporated into science teaching at colleges and high schools: 'One hundred years without Darwin are enough.'[23] But the fact that Huxley's broadcast address to the Chicago meeting had attracted significant criticism made it clear that even in its most sanitized form, the modern synthesis was not going to go down well with conservative religious believers.

[22] Simpson, *This View of Life*, chap. 13, see p. 267 for the quotation.

[23] Muller, 'One Hundred Years without Darwin Are Enough', *The Humanist*, 19 (1959): 139–49.

10 Toward the Modern World

In the late twentieth century there was continued expansion among the media through which science could be disseminated and debated. Even in print, the availability of cheap, high-quality colour photography transformed magazines such as *Scientific American* and *National Geographic*, sometimes at the risk of subordinating the text to the image. Coffee-table books proliferated, some on topics that included material on evolution, while paperbacks were passing beyond the age of cheap reprints. The most significant transformation was the rise of television, which replaced radio as the means by which most people accessed news and current affairs. The moving image was even more effective at presenting the immediacy of the natural world and the process of discovery, especially when it too became coloured. Seeing the harshness of life in the wild, as well as its more cuddly aspects, could help spread awareness of the struggle for existence. More formal documentaries such as the BBC's *Horizon* series (later transferred to ITV) attracted significant audiences. But it was blockbuster serials fronted by high-profile presenters that had the greatest impact, spreading the idea that evolutionism could be applied to the cosmos itself as well as in biology. Until the emergence of the internet, this was now the public face of science.

As the sophistication of the media expanded, so did the power of those who were in control to shape what was presented. It was no longer possible for the scientists to dictate what the producers and presenters served up. They had always feared oversimplification and trivialization, but they had to accept that if they wanted to keep the public's interest, they had to work with those who knew what would hold an audience. The two areas sometimes worked well together, as when the BBC encouraged Huxley's humanistic approach, although in the United States the proliferation of rival channels generated an array of competing views.

Scientists who could adapt to the media became celebrities, as in the case of the primatologists Jane Goodall and Dianne Fossey, who presented their lives with the apes. David Attenborough on natural history, Carl Sagan on cosmology and Jacob Bronowski on science and culture all became famous. Others became celebrities through writing books and magazine articles, especially when their work attracted wider press and media attention. Stephen Jay Gould and Richard Dawkins fall into this category. Their work was paralleled by science writers who came into the profession with a substantial scientific training. Matt Ridley, for instance, gained an Oxford PhD in zoology before working for *The Economist* and going on to write best-selling books. There could be tensions when the scientific community was divided: some did not endorse Huxley's optimistic version of Darwinism, and Gould faced opposition for using his high profile to present unorthodox views. Dawkins became famous for promoting his gene-centred version of the selection theory but annoyed many scientists by his efforts to link Darwinism with atheism.

Some new areas of science emerged directly into the media spotlight and were shaped by the resulting public attention. Elizabeth Jones' study of the attempts to recover ancient DNA coins the term 'celebrity science' to denote this phenomenon. In this case the technique held out the hope of reconstructing extinct species such as the mammoth, leading to speculation that it might even be extended to the dinosaurs, as highlighted by the movie *Jurassic Park*.[1] The scientists involved had to learn how to cope with the public expectations generated by these rather implausible hopes. Other areas of science relevant to evolution also emerged with the same blaze of publicity, including the discovery of evidence that the age of the dinosaurs most likely was ended by an asteroid impact.

For writers and producers with less commitment to science, the prospect of the scientists falling out among themselves offered an opportunity, especially when the rival positions impinged on beliefs about human nature and society. It was also news if it seemed that a major theory such as Darwinism was coming under fire, either from rival biologists or from external sources inspired by philosophical or religious positions. Many readers and viewers actively welcomed the appearance of ideas that challenged scientific orthodoxy. In the counter-culture, it became fashionable to take on not just Darwinism but also the whole scientific view of prehistory. Cosmic catastrophes and visiting aliens were perhaps the most startling alternatives, each appealing to the same anti-Darwinian arguments to promote their mutually incompatible positions. It was pure coincidence that catastrophism eventually re-entered mainstream science when the possibility of asteroid impacts emerged. The most powerful assault was the growing hostility of evangelical Christians, driven to renew the opposition that had hit the headlines in the 1920s. A series

[1] Elizabeth D. Jones, *Ancient DNA*.

of challenges emerging from rival positions within the evangelical community has created what looks like a coordinated campaign from the evolutionists' perspective.

To those who reject evolutionism altogether, disagreements within the scientific community can be presented as evidence that the whole framework is unsound. Media presenters looking for good stories also find it easy to sell news of arguments among the experts to a public increasingly wary of those who claim authority. The biologists certainly grew hot under the collar as the consensus of the modern synthesis began to unravel in the 1960s and the rival positions broke out into open conflict in the next decade. The whole episode became known as the 'Darwin wars', fought not just between Darwinians and outright opponents but also between rival factions within the Darwinian camp. The sociobiology of E. O. Wilson and Richard Dawkins' notion of the 'selfish gene' were attacked as forms of an 'ultra-Darwinism' that has lost sight of the organism as a valid entity. When the positions could be linked to debates about human nature and society, all sides became excited and the mass media began to pay attention. There were claims of renewed efforts to promote eugenics and social Darwinism, countered by charges of Marxist infiltration and wooly-minded anti-materialism. Early efforts to revive Lamarckism were soon dismissed, but the emergence of evolutionary developmental biology, or 'evo-devo', has made it possible for biologists to suggest that elements of older research programs might still have something to offer.

Alongside these debates over the mechanism of evolution there were further developments in some of the associated areas of interest to the wider public. The flow of fossil discoveries continued unabated, creating disagreements among the experts trying to understand their significance. The dinosaurs recovered from a temporary loss of public interest, confirming their position as a once-successful branch of the tree of life cut down suddenly to make room for the mammals. Perhaps they had even been warm-blooded. Toward the end of the century it became clear that some of them had feathers, leading to the realization that the birds were, in effect, living dinosaurs. Evidence for an asteroid impact as the cause of the last mass extinction added a new dimension to the apparent unpredictability of the overall sequence of events in the history of life. Hominid fossil discoveries retained their fascination and confirmed the realization that our ancestors had stood upright before gaining bigger brains, although the precise sequence of events remains obscure. The process by which species are classified became controversial as the most extreme advocates of the new technique of cladism argued that it is impossible to recognize ancestor-descendant relationships. Amidst all this excitement the scientific community as a whole looked on in amazement as the advocates of young-earth creationism rejected the framework erected to understand the earth's history, restoring Noah's flood to its traditional role.

Life in an Uncertain World

Whatever the disagreements emerging in the scientific community, by the 1970s evolutionism was routinely presented to the public within a broadly Darwinian framework. Jacob Bronowski's 1973 TV series *The Ascent of Man* opened with the latest views on the origin of humanity in Africa and went on to include material on the mechanism of evolution that, unusually, devoted more time to Wallace than to Darwin. Relics of older ways of thinking persisted, however: Bronowski's general chapter on evolution was titled 'The Ladder of Creation', and when David Attenborough's *Life on Earth* serial used the latest techniques of wildlife photography to illustrate the diversity of life, it too used a linear sequence of living species reminiscent of the chain of being. There was, at least, a warning that no living creature could be seen as an exact equivalent of earlier stages in evolution. The final chapter presented the separating of humans from apes as a process initiated by the adoption of an upright posture and warned against seeing ourselves as 'the ultimate triumph of evolution'. It suggested that if we were to disappear, 'There is a modest, unobtrusive creature somewhere that could develop into a new form and take our place.'[2]

A more direct vehicle for popularizing Darwinism was the appeal to Darwin himself, especially his visit to the Galápagos Islands and the evidence they provided for adaptive diversification. The scientific community and the general public had begun to take Galápagos biogeography – and especially 'Darwin's finches' – seriously at the time of the Darwin centenary in 1959. Encouraged by Huxley, UNESCO created the Charles Darwin Foundation there, initiating a flow of research and publicity. The Ecuadorian government established a national park and encouraged tourism. Alan Moorhead's profusely illustrated *Darwin and the Voyage of the Beagle* appeared in 1969 and a seven-part TV docu-drama *The Voyage of Charles Darwin* in 1978. In 1973 Peter and Rosemary Grant began their long-running research on the finches, hailed

[2] I have used the coffee-table books that were spinoffs from the series; Bronowski, *The Ascent of Man*, see chap. 1 and chap. 8, pp. 291–309; Attenborough, *Life on Earth*, quotations from p. 308. See Hall, *Evolution on British Television and Radio*, and on Attenborough, see Morgan Richards, 'Wild Visions'.

in Jonathan Weiner's *The Beak of the Finch* in 1995. Back in Britain, Darwin's home, Down House in Kent, had become a place of pilgrimage. Children were encouraged to think about the finches and their implications – they even appeared in a board book for the youngest readers, with a text noting Darwin's realization that only the best-adapted survive.[3]

For those more interested in the overall development of life on earth, the fossil record was still the main focus of attention, eventually supplemented by the efforts to recover ancient DNA in the hope of actually recreating some extinct species – including (rather fancifully) the dinosaurs. Stephen Gould recalled that when he was growing up in the 1940s he had been the only one in his class to be interested in dinosaurs – the wave of enthusiasm that peaked at the turn of the century had abated.[4] In the 1970s the situation changed again, driven in part by new discoveries but also by a revision of the dinosaurs' capacities and behaviour. For all the excitement generated by the discoveries of the late nineteenth century, the huge herbivorous dinosaurs had still been seen as lumbering, slow-witted beasts held in check by their cold-bloodedness. Now palaeontologists such as Robert Bakker began to argue that they were much more active and intelligent, fully deserving their place as the dominant life form of their time. Adrian Desmond's *The Hot-Blooded Dinosaurs* of 1975 became a bestseller by challenging their status as reptiles in the modern sense. The theory that the birds were evolved from dinosaur ancestors was revived, and it began to be suggested that the birds were little more than specialized dinosaurs, a view that would be endorsed at the end of the century by discoveries of feathered dinosaurs in China.

This more positive view began to make its way into general surveys of the fossil record, including *The Book of Life* of 1995, a well-illustrated popular text edited by Gould. This highlighted the diversity of life in each geological period, although Michael Ruse notes that for all Gould's hostility to the idea of progress the book did end up with humanity as the triumphant outcome. The new vision of the dinosaurs reached an even wider public through high-quality movie representations such as the TV series *Walking with Dinosaurs* (1999) and, of course, *Jurassic Park* (1997). The possibility that intelligence could emerge in non-human

[3] See Larson, *Evolution's Workshop*, chaps. 8 and 9. The board-book (in the author's collection) is Harry Karlinsky, *The Origin of Species by Charles Darwin*, which contains only images apart from a few sentences of text (presumably for the parents) on the back cover.

[4] Gould, *Bully for Brontosaurus*, pp. 95–7; for examples of popular interest in dinosaurs, see Mitchell, *The Last Dinosaur Book*, and Rieppel, *Assembling the Dinosaur*.

forms, hinted at earlier in the century, now moved centre-stage: science fiction author Harry Harrison imagined a world with intelligent reptiles (in his case evolved from mosasaurs, not dinosaurs), while biologist Dale Russell's depiction of a humanoid dinosaur appeared in the press and museum displays. Progress might still be assumed, but it was an open-ended form of progress with no predictable outcome.[5]

If the dinosaurs were such active creatures, why did they die out completely at the end of the Cretaceous period? The previous generation of palaeontologists had been aware of their relatively abrupt disappearance and had attributed it to an environmental transition, drastic in its long-term effects but far less dramatic than the catastrophes imagined by early nineteenth-century geologists. That form of catastrophism had been reintroduced in Immanuel Velikovsky's *World's in Collision* of 1950, which postulated astronomical events to explain the upheavals recorded in the Bible. In a follow-up, he extended his views into the more distant past, invoking rapid mass extinctions followed by bursts of evolution triggered by radiation. The counter-culture of the time lambasted the scientists for attempting to silence Velikosky, but there were few converts and his revival of catastrophism gradually faded from sight even among those drawn to unorthodox views.[6]

The situation was reversed in 1980 when Luis and Walter Alvarez startled the scientific community with their theory that the Age of Reptiles had been brought to an end by the impact of a large asteroid. Evidence that the end of the dinosaurs had been quite abrupt had been mounting, and the Alvarez team (father and son) showed that a crucial layer in the rocks marking the end of the Cretaceous contained high levels of iridium, an element known to be abundant in meteorites. The resulting controversy made the cover of *Time* magazine, and the idea even found its way into a 1982 episode of Doctor Who. Some palaeontologists warned that the dinosaurs were already in decline in the late Cretaceous, while geologists pointed to the massive volcanic upheavals that formed the Deccan traps in India at the end of the period. By the 1990s it was widely accepted that an asteroid impact had played a significant role, especially once the impact crater was located off the

[5] See Ruse, *Monad to Man*, on Gould's progressionism pp. 505–7 and on Russell's intelligent dinosaur p. 528. Mike Benton describes the making of *Jurassic Park* and *Walking with Dinosaurs* in his *The Dinosaurs Rediscovered*, pp. 134–6 and 249–52. Harrison's book is *West of Eden*, although it rather spoils the point by also having something very like modern humans evolved from the New World monkeys; see my *Progress Unchained*, pp. 218–21.

[6] See Velikovsky, *Earth in Upheaval*, chap. 15; see A. De Grazia, *The Velikovsky Affair*.

Yucatán Peninsula in Mexico. Here was clear evidence that whatever the laws governing the biological processes of evolution, unpredictable events could deflect the course of history and reshape the tree of life.[7]

The birds and mammals, of course, survived but the suggestion that the birds are actually modified dinosaurs provided a window into a controversy that had broken out among taxonomists (the biologists who specialize in classification) during the 1970s. The technique that came to be known as cladism originated within the evolutionary paradigm but had now begun to challenge existing assumptions about relationships. A clade is a group of species united by a shared character inherited from a common ancestor (on the assumption that it is unlikely to have been invented more than once). The arrangements used to display these relationships – cladograms – look like evolutionary trees but do not recognize a time element and cannot be used to infer ancestor-descendant links, even when fossil species are included. They just show degrees of similarity. The more extreme cladists began to reject the whole programme of reconstructing the history of life, prompting outrage among palaeontologists who were used to doing just that and were committed to projecting their interpretations to the public. A debate conducted mostly in the academic literature erupted into the public domain in 1980 when the Natural History Museum in London, now moving into an era of push-button displays, mounted exhibitions on the dinosaurs and on human fossils in which visitors could choose between rival schemes of classification. The maverick palaeontologist Beverly Halsted conducted a campaign in the pages of *Nature* against the Museum linking cladism to theories of discontinuous evolution (discussed below) and Marxism. In 1999 the *Nature* journalist Henry Gee undertook a lecture tour in the United States to promote his book *Deep Time* in which he ridiculed the 'just so' stories of how new types evolve and argued that science would be better off if it abandoned the attempt to define evolutionary links.[8]

These attempts to deflect interest from efforts to understand the course of evolution had little long-term effect, and the continuing flood of books, popular articles and press reports devoted to fossils and their implications remains unabated. The expansion of information available,

[7] Walter Alvarez provided an account of these developments in his *T. rex and the Crater of Doom*; for modern accounts, see Steve Brusatte, *The Rise and Fall of the Dinosaurs*, and Mike Benton, *The Dinosaurs Rediscovered*, both of which discuss earlier ideas. The Doctor Who reference is in the serial 'Earthshock' of March 1982; my thanks to James Sumner for this information.

[8] Gee's *Deep Time* explores the rise of cladism; see also David Hull, *Science as a Process*, chap. 6, and on the Natural History Museum episode, pp. 265–70.

through new techniques of analysis as well as new fossil discoveries, has led to many debates about the precise details of how the main animal types appeared and (in many cases) disappeared. Dinosaurs continue to fascinate children as well as adults, and there is no let-up in the press attention paid to the latest developments. The one thing that does emerge from any survey of this material is the absence of theories that assume the inevitability of all the stages in the emergence of the modern world of life. The growing concerns about the possibility of rapid climate change is serving only to increase the plausibility of the Darwinian vision of an evolutionary process that has no single goal and is constantly impinged upon by external forces.

Apes and Aggression

One of the Museum's displays criticized by Halstead dealt with human origins and encouraged visitors to choose between various interpretations of the relationships between fossil hominids. It seemed to question what had become the generally accepted view that the Australopithecines had given rise eventually to the first members of the genus *Homo* and then to modern humans. In fact, there were alternative views already in circulation among palaeoanthropologists, a community always divided over how to make sense of the few available fossils. By the 1960s Louis Leakey had emerged as an internationally known fossil hunter who championed his own idiosyncratic views. Articles in *National Geographic* and well-publicized lecture tours focused the public's attention, first on his revelation in 1959 of *Zinjanthropus boisei*, which he claimed as a distant ancestor of modern humans but which most authorities saw as another Australopithecine. For Leakey, the latter were a side branch of the human family tree, not the original trunk, a view he also imposed on his discovery of *Homo habilis* in 1964, a slightly larger-brained form associated with very primitive stone tools. Everyone else saw it as the first step on the way from the Australopithecines toward the later, more intelligent, forms of *Homo*. In later decades, Leakey's son Richard and second wife Mary also achieved fame with their own finds, including the '1470' skull and fossil footprints of early humans walking fully upright.[9]

The new fossils did not make it easy to trace a path through to modern humans, even without Leakey's unorthodox theories. Donald Johannsen achieved fame for his discovery of 'Lucy' – later given the formal name

[9] For details of the publicity surrounding these finds, see Reader, *Missing Links*, and Roger Lewin, *Bones of Contention*.

Australopithecus afarensis – in 1974 and his claim that this was the starting point for all later developments. By the end of the century there were several more ancient species, showing that whatever the links, the hominid family tree was very bush-like, with numerous extinctions and no sign of a central main line of development. The status of the Neanderthals also fluctuated as alternatives to the old image of them as brutal ape-men came to the fore. Most authorities (Leakey excepted) now treated them as a subspecies of *Homo sapiens*, perhaps adapted to ice-age conditions. The discovery of a burial with the remains of flowers led Ralph Soleki to name them the 'first flower people' in 1971, evoking an image of peaceful and cooperative tribes also featured in the popular novels of Jean Auel. Later studies have suggested that they were a separate development from *Homo erectus*, but close enough to ourselves for there to have been at least limited interbreeding – although the suspicion that our ancestors might have played a role in their disappearance remained.

The debates were driven in part by new techniques in molecular biology providing evidence that challenged traditional ideas of dating and relationships. The timing of the split between the human and ape lineages now seemed to be more recent than the palaeoanthropologists had believed. It became possible to argue that there was a comparatively limited degree of genetic separation between the ape and human lines. The possibility of hybridization with the Neanderthals was another product of this revolution.

Most controversial of all was the new techniques' impact on ideas about how modern humans had emerged from *Homo erectus*. It had been known from the time of the Java finds of 'Pithecanthropus' in the 1890s that this type had spread across the Old World, and it was now taken for granted that its origin had been in Africa. Many palaeoanthropologists believed that its scattered populations had then given rise to the various races of modern humanity. In the 1980s evidence from mitochondrial DNA was used to challenge this position, which could be dismissed as a relic of older ideas of parallel evolution. The 'out of Africa' thesis maintained that a second wave of emigration from that continent had spread *Homo sapiens* across the world, totally displacing the earlier populations of *H. erectus*. The concept of a 'mitochondrial Eve' – a single female from which all later humans could trace descent – generated huge publicity. This position could be used to claim that all living humans share a close genetic heritage, while serving as an uncomfortable reminder that in a Darwinian world, new innovations confer an adaptive advantage in the struggle for existence. Supporters of the multi-regional hypothesis have continued to defend their position, and it is still favoured by Chinese authorities.

The claim that we share a close relationship with the apes drew attention to primatology as a source of information on the origins of human behaviour. If our capacities and instincts are not as far removed from those of the apes as once assumed, the study of their lifestyles might throw light on our own predispositions. This expectation encouraged efforts to play down the differences by showing that the apes have rudimentary versions of capacities once thought to be uniquely human. Language had always been seen as a distinguishing feature, a position still defended by the eminent linguist Noam Chomsky, who insisted that it was a capacity that could only have emerged fully formed by a single evolutionary jump. To undermine this assumption, efforts were made to teach chimpanzees American sign language, and in late 1969 it was claimed that these were achieving some success. The October 1972 edition of *Scientific American* featured the programme as its lead story. A young chimp who seemed to have some awareness of grammar as well as word recognition was introduced to the public as 'Nim Chimpsky' to drive home the point. In the end, though, these claims were much disputed, and it soon became clear that all attempts to raise infant chimps as human children were doomed to fail as the animals matured.[10]

Even more controversial were suggestions that the more aggressive aspects of human behaviour might be locked in by instincts inherited from our distant forbears or even from our animal ancestors. The claim that early humans had been differentiated from the apes by their adoption of a lifestyle based on hunting on the open plains had become popular in the post-war years. Raymond Dart, who had discovered the first Australopithecine fossil back in 1924, began to argue that these early hominids had already begun to acquire violent instincts arising from hunting and scavenging. This claim soon became incorporated into a wave of popular literature promoting what was called the 'anthropology of aggression' by those who feared that it marked the re-emergence of something resembling the old social Darwinism. Resonances with Cold War anxieties and the counter-culture's reactions were obvious to many commentators.

In the late 1960s Konrad Lorenz, Desmond Morris and Robert Ardrey published bestselling books promoting various interpretations of the belief that aggression was an innate instinct. Lorenz's *On Aggression* argued that the instinct to compete ran throughout the animal kingdom, with each species developing its own way of coping with it, depending on

[10] See Greg Radick, *The Simian Tongue*, chaps. 8 and 9; the cover of the October 1972 *Scientific American* is reproduced on p. 317; also Donna Haraway, *Primate Visions*, esp. pp. 140–9. An almost contemporary survey is Adrian Desmond, *The Ape's Reflection*.

its lifestyle. Morris's *The Naked Ape* linked human behaviour more specifically to our ape ancestors. In a series of books Ardrey expanded Dart's vision of our early forbears in Africa adopting a hunting lifestyle with the attendant development of territoriality and aggression between individuals and groups. The image of an Australopithecine hammering skulls with a bone tool in Stanley Kubrick's 1968 movie *2001: A Space Odyssey* brought this home to a very wide audience indeed. The anthropology of aggression was not without its critics, however. Elaine Morgan challenged the whole idea of a move onto the open plains with her theory of the 'aquatic ape', according to which our ancestors had for a considerable period spent most of their lives living in lakes. She also joined a chorus of female critics of the hunting hypothesis who stressed the role of what Sally Linton called 'woman the gatherer'.[11]

The focus on aggressive instincts had a complex relationship with Darwinian theories of biological evolution. If we had such instincts, were they derived from a distant animal past or from the fairly recent adoption of a hunting lifestyle? In the former case especially, it might seem inevitable that in a world subject to the struggle for existence competition between individuals and groups would drive the formation of violent behaviour patterns. Yet the authors who promoted the anthropology of aggression had very little contact with the new Darwinism that had emerged with the Modern Synthesis, despite their constant references to competition. Links did eventually emerge: in 1987 the science writer John Gribben produced a radio series, 'Being Human', introducing listeners to both the debate on aggression and sociobiology.[12] But the debate over the origin of aggressive instincts was at first conducted with weapons derived from the observation of human and ape behaviour, not from the emerging neo-Darwinism of the 'selfish gene'.

Louis Leakey had hoped that the study of ape behaviour would throw light on the origins of human nature, and in the 1960s he encouraged several female primatologists to begin observations in the wild. Jane Goodall and Dian Fossey became global celebrities on the strength of their studies, hailed in articles in magazines such as *National Geographic*, in books, TV programmes and eventually films. Fossey's *Gorillas in the Mist* of 1983 stressed the peaceful life of the mountain gorillas, demolishing the traditional image of the 'killer ape'. After her death at the hands of poachers two years later her work was immortalized in a classic movie. Goodall's work had a more complex legacy. Her work with chimpanzees

[11] There is a huge literature on this episode and its popular impact, including Haraway's *Primate Visions*, esp. chap. 11, and Matt Cartmill, *A View to a Death in the Morning*.

[12] Gribben's radio series was reviewed in *New Scientist*, 5 March 1987, p. 60.

revealed their social hierarchies and gender relationships, emphasized in scenes that created a sense of continuity with our own family lives. Later observations, though, revealed more disturbing aspects of their behaviour, including occasional hunting expeditions and even violent conflicts between rival groups. The complex interplay between science, publicity and rival ideologies in these efforts to reveal the biological origins of human nature is explored in Donna Haraway's classic *Primate Visions*.

Darwin Wars

Among those whose main concern was the actual mechanism of evolution, the loose consensus known as the Modern Synthesis began to unravel around 1970. New insights into the operations of natural selection began to extend the theory in ways that led to controversy when they seemed to threaten traditional values relating to altruism. What the innovators saw as an extension of the inner logic of the mechanism looked like a revival of social Darwinism to those who worried about the implications for human nature. The anxiety was perhaps inevitable given the existing debate over the anthropology of aggression, which had revived fears that a biological explanation of human behaviour would encourage racism and eugenics. Even when the biologists involved repudiated such implications, their opponents saw their work as implicit support for the ideologies they found so offensive. The confrontations would eventually die down as it became recognized that the polarized debate was obstructing the possibility of a more sophisticated approach to the relationship between nature and nurture. In the meantime, though, the heat of debate led to challenges that temporarily rejected the Darwinian focus on adaptation and a revival of the theory of evolution by jumps.

The innovations were in part inspired by the emergence of molecular biology following the discovery of the structure of DNA and the research that sought to uncover how it shapes the development of the organism. At one level, this revolution created a major challenge to evolutionary biology's status by focusing public attention on the biomedical sciences and the hope of curing inherited diseases. Here was yet another source of concern for those who worried about hereditarian attitudes – it seemed a small step from eliminating harmful genes to choosing those that should be incorporated into the next generation. This would be analogous to artificial rather than natural selection, but the same research also generated techniques that led to new ways of understanding the genetic relationships between organisms and reshaped understanding of the tree of life. It also focused attention onto the gene as the key agent involved in natural selection, displacing the traditional Darwinian view in which the organism as a whole interacts with the environment. A new generation of biologists realized that when the environment includes other organisms of the same species, the consequences of the interactions can include instincts that seem to mimic altruism. What helps the gene to replicate itself is not necessarily what is good for the organism.

The breakdown of the Modern Synthesis took different paths on either side of the Atlantic. The two books that became icons of what became known as 'ultra-Darwinism' – E. O. Wilson's *Sociobiology: The New Synthesis* and Richard Dawkins' *The Selfish Gene* – were published in 1975 and 1976, respectively, but had different agendas and audiences. Wilson's was a mammoth survey of the social behaviour of animals based partly on his own work with ants and concluded with a chapter openly suggesting ways in which his views could be applied to human society. Dawkins' was intended to introduce the general reader to the gene-centred approach to evolution (although he did not focus on individual genes) and explicitly ruled out humans on the grounds that culture could evade whatever instincts might be imposed by biology. He later argued that ideas and values could serve as replicators within culture equivalent to the genes in biology. Both books generated huge controversy in their authors' home countries, the United States and Britain, respectively, but only gradually became seen as playing similar roles in a wider debate.

Both exponents started from the same insight, an extension of the theory of natural selection that challenged the accepted way of explaining altruism. In biology, this means behaviour that seems to benefit other members of the species while putting the individual at risk, traditionally seen as incompatible with the 'survival of the fittest'. According to V. C. Wynne-Edwards, the explanation was that natural selection could work at the level of groups rather than individuals – as Darwin had hinted, those groups whose members cooperated were the most successful. The classic example was the warning cry given when a predator is spotted, which alerts the group but may draw attention to the individual who cries out. Opponents of the new way of thinking pointed out that an individual with a mutation leading them to cheat on the system would benefit even more and would soon swamp the gene for cooperation. The question then became: How can apparently altruistic behaviour evolve in a world in which reproductive success is the only driving force?

The answer was to think in terms of genes, or gene complexes, rather than the organisms themselves. One sign of the new interest was an outburst of interest in sexual selection, a topic that had been largely neglected since Darwin had suggested the idea. It made sense to recognize that the breeding success of the most colourful peacocks outweighed the risk they ran by carrying such an unwieldy tail around. Extending this insight into social behaviour would undermine any sense that altruism was developed for the good of the whole group. As Dawkins explained, the best way for the non-specialist to imagine what is going on is to think of the genes acting for their own benefit, even though the process is actually a mechanistic application of reproductive success.[13]

Andrew Brown's book *The Darwin Wars* drove home the impact of this approach by opening with the suicide in 1974 of George Price, an American biologist living in London. Along with John Maynard Smith, he had worked out the consequences of an analysis first proposed by Bill Hamilton in 1964 leading to the theory of 'kin selection'. Natural selection can generate an instinct that leads an animal to sacrifice its own interests for those of another provided they share some of the same genes, especially if the chance of the beneficiary reproducing is better than that of the one who makes the sacrifice. What looks like altruism in the organism is actually driven by the advantage given to the 'selfish gene' – which does not really care about the individual who carries it. Price was so upset by the implications of this for human morality that he turned to religion, gave all his goods to the poor and committed suicide in poverty.[14]

Hamilton's work was the starting point for Maynard Smith, Dawkins and Wilson. In Britain, Smith added a chapter on kin selection to the 1966 reissue of his paperback *The Theory of Evolution*, stressing that it undermined the case for group selection and the whole idea of sacrifice for the good of the community. He had applied game theory to understand how apparently disadvantageous characteristics could be maintained in a population by selection. If pushed too far, the apparently more advantageous characteristic loses its benefits and creates room for the less obviously positive one to retain a foothold. As a presenter for the popular *Horizon* television series, Smith was in a good position to alert the audience to new developments and introduced them to the concept of the selfish gene following the publication of Dawkins' book. He also edited a collection of papers sponsored by *Nature* to make the detailed arguments available for a wider readership.[15]

[13] Helena Cronin's *The Ant and the Peacock* focused on the renewed interest in sexual selection.

[14] Brown's *The Darwin Wars* was an early survey of these developments; for later analyses, see especially Ullica Segerstråle, *Defenders of the Truth*; also Hull, *Science as a Process*, chap. 6, and Ruse, *Monad to Man*, chap. 12.

[15] See Maynard Smith, ed., *Evolution Now*. On his work as a populariser, see Helen Piel, 'Scientific Broadcasting as a Social Responsibility'.

In Britain, Dawkins' book soon became the centre of a fierce controversy. Although he excluded human behaviour, his hard-hitting demolition of the case for genuinely altruistic instincts in animals was assumed to have wider applications. To religious believers, moralists and left-wing intellectuals alike this looked like an attempt to show that selfishness was the only foundation for the interaction between living things, including humans. Activists such as Stephen Rose accused Dawkins of justifying not only an ideology based on the inevitability of competition but also more specific behaviour patterns including male philandering. Dawkins' protestations that his explanations of particular behaviours (such as male infidelity) were linked to the lifestyles of the particular species involved went unheeded. It did not help matters that he was also emerging as an outspoken critic of religion, the country's most visible atheist. Less aggressive supporters of Darwinism thought he was doing the theory a disservice by seeking to identify it with rejection of even the most liberal forms religious belief.[16]

Meanwhile in the United States, Wilson had taken up Hamilton's work to solve the long-standing puzzle posed by the sterile castes found among ants, bees and wasps. Because their peculiar reproductive system leads to a closer relationship between sisters than between mothers and daughters, it is advantageous (from the genes' viewpoint) for a female worker to give up her reproductive potential to help raise the offspring of a sister, the queen of the nest or hive. *Sociobiology* extended the model to a whole range of species with different lifestyles, showing how selection could explain the instincts that govern both reproductive and other forms of behaviour. Unlike Dawkins, Wilson was not committed to kin selection and was willing to consider cases where group selection might apply. In the end, though, his analysis stressed the power of natural selection to impose instincts that would result in the maximization of the genes' ability to reproduce themselves, whatever the consequences for our sense of right and wrong. By including a chapter on the possibility that this might apply to humans, he left himself open to the same array of critics as that facing Dawkins in Britain, all charging him with endorsing a conservative, sexist ideology based on the assumption that certain aspects of our lives are predetermined by heredity and cannot be modified by education or social conditioning.

Wilson had repudiated the anthropology of aggression and made it clear that he did not see the genes as absolute determinants of human nature. He later conceded that he had been politically naïve and failed to

[16] On Dawkins' popular image, see Fern Elsdon-Baker, *The Selfish Genius*.

realize that his position would inflame the left-wing opponents of all forms of social Darwinism. Curiously, having turned his back on a fundamentalist upbringing, he was quite happy to challenge the traditional Christian vision of human nature. But to his ideological opponents any suggestion of a genetic basis for instincts that would block the route to social reform was anathema. They were already alarmed at the revival of hereditarianism by figures such as Arthur Jensen, who in 1969 had claimed that the races have different levels of IQ. A major conference on 'Man and Beast' had brought together scientists and social thinkers from a variety of backgrounds to debate the issues. Following a positive review of *Sociobiology* in the *New York Review of Books* and front-page coverage in the *New York Times*, left-wing scientists and activists rushed to attack Wilson and formed the Sociobiology Study Group to coordinate their efforts. They included geneticist Richard Lewontin and palaeontologist Stephen Jay Gould, the latter already well-known for his popular-science writing. At the 1979 meeting of the American Association for the Advancement of Science a group of hecklers rushed to the front and poured iced water on Wilson's head.[17]

Gould already had a reputation for rocking the Darwinian boat and was well-positioned to publicize his views thanks to his regular contributions to the magazine *Natural History*, subsequently collected in a series of books. He was fully onboard with the open-ended model of the history of life, and had attacked the whole concept of progress in evolution. Later on, his *Wonderful Life* of 1989 used the strange fossils of the Burgess Shale in Canada to argue that the earliest animals had been formed with a wide range of body plans, most of which had disappeared at an early stage due to unpredictable events. If the 'tape of life' could be rewound and played again, the end result might be quite different to what we see today. But this emphasis on contingency arose from a long-standing hostility to the ultra-Darwinian claim that natural selection is the only factor involved, which Gould took to imply that those early animals that survived must have been better adapted than those that died.

The earliest indication of these doubts about the conventional Darwinian view came in 1972 when Gould and Niles Eldredge introduced their theory of punctuated equilibrium (often abbreviated to 'punk eek' in the press). They argued that the fossil record did not support the idea that evolution is a process of gradual, cumulative change. On the contrary, many species remained almost unchanged over long periods and were then replaced abruptly by another equally static form. Some

[17] A series of contributions to this debate are reprinted in Arthur L. Caplan, ed., *The Sociobiology Debate*, part 5.

level of discontinuity was accepted by the Darwinians – Ernst Mayr acknowledged that innovations would often appear in small, isolated populations at the edge of a species' range, which would take over the whole range if they turned out to have an advantage over the main population. The transition would be quite rapid by geological standards and would look abrupt in the fossil record.

The real problem was the claim that species remain essentially static over their whole lifespan, which seemed unlikely if they were having to adapt to the inevitable changes in the local environment that would happen over such a long period. Gould argued that such small modifications could not occur because species acquired developmental constraints that restricted individual variation and could only be uncoupled in unusual circumstances, such as a small population exposed to an untypical environment. To his opponents this was beginning to look like the non-Darwinian ideas popular around 1900, and indeed Gould briefly flirted with the possibility that the punctuations were caused by large-scale saltations generated by internal transformations of development.

He did not follow up the suggestion, but he did launch a full-scale assault on the adaptationist programme of classical Darwinism, the assumption that every character of every species has been shaped by adaptation either in the past or recently. He now argued that many of the visible features of an organism are not 'characteristics' in the sense understood by Darwinism – they are merely essential by-products of the process of individual development. A much-debated 1979 paper with Lewontin drew a comparison with the highly decorated spandrels of the San Marco cathedral in Venice. The spandrels were not designed to provide a convenient space for painters to work on because they are necessary structures required in order to place a round dome on top of a square building. By analogy, the organic structures that are modified by adaptation have their origin not in previous adaptive modifications but in the developmental constraints needed to produce that kind of organism. Gould was now arguing that a new evolutionary synthesis was needed that might include elements of Darwinism along with a recognition of the role played by developmental forces. In effect, aspects of the previous 'eclipse of Darwinism' were being rehabilitated in modern form.

Despite their dislike of hereditarianism, the radicals were not tempted by the Lamarckian alternative in which environmental pressures directly modify the development of the organism and eventually the whole species. This had always been the favoured position of anti-Darwinians anxious to give the organism an active role in choosing its own lifestyle. It still appealed to those hostile to the materialism associated with the theory of natural selection. In the 1970s the well-known writer Arthur

Koestler attracted some attention with his efforts to arouse renewed interest in the theory, and his book *The Case of the Midwife Toad* tried to rehabilitate the experimental programme of the Lamarckian biologist Paul Kammerer. Most biologists remained hostile, in part because they had spent the previous decades arguing against the similar programme advanced by the plant-breeder T. D. Lysenko in the Soviet Union. The one high-profile line of support came from a group led by Ted Steele, which published evidence suggesting that acquired immunity in mice could be transmitted to their offspring, by a process that allowed their DNA to be infected by characteristics derived from the immune system. While the experimental evidence for this process remained controversial, wider developments did gradually begin to undermine the rigid assumption that the genes are the only factor to be taken into consideration.

The call for a new synthesis was reinvigorated toward the end of the century with the emergence of evolutionary developmental biology. This built on the suggestion by Gould and others that the ultra-Darwinian focus on the gene had sidelined any role for other processes that might also shape the development of the organism. One sign of a more broadly based approach was an increased willingness to allow for gene interactions and an input from the environment. Wilson and Dawkins had both insisted that they did not subscribe to a rigid hereditarianism in which human characteristics are completely predetermined, and eventually it became widely accepted that environmental influences do shape the development of the organism by modifying or suppressing the way genes operate. Publication in 2001 of the news that the Human Genome Project had found that we only have 30,000 genes, far fewer than most experts had predicted, generated headlines speculating that the genome was too small to rigidly predetermine structure and, more importantly, behaviour. Science writer Matt Ridley proclaimed that the old debate between nature and nurture had been resolved – it was now a case of 'nature via nurture', the title of his bestselling 2003 book.

Evolutionary developmental biology built on growing evidence that environmental influences not only shape the development of the individual but might also be transmitted to the organism's offspring and even later generations. It expanded the claim for a more flexible understanding of the genes' activity by arguing that constraints on how the embryo is formed are responsible for many of the fundamental characteristics the Darwinians ascribe to the accumulated effects of selection.[18] The more enthusiastic supporters of this approach openly proclaim that it has

[18] See, for instance, Snait B. Gissis and Eva Jablonka, eds., *Transformations of Lamarckism*, and Manfred D. Laublichler and Jane Maienschein, eds., *From Embryology to Evo-Devo.*

reintroduced elements of the old Lamarckism, and some even claim that the time has come to replace the whole edifice of Darwinism. More cautious supporters see the move as a modification of Darwinism (indeed a revival of Darwin's own position), showing that the theory can itself evolve. Some older versions of the focus on development have disappeared from sight, most obviously the recapitulation theory. But others remain in a modernized form and are now reconstituting the evolutionary synthesis. The ongoing debates sparked by these initiatives are beyond the scope of this historical account. Yet it should be remembered that Darwinism survives, and continued public interest in traditional topics such as the Galápagos finches suggests that the role of adaptation is still widely appreciated.

Creationism Revived

Whatever his doubts about the selection theory, Gould played an active role in the defence of Darwinism against its critics in the wider world. The movement known as 'creationism' was actually an assemblage of different positions favoured by rival factions within the more fundamentalist churches, unified only by their hatred of Darwinian materialism. To the evolutionists, though, the series of challenges launched in the late twentieth century looked like a concerted campaign that merely changed its tactics to suit the prevailing situation. The creationists relished the disunity of the evolutionists because they could present it as evidence that the scientific position was only guesswork – in their simplified model of the scientific method the product should be demonstrable truths, while theories are mere speculation. Centres to develop an alternative 'creation science' were founded, but it proved hard to generate positive evidence of divine action, and most campaigns exploited old arguments against Darwinism reinforced by appeals to the modern disagreements on the history of life. These were skilfully presented to the public in debates, printed material and TV programmes. The scientists often found it hard to respond without getting bogged down in details that looked like hair-splitting to the audiences. As a result many people with evangelical religious beliefs got (and are still getting) their information about evolutionism from these negative presentations.

Support comes mostly from conservative sections of evangelical Protestantism, where rival churches compete for influence using their different stances on the biblical story of creation to define their faith. Their interpretations of Genesis reflect a spectrum of opinion depending on the degree of literalism applied to the sacred text. The most extreme position, known as Young Earth Creationism, returns to an interpretation abandoned by science in the eighteenth century: the earth is only a few thousand years old and the earth's crust was completely reshaped by Noah's flood. There is no long sequence of creations, and the fossil record was built up by the settling of the sediments after the flood. In principle, Young Earth creationists should not need to attack Darwinism since their model rules out any form of evolutionism, but they find the materialism of natural selection too tempting a target. Less extreme positions accept the evidence for a sequence of geological periods but focus on the discontinuities to undermine any theory of slow, gradual change, especially Darwinism. Curiously, the same tactic was used by Erich von Däniken in the 1970s to argue for episodes of genetic engineering by visiting aliens.[19] But for the followers of Intelligent Design, the alternative is supernatural action since this is the only way the complexity of living things could be achieved. In effect this reinstates the argument from design as applied by William Paley in 1802 using updated evidence. Some liberal creationists are even willing to allow limited amounts of theistic evolutionism, although this is anathema to most.

There is a substantial academic literature on the rise of, influence of and opposition to the creationist movement in the United States, so the key events are only summarized here.[20] What looked like a coherent campaign that changed its tactics to suit the times was actually a series of initiatives by rival groups within the evangelical community, each seeking to expand its following. The issue had lain dormant for several decades after the Scopes trial, in part because textbook publishers avoided the term 'evolution' if not the topic. The revived interest in Darwinism during the post-war decades led biologists to regain the confidence needed to demand more explicit coverage, which soon provoked a response. The first wave of opposition followed the publication of John C. Whitcomb and Henry M. Morris' book *The Genesis Flood* in 1961, which revived the 'flood geology' developed by George McCready Price in the 1920s. Renamed 'creation science', this retained a Young Earth position based on the claim that the geological strata were laid down as sediments in the biblical flood. The apparent progress shown by

[19] Von Däniken became famous for his *Chariots of the Gods?* in 1970 and produced his extended model of earth history in his *According to the Evidence* seven years later.

[20] The classic analysis is Ronald Numbers, *The Creationists.* On the equal-time movement, see Dorothy Nelkin, *Science Textbook Controversies and the Politics of Equal Time* and *The Creation Controversy*, and Adam Laats and Harvey Segal, *Teaching Evolution in a Creation Nation.* On the diversity of responses, see Jeffrey P. Moran, *American Genesis*, and on the rivalries within the evangelical movement, Benjamin Huskinson, *American Creationism, Creation Science, and Intelligent Design in the Evangelical Market.*

the fossil record was an artefact of the process by which the bones of animals wiped out in the flood settled to the bottom. In fact, it was claimed, all the extinct species had co-existed with the early descendants of Adam and Eve.

The Creation Research Society was founded in 1963, followed by the Institute for Creation Research in 1972. Arguments against the orthodox scientific view of the past were promoted by figures such as Duane T. Gish, an effective debater and popular writer, author of *Evolution: The Fossils Say No* and a host of widely circulated pamphlets. I can personalize the account at this point because I played a minor role in the controversies and still have a collection of ephemera from the period. It includes the pamphlet *Gish Answers Faculty*, which has an order form on its back page offering quantities up to 10,000 (price $550) for those anxious to spread the word. I also have dog-eared copies of magazines published by various churches attacking evolution, including the Seventh-Day Adventists' *Signs of the Times* and the Jehovah's Witnesses' *Watchtower* and *Awake!* In 1979 I moved to Northern Ireland, the most religiously conservative part of the United Kingdom, and was sometimes called in to confront visiting creationists on local radio. I even debated Gish in our Students' Union and held my own by pointing out that he did not actually believe the interpretation of the fossil record he presented. His talk implied a series of creations in geological time, although his book reveals his belief that all the strata were laid down in Noah's flood.

The Young Earth interpretation of the geological sequence continues to be widely defended in creationist presentations. On 1 June 1996 a 'Creation Discovery Day' was held at a Belfast Presbyterian church (see Fig. 10.1) and featured the claim that humans and dinosaurs coexisted, while dismissing 'ape-men' as a myth. Similar meetings were advertised as taking place in local schools. Local and national newspapers reacted when ministers in the Northern Ireland government lobbied to block both the teaching of evolution and displays favouring it in the Ulster Museum. In 2012 there was again press controversy when the National Trust's display in the Giant's Causeway Centre admitted that many local people didn't believe in the scientists' interpretation of the geological record.

In the United States, of course, the pressure has been far more intense. In the 1970s there was a major effort to block the teaching of evolution in the schools, which sidestepped the topic of flood geology and claimed that there was scientific evidence for creation, mostly arguments against evolution that were assumed to leave creation as the only alternative. There were demands that this alternative position should be given at least equal time in the science classes, resulting in laws passed in a number of states. Several organizations from within the scientific community joined

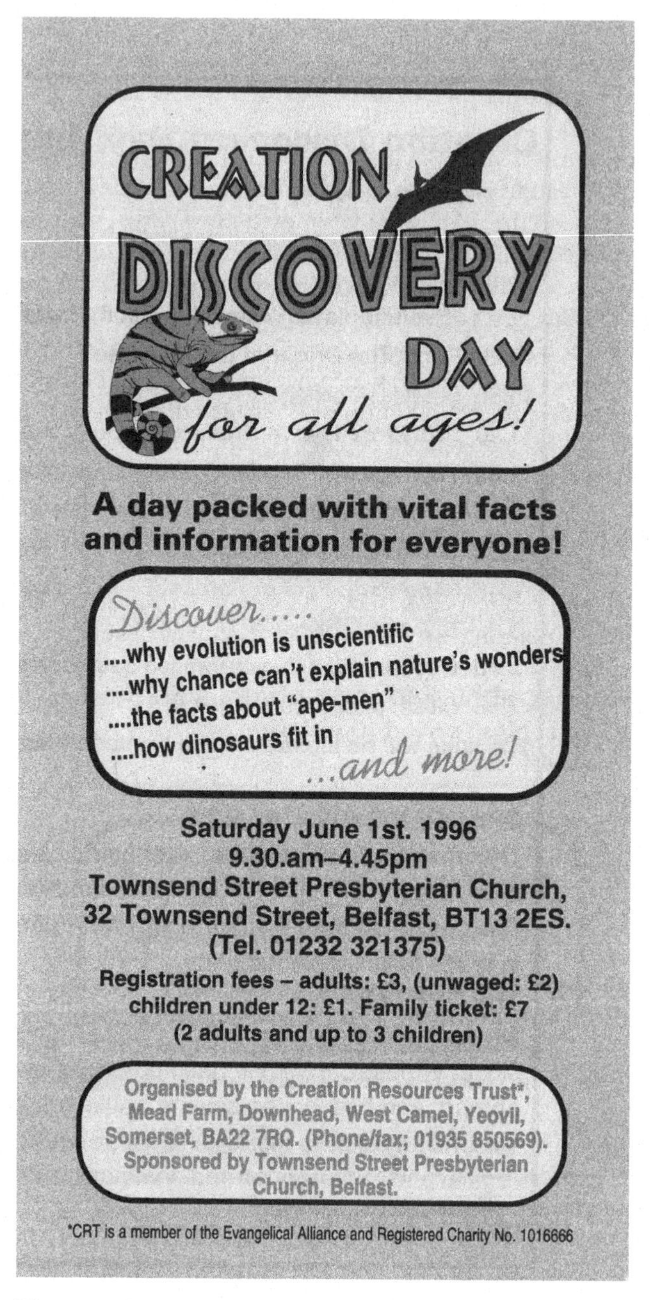

Fig. 10.1 Flyer advertising a meeting organized by the Creation Resources Trust in Belfast on 1 June 1996.

the American Civil Liberties Union in challenging these laws on the grounds that 'creation science' is based on religion rather than valid scientific research and is thus excluded by the US Constitution, a position upheld in a court in Arkansas in 1981 following evidence given by Gould and numerous other experts.

Another wave of attacks began a decade later with the emergence of the Intelligent Design movement. High-profile books by lawyer Philip Johnson (*Darwin on Trial*, 1991) and biochemist Michael Behe (*Nature's Black Box*, 1998) proclaimed the inability of natural processes to generate the complex structures and functions shown by living things. In principle the supporters of Intelligent Design did not invoke miraculous creation, although in private one suspected that many were closet creationists. Some even accepted the Young Earth position but, as I discovered in debates, were extremely reluctant to admit that in public. In the United States the equal-time movement renewed its activities, again blocked by a court decision in Dover, Pennsylvania, in 2005. Complaints against the decision by Behe and many others followed, and the issues simmer on to this day, the debates becoming all the more complex and virulent in the age of social media.

Surveys suggest that nearly half the American population believes in some element of supernatural creation for either species in general or the human race in particular. There are real issues involved in the education debate: Should students have to learn about something they don't actually believe in? How should any attempt to give creationism equal time handle the diversity of opinions among the anti-evolutionists? Is it valid to teach only their negative arguments against the theory? The different positions of the Young Earth creationists and the Intelligent Design movement reflect the diversity of opinions and attitudes involved, but are often intermingled in practice. There are regional differences, the Southern states still being more active. There are also important variations in the response of different communities, religious affiliation being obviously relevant, but with both race and gender also affecting how people respond. This is a movement driven by faith groups who share a common hostility to Darwinism, but differ widely on the alternative to be offered depending on the level of commitment to a literal reading of the Bible. Which group is most active in a given place and time determines the ideas and the tactics to be used.

2009: A Darwin Odyssey

A good illustration of evolutionism's enduring popularity can be seen in the Darwin celebrations of 2009, marking 200 years since his birth and

150 since the *Origin of Species* was published. Museums around the world put on special displays; there was a travelling circus of well-publicized conferences; news reports abounded in the press and popular magazines; and a plethora of books and pamphlets gave even the least science-conscious reader a grasp of the basic ideas involved. Darwin's face appeared on everything from T-shirts to coffee cups, along with images of finches, dinosaurs and other icons of evolution. Not all of this material was thoroughly in tune with the latest thinking – Rudolph Zallinger's much-copied image 'The Road to Homo Sapiens' represented an ape turning into a human as the figures walk across the page, retaining a linear model of the ascent of life. (Most derivatives are a cut-down version of the original, published in 1965 for a *Time-Life* volume on early humans.)[21] Whatever the level of comprehension, few could have escaped an appreciation of the jamboree's central purpose: to drive home the theory's impact on the way we think and shape our societies.[22] There has been nothing on this scale since 2009, but the subject remains as fascinating and as controversial to this day.

Evolutionism thus retains a significant role in the public imagination into the twenty-first century. The dinosaurs are still getting regular workouts on television and in popular books, while creationism and other anti-Darwinian campaigns remain active. The media involved in dissemination and debate have expanded in the age of digitalization and the internet. Evolution in the most general sense finds its way into websites on a whole range of topics beyond biology. It is widely appealed to in advertising, where its implication of inevitable change (usually assumed to be beneficial) finds ready applications. There are computer games that also feature transformations at least superficially reminiscent of the biological world. The possibilities of genetic modification of animals and plants has become hugely controversial, as has the potential for tinkering with the human genome, all reminiscent of the engagement of evolutionism with the eugenics movement. It is not always easy to convince people that there is not necessarily a single gene controlling each character, a legacy of earlier ideas that gained too great a hold on the imagination.

All of these issues are debated endlessly and often acrimoniously on social media platforms, and anyone putting up material on evolutionism is likely to face attacks from trolls. Fern Elsdon-Baker records that while

[21] Gowan Dawson notes that the image has its origins in the frontispiece to Huxley's *Man's Place in Nature*; see his 'A Monkey into a Man'. Dawson is working on a more extensive study of this icon.

[22] For a survey of the material generated by the 2009 event, including a vast array of images, see Schwarz, ed., *Streitfall Evolution*, part 7.

she was writing her book on Richard Dawkins, news of her criticism of his stance on religion elicited online abuse on the assumption that she herself had an anti-Darwinian agenda.[23] The transition to the age of social media obviously represents a major turning point in the way evolutionism and other controversial issues in science are debated. However, it also marks a point beyond which the author of this survey does not venture. At the age of seventy-nine I have decided that life really is too short to get involved in a new level of activity, so my ability to comment is in any case far too restricted.

[23] Elsdon-Baker, *The Selfish Genius*, p. 268.

Bibliography

References to ephemeral material in newspapers are provided in the footnotes. Where later editions are cited, the date of first publication is given in brackets.

Adams, Mark. *The Evolution of Theodosius Dobzhansky*. Princeton, NJ: Princeton University Press, 1994.

Adelman, Juliana. 'Evolution on Display: Irish Natural History and Darwinism at the Dublin Science and Art Museum'. *British Journal for the History of Science*, 38 (2005): 411–36.

Agassiz, Louis. *An Essay on Classification*. London: Longman, Brown, 1860.

Agassiz, Louis, and Augustus A. Gould. *Outlines of Comparative Physiology, Touching the Structure and Development of the Races of Animals, Living and Extinct*. Revised ed. London: H. G. Bohn, 1851.

Allen, Garand E. *Life Science in the Twentieth Century*. New York: Wiley, 1975.

Allen, Grant. 'Evolution'. *Cornhill Magazine*, 55 (1888): 34–47.

The Evolutionist at Large. London: Chatto and Windus, 1881. Reprinted New York: J. Fitzgerald (Humboldt Library), 1881.

Vignettes from Nature. New York: J. Fitzgerald (Humboldt Library), 1882.

Alvarez, Walter. *T. rex and the Crater of Doom*. Princeton, NJ: Princeton University Press, 1997.

Appel, Toby. *The Cuvier-Geoffroy Debate: French Biology in the Decades before Darwin*. Oxford: Oxford University Press, 1987.

Archibald, J. David. 'Edward Hitchcock's Pre-Darwinian (1840) "Tree of Life"'. *Journal of the History of Biology*, 42 (2009): 561–92.

Ardrey, Robert. *The Territorial Imperative: A Personal Inquiry into the Animal Origins of Property and Nations*. New York: Atheneum, 1966.

Argyll, George Douglas Campbell, 5th Duke of. 'On Variety as an Aim in Nature'. *Contemporary Review*, 17 (1871): 153–60.

The Reign of Law. 5th ed. London: Alexander Strahan, 1868.

Attenborough, David. *Life on Earth: A Natural History*. London: Collins/BBC, 1979.

Auffermann, Bårbel, and Gerd-Christian Weniger. 'Von Wilden Månnern und Frauen'. In Angela Schwarz, ed., *Streitfall Evolution: Eine Culturgeschichte*. Cologne: Böhlau Verlag, 2017, pp. 141–52.

Aveling, Edward B. *The Student's Darwin*. London: Progressive Publishing, 1881.

Bagehot, Walter. *Physics and Politics, or Thoughts on the Application of the Principles of 'Natural Selection' and 'Inheritance' to Political Society*. London: H. S. King, 1872.

Ballantyne, R. M. *The Gorilla Hunters: A Tale of the Wilds of Africa*. New ed. London: Nelson, 1897 [1861].

Bannister, Robert C. *Social Darwinism: Science and Myth in Anglo-American Social Thought*. Philadelphia: Temple University Press, 1979.

Barkan, Elazar. 'The Dynamics of Huxley's Views on Racism and Eugenics'. In Waters and Van Helden, eds., *Julian Huxley: Biologist and Statesman of Science*, pp. 231–7.

The Retreat of Scientific Racism: Changing Concepts of Race in Britain and the United States between the World Wars. Cambridge: Cambridge University Press, 1992.

Bartholomew, Michael. 'Huxley's Defence of Darwinism'. *Annals of Science*, 32 (1975): 525–35.

Barton, Ruth. 'John Tyndall, Pantheist: A Rereading of the Belfast Address'. *Osiris*, 3 (1987): 111–34.

'Just before *Nature*: The Purposes of Science and the Purposes of Popularization in Some English Popular Science Journals of the 1860s'. *Annals of Science*, 55 (1998): 1–33.

The X Club: Power and Authority in Victorian Science. Chicago: University of Chicago Press, 2018.

Bashford, Alison. *An Intimate History of Evolution: The Huxleys in Nature and Culture*. London: Allen Lane, 2022.

Bashford, Alison, and Philippa Levine, eds. *The Oxford Handbook of the History of Eugenics*. Oxford: Oxford University Press, 2010.

Bateson, William. *Problems of Genetics*. Reprinted New Haven, CT: Yale University Press, 1979 [1913].

Beddoe, John. *The Races of Britain: A Contribution to the Anthropology of Western Europe*. Reprinted London: Hutchinson, 1971 [1885].

Beecher, Henry Ward. *Evolution and Religion*. London: James Clarke; New York: Fords, Howard and Hulbert, 1885. 2 vols.

Beer, Gillian. *Darwin's Plots: Evolutionary Narrative in Darwin, George Eliot and Nineteenth-Century Fiction*. London: Routledge and Kegan Paul, 1983.

Behe, Michael. *Nature's Black Box: The Biochemical Challenge to Evolution*. New York: Simon and Schuster, 1996.

Belknap, Geoffrey. 'Natural History Periodicals and Changing Conceptions of the Naturalist Community, 1828–1865'. In Gowan Dawson and Bernard Lightman, eds., *Science Periodicals in Nineteenth-Century Britain*. Chicago: University of Chicago Press, 2020, pp. 172–204.

Belloc, Hilaire. *A Companion to Mr. Wells' Outline of History*. London: Sheed and Ward, 1926.

The Crisis of Our Civilization. London: Cassell, 1937.

Bellomy, Donald C. 'Social Darwinism Revisited'. *Perspectives in American History*, n.s., 1 (1984): 1–129.

Benton, Mike. *The Dinosaurs Rediscovered: How a Scientific Revolution Is Rewriting History*. London: Thames and Hudson, 2019.

Berkowitz, Carin, and Bernard Lightman, eds. *Science Museums in Transition: Cultures of Display in Nineteenth-Century Britain and America*. Philadelphia: University of Pennsylvania Press, 2017.

Betts, John. 'P. T. Barnum and the Popularization of Natural History'. *Journal of the History of Ideas*, 20 (1959): 353–68.

Bishop, Rebecca. 'Evolution im Zoo: Die Veranschaulichung einer Theories am lebenden Objekt'. In Angela Schwarz, ed., *Streitfall Evolution: Eine Kulturgeschichte*. Cologne: Böhlau, 2017, pp. 103–204.

Blinderman, Charles. *The Piltdown Inquest*. Buffalo, NY: Prometheus Books, 1986.

Boon, Timothy. *Films of Fact: A History of Science in Documentary Films and Television*. London: Science Museum, 2007.

Boule, Marcellin. *Fossil Men: Elements of Human Palaeontology*. Edinburgh: Oliver and Boyd, 1923.

Bowler, Peter J. 'American Paleontology and the Reception of Darwinism'. *Centaurus*, 66 (2017): 3–7.

'Bonnet and Buffon: Theories of Generation and the Problem of Species'. *Journal of the History of Biology*, 6 (1973): 259–81.

'E. W. MacBride's Lamarckian Eugenics and Its Implications for the Social Construction of Scientific Knowledge'. *Annals of Science*, 41 (1984): 245–60.

The Eclipse of Darwinism: Anti-Darwinian Evolution Theories in the Decades around 1900. Baltimore: Johns Hopkins University Press, 1983.

Evolution: The History of an Idea. 25th anniversary edition. Berkeley: University of California Press, 2009 [1984].

'Evolutionism in the Enlightenment'. *History of Science*, 12 (1974): 159–83.

Fossils and Progress: Paleontology and the Idea of Progressive Evolution in the Nineteenth Century. New York: Science History Publications, 1976.

'Herbert Spencer and "Evolution" – An Additional Note'. *Journal of the History of Ideas*, 36 (1975): 367.

'Herbert Spencer and Lamarckism'. In Mark Francis and Michael Taylor, eds., *Herbert Spencer: Legacies*. London: Routledge, 2015, pp. 203–21.

A History of the Future: Prophets of Progress from H. G. Wells to Isaac Asimov. Cambridge: Cambridge University Press, 2017.

The Invention of Progress: The Victorians and the Past. Oxford: Basil Blackwell, 1989.

Life's Splendid Drama: Evolutionary Biology and the Reconstruction of Life's Ancestry, 1860–1940. Chicago: University of Chicago Press, 1996.

The Mendelian Revolution: The Emergence of Hereditarian Concepts in Modern Science and Society. London: Athlone; Baltimore: Johns Hopkins University Press, 1989.

Monkey Trials and Gorilla Sermons: Evolution and Christianity from Darwin to Intelligent Design. Cambridge, MA: Harvard University Press, 2007.

The Non-Darwinian Revolution: Reinterpreting a Historical Myth. Baltimore: Johns Hopkins University Press, 1988.

Progress Unchained: Ideas of Evolution, Human History and the Future. Cambridge: Cambridge University Press, 2021.

Reconciling Science and Religion: The Debate in Early Twentieth-Century Britain. Chicago: University of Chicago Press, 2001.

Science for All: The Popularization of Science in Early Twentieth-Century Britain. Chicago: University of Chicago Press, 2009.

'The Specter of Darwinism: The Popular Image of Darwinism in Early Twentieth-Century Britain'. In Abigail Lustig, Robert J. Richards and Michael Ruse, eds., *Darwinian Heresies*. Cambridge: Cambridge University Press, 2004, pp. 48–68.

Theories of Human Evolution: A Century of Debate, 1844–1944. Baltimore: Johns Hopkins University Press; Oxford: Basil Blackwell, 1986.

Bowler, Peter J., and Iwan Rhys Morus. *Making Modern Science: A Historical Survey*. Chicago: University of Chicago Press, 2020 [2005]. 2nd ed.

Brinkman, Paul D. *The Second Jurassic Dinosaur Rush: Museums and Paleontology in America at the Turn of the Twentieth Century*. Chicago: University of Chicago Press, 2010.

Broca, Paul. *On the Phenomena of Hybridity in the Genus Homo*. London: For the Anthropological Society, Longmans, Green, 1864.

Broks, Peter. *Media Science before the Great War*. London: Macmillan, 1996.

'Science, Media and Culture: British Magazines, 1890–1914'. *Public Understanding of Science*, 2 (1993): 123–39.

Understanding Popular Science. Maidenhead: Open University Press, 2006.

Bronowski, Jacob. *The Ascent of Man*. London: Book Club Associates, 1973.

Broom, Robert. *The Coming of Man: Was It Accident or Design?* London: H. and F. Witherby, 1933.

Finding the Missing Link. London: Watts, 1950.

Brown, Andrew. *The Darwin Wars: How Stupid Genes Became Selfish Gods*. New York: Simon and Schuster, 1999.

Browne, Janet. *Charles Darwin: The Power of Place*. London: Jonathan Cape, 2002.

Charles Darwin: Voyaging. London: Jonathan Cape, 1995.

Bruce, R. V. *The Launching of Modern American Science, 1844–1876*. New York: Knopf, 1988.

Brusatte, Steve. *The Rise and Fall of the Dinosaurs*. London: Picador, 2019.

Buckland, William. *Geology and Mineralogy Considered with Respect to Natural Theology* [*Bridgewater Treatise*, vol. 6]. 2nd ed., London: Pickering, 1837 [1836]. 2 vols.

Buckley, Arabella B. *Winners in Life's Race, or The Great Backboned Family*. London: Macmillan, 1903 [1882].

Buffon, George Louis Leclerc, comte de. *Barr's Buffon: Buffon's Natural History Containing a Theory of the Earth, a General History of Man, of the Brute Creation, and of Vegetables and Minerals*. London: The Proprietor, 1797. 10 vols.

Barr's Buffon: Natural History of Birds, Fish, Insects and Reptiles. London: The Proprietor, 1798.

Natural History, General and Particular. Trans. William Smellie. 2nd ed., London: W. Strahan and T. Cadell, 1785. 18 vols.

Burchfield, Joe D. *Lord Kelvin and the Age of the Earth*. New York: Science History Publications, 1975.

Burnham, John C. *How Superstition Won and Science Lost: Popularizing Science and Medicine in the United States*. New Brunswick, NJ: Rutgers University Press, 1987.

Burroughs, Edgar Rice. *At the Earth's Core*. London: Methuen, 1923.
The Return of Tarzan. London: Methuen, 1920 [1918].
Butler, Samuel. *Erewhon, or Over the Range*. Reprinted Harmondsworth: Penguin, 1954 [1872].
Essays on Life, Art and Science. London: A. C. Fifield, 1908.
Evolution, Old and New, Or The Theories of Buffon, Dr. Erasmus Darwin, and Lamarck, as Compared with That of Mr. Charles Darwin. London: Hardwicke and Bogue, 1879.
Cain, A. J. *Animal Species and Their Evolution*. New York: Harper, 1960.
Cain, Joe. 'Julian Huxley, General Biology, and the London Zoo, 1935–42'. *Notes and Records of the Royal Society*, 64 (2010): 359–78.
'Publication History for *Evolution: A Journal of Natural History*'. *Archives of Natural History*, 30 (2003): 68–71 and 298.
Campbell, Reginald. *The New Theology*. London: Chapman and Hall, 1907.
Cantor, Geoffrey, Gowan Dawson, Graeme Gooday, Richard Noakes, Sally Shuttleworth and Jonathan R. Topham. *Science in the Nineteenth-Century Periodical*. Cambridge: Cambridge University Press, 2004.
Caplan, Arthur L., ed. *The Sociobiology Debate: Readings on Ethical and Scientific Issues*. New York: Harper and Row, 1978.
Carpenter, William Benjamin. *Nature and Man: Essays Scientific and Philosophical: With an Introductory Memoir by J. Estlin Carpenter*. New York: Appleton, 1889.
Cartmill, Matt. *A View to a Death in the Morning: Hunting and Nature through History*. Cambridge, MA: Harvard University Press, 1993.
Chamberlain, Alexander. *The Child: A Study in Evolution*. London: Walter Scott, 1900.
Chamberlin, J. Edward, and Sander L. Gilman, eds. *Degeneration: The Dark Side of History*. New York: Columbia University Press, 1985.
Chambers, Robert. *Explanations: A Sequel to 'Vestiges of the Natural History of Creation'*. 2nd ed. London: Churchill, 1846 [1845].
Vestiges of the Natural History of Creation. 5th ed. London: John Churchill, 1846; 11th ed., London: John Churchill, 1860 [1844].
Vestiges of the Natural History of Creation and Other Evolutionary Writings. Ed. James Secord. Chicago: University of Chicago Press, 1994.
Chesterton, G. K. *As I Was Saying: A Book of Essays*. London: Methuen, 1936.
Eugenics and Other Evils. London: Cassell, 1922.
The Everlasting Man. London: Hodder and Stoughton, 1925.
Clark, Constance Areson. *God – Or Gorilla: Images of Evolution in the Jazz Age*. Baltimore: Johns Hopkins University Press, 2008.
Clark, Ronald W. *The Huxleys*. London: Heinemann, 1968.
[Clark, Samuel]. *Peter Parley's Wonders of the Earth, Sea and Sky*. Ed. Rev. T. Wilson. Facsimile reprint ed. Aileen Fyfe. Bristol: Thoemmes, 2003 [1837].
Clark, W. E. Le Gros. *The Antecedents of Man*. New York: Harper, 1963.
Clodd, Edward. *The Childhood of the World: A Simple Account of Man in Early Times*. London: Macmillan, 1872.
A Primer of Evolution. New York: Longman, Green, 1895.
The Story of Creation: A Plain Account of Evolution. New ed. London: Longman, Green, 1909 [1888].

Cohen, Claudine. *The Fate of the Mammoth: Fossils, Myth and History*. Chicago: University of Chicago Press, 2002.

Colbert, Edwin H. *Men and Dinosaurs: The Search in Field and Laboratory*. Reprinted Harmondsworth: Penguin, 1971.

Conklin, E. Grant. *The Direction of Human Evolution*. London: Humphrey Milford/Oxford University Press, 1921.

Conrad, Joseph. *Heart of Darkness*. London: Penguin Classics, 2007 [1899].

Cooke, Bill. *A Rebel to His Last Breath: Joseph McCabe and Rationalism*. Amherst, NY: Prometheus Books, 2001.

Coombe, George. *The Constitution of Man Considered in Relation to External Objects*. 3rd ed. Edinburgh: John Anderson, Jr., 1835 [1828].

Cooter, Roger. *The Cultural Meaning of Popular Science: Phrenology and the Organization of Consent in Nineteenth-Century Britain*. Cambridge: Cambridge University Press, 1985.

Cope, Edward Drinker. 'The Genealogy of Man'. *American Naturalist*, 27 (1893): 321–35.

'On the Origin of Genera'. *Proceedings of the Academy of Natural Sciences, Philadelphia*, 20 (1867): 242–300.

The Origin of the Fittest. New York: Macmillan, 1887.

The Primary Factors of Organic Evolution. Chicago: Open Court, 1904 [1896].

The Theology of Evolution: A Lecture. Philadelphia: Arnold & Co., 1887.

Cornish, Caroline. 'Botany behind Glass: The Vegetable Kingdom on Display at Kew's Museum of Economic Botany'. In Carin Berkowitz and Bernard Lightman, eds., *Science Museums in Transition: Cultures of Display in Nineteenth-Century Britain and America*. Philadelphia: University of Pennsylvania Press, 2017, pp. 188–213.

Corsi, Pietro. *The Age of Lamarck: Evolutionary Theories in France, 1790–1830*. Berkeley: University of California Press, 1988.

"Edinburgh Lamarckians? The Authorship of Three Anonymous Papers (1826–1829)." *Journal of the History of Biology*, 54 (2021): 345–74.

Cravens, Hamilton. *The Triumph of Evolution: American Scientists and the Heredity-Environment Controversy, 1990–1941*. Philadelphia: University of Pennsylvania Press, 1978.

Cronin, Helena. *The Ant and the Peacock: Altruism and Sexual Selection from Darwin to the Present*. Cambridge: Cambridge University Press, 1991.

Crook, Paul. *Darwinism, War and History: The Debate over the Biology of War from the 'Origin of Species' to the First World War*. Cambridge: Cambridge University Press, 1994.

Cross, J. W., ed. *George Eliot's Life: As Related in Her Letters and Journals*. Edinburgh: William Blackwood, 1885. 3 vols.

Curry, Helen Anne. *Evolution Made to Order: Plant-Breeding and Technological Innovation in Twentieth-Century America*. Chicago: University of Chicago Press, 2016.

Curtis, Lewis P., Jr. *Apes and Angels: The Irishman in Victorian Caricature*. Newton Abbot: David and Charles, 1971.

Czerkas, Sylvia Massey, and Donald E. Glut. *Dinosaurs, Mammoths and Cavemen: The Art of Charles R. Knight*. New York: Dutton, 1982.

Däniken, Erich von. *According to the Evidence: My Proof of Man's Extraterrestrial Origins*. London: Souvenir Press, 1977.

Darwin, Charles. *The Correspondence of Charles Darwin*. Ed. Frederick Burkhardt et al. Cambridge: Cambridge University Press, 1985–2022. 30 vols.

The Descent of Man: and Selection in Relation to Sex. London: Murray, 1871, 2 vols.; 2nd edn., London: Murray, 1874.

On the Origin of Species by Means of Natural Selection. London: John Murray, 1859. Facsimile reprint with introduction by Ernst Mayr. Cambridge, MA: Harvard University Press, 1964.

The Voyage of the Beagle. Reprinted London: Wordsworth Editions, 1997.

Darwin, Erasmus. *The Botanic Garden: A Poem in Two Parts*. London: J. Johnson, 1791.

The Temple of Nature, or The Origin of Society. London: J. Johnson, 1803.

Zoonomia, or The Laws of Organic Life. London: J. Johnson, 1794–96, 2 vols. 3rd ed. Boston: Thomas and Andrews, 1803, 2 vols.

Darwin, Leonard. *What Is Eugenics?* London: Watts, 1928.

Davis, Edward B. 'Science and Religious Fundamentalism in the 1920s'. *American Scientist*, 93 (2003): 253–60.

Dawkins, Richard. 'Progress'. In Evelyn Fox Keller and Elizabeth A. Lloyd, eds., *Keywords in Evolutionary Biology*. Cambridge, MA: Harvard University Press, 1992, pp. 263–72.

The Selfish Gene. Oxford: Oxford University Press, 1976.

Dawkins, W. Boyd. *Cave Hunting: Researches on the Evidence of Caves Respecting the Early Inhabitants of Europe*. London: Macmillan, 1874.

Dawson, Gowan. 'The *Cornhill Magazine* and Shilling Monthlies in Mid-Victorian Britain'. In Geoffrey Cantor et al., eds., *Science in the Nineteenth-Century Periodical*. Cambridge: Cambridge University Press, 2004, pp. 123–50.

Darwin, Literature and Victorian Respectability. Cambridge: Cambridge University Press, 2007.

'"A Monkey into a Man": Thomas Henry Huxley, Benjamin Waterhouse Hawkins, and the Making of an Evolutionary Icon'. In Ian Hesketh, ed., *Imagining the Darwinian Revolution*. Pittsburgh: Pittsburgh University Press, 2022, pp. 80–99.

'The *Review of Reviews* and the New Journalism in Late Victorian Britain'. In Geoffrey Cantor et al., eds., *Science in the Nineteenth-Century Periodical*. Cambridge: Cambridge University Press, 2004, pp. 179–95.

Show Me the Bone: Reconstructing Prehistoric Monsters in Nineteenth-Century Britain and America. Chicago: University of Chicago Press, 2016.

Dawson, Gowan, Bernard Lightman, Sally Shuttleworth and Jonathan R. Topham, eds. *Science Periodicals in Nineteenth-Century Britain: Constructing Scientific Communities*. Chicago: University of Chicago Press, 2020.

De Beer, Gavin. *Charles Darwin: Evolution by Natural Selection*. London: Nelson, 1963.

A Handbook on Evolution. 2nd ed. London: British Museum (Natural History), 1959.

De Grazia, A. *The Velikovsky Affair*. New Hyde Park, NY: University Books, 1966.

De Quatrefages, Armand. *The Human Species*. London: Kegan Paul, 1879.

Dean, Dennis R. *Gideon Mantell and the Discovery of Dinosaurs*. Cambridge: Cambridge University Press, 1999.

Delisle, Richard G. 'Natural Selection as a Mere Auxiliary Hypothesis (sensu stricto I. Lakatos) in Charles Darwin's *Origin of Species*'. In Richard G. Delisle, ed., *Natural Selection: Revisiting Its Explanatory Role in Evolutionary Biology*. Cham: Springer, 2021, pp. 73–104.

Dennert, Eberhart, ed. *At the Deathbed of Darwinism*. Burlington, IA: German Literary Board, 1904.

Desmond, Adrian. *The Ape's Reflection*. London: Blond and Briggs, 1979.

Archetypes and Ancestors: Palaeontology in Victorian London, 1850–1875. London: Blond and Briggs, 1982; reprinted Chicago: University of Chicago Press, 1984.

'Artisan Resistance and Evolution in Britain, 1819–1848'. *Osiris*, 2nd series 3 (1987): 77–110.

The Hot-Blooded Dinosaurs: A Revolution in Palaeontology. Reprinted London: Futura, 1977 [1976].

Huxley: The Devil's Disciple. London: Michael Joseph, 1994.

Huxley: Evolution's High Priest. London: Michael Joseph, 1997.

The Politics of Evolution: Morphology, Medicine and Reform in Radical London. Chicago: University of Chicago Press, 1989.

Desmond, Adrian, and James R. Moore. *Darwin*. London: Michael Joseph, 1991.

Darwin's Sacred Cause: London: Allen Lane, 2009.

[D'Holbach, Baron]. *The System of Nature, or The Laws of the Moral and Physical World*. London: T. Davison and R. Helder, 1820–21. 3 vols.

Dickens, Charles. *Bleak House*. Revised ed. reprinted London: Hazel, Watson and Viney, n.d. [1868; original publication 1852–3].

Dixon, Thomas. *The Invention of Altruism: Making Moral Meanings in Victorian Britain*. Oxford: Oxford University Press, 2008.

Dobzhansky, Theodosius. *The Biology of Ultimate Concern*. Reprinted London: Fontana, 1971 [1967].

Genetics and the Origin of Species. New York: Columbia University Press, 1937.

Donald, Diana, and Jane Munro, eds. *Endless Forms: Charles Darwin, Natural Science and the Visual Arts*. New Haven, CT: Yale University Press, 2009.

Doyle, Sir Arthur Conan. *The Lost World*. London: John Murray, 1960 [1912].

Draper, John William. 'Dr. Draper's Lectures on Evolution: Its Origin, Progress, and Consequences'. *Popular Science Monthly*, 12 (1877–78): 175–92.

History of the Conflict between Science and Religion. London: Kegan Paul, 1882 [1874].

Drummond, Henry. *The Ascent of Man*. 13th ed. New York: James Pott, 1904 [1894].

Dunn, Leslie, and Theodosius Dobzhansky. *Heredity, Race and Society*. New York: New American Library, 1946.

Dupree, A. Hunter. *Asa Gray: 1810–1888*. Reprinted New York: Atheneum, 1968.

Eimer, Gustav Heinrich Theodor. *On Orthogenesis and the Impotence of Natural Selection in Species Formation*. Chicago: Open Court, 1898.

Organic Evolution as the Result of the Inheritance of Acquired Characters According to the Laws of Organic Growth. Trans. J. T. Cunningham. London: Macmillan, 1890.

Eiseley, Loren. *Darwin's Century: Evolution and the Men Who Discovered It.* New York: Doubleday, 1958.

Ellegård, Alvar. *Darwin and the General Reader: The Reception of Darwin's Theory of Evolution in the British Periodical Press, 1859–1871.* Reprinted Chicago: University of Chicago Press, 1990 [1958].

Elliott, Paul. 'Erasmus Darwin, Herbert Spencer, and the Origins of the Evolutionary Worldview in British Provincial Scientific Culture, 1770–1850'. *Isis* 94 (2003): 1–29.

Elsdon-Baker, Fern. *The Selfish Genius: How Richard Dawkins Rewrote Darwin's Legacy.* London: Icon, 2009.

Endersby, Jim. 'Mutant Utopias: Evening Primroses and Imagined Futures in Early Twentieth-Century America'. *Isis*, 104 (2013): 471–503.

Fara, Patricia. *Erasmus Darwin: Sex, Science and Serendipity.* Oxford: Oxford University Press, 2012.

Farber, Paul L. 'Buffon and the Problem of Species'. *Journal of the History of Biology*, 5 (1972): 259–84.

Fay, Margaret A. 'Did Marx Offer to Dedicate *Capital* to Darwin?' *Journal of the History of Ideas*, 39 (1978): 133–46.

Fichman, Martin. 'Ideological Factors in the Dissemination of Darwinism in England, 1860–1900'. In E. Mendelsohn, ed., *Transformation and Tradition in the Sciences: Essays in Honor of I. Bernard Cohen.* Cambridge: Cambridge University Press, 1984, pp. 471–84.

Figuier, Louis. *The World before the Deluge.* London: Chapman and Hall, 1865; new ed. London: Cassell, Petter and Galpin, 1872.

Finnegan, Diarmid. *The Voice of Science: British Scientists on the Lecture Circuit in Gilded Age America.* Pittsburgh: Pittsburgh University Press, 2021.

Fisher, Ronald Aylmer. *The Genetical Theory of Natural Selection.* Oxford: Clarendon Press, 1930.

Fiske, John. *The Destiny of Man Viewed in the Light of His Origin.* Boston: Houghton Mifflin, 1884.

Outlines of Cosmic Philosophy: Based on the Doctrine of Evolution. 18th ed. Boston: Houghton Mifflin, 1898 [1874]. 2 vols.

Fosdick, Harry Emerson. *Christianity and Progress.* London: Nisbett, 1922.

Fossey, Diane. *Gorillas in the Mist.* Harmondsworth: Penguin, 1983.

Foster, Roy F. *Paddy and Mr Punch: Connections in Irish and English History.* London: Allen Lane, 1993.

Francis, Mark. *Herbert Spencer and the Invention of Modern Life.* Durham, NC: Acumen, 2007.

Fyfe, Aileen. 'Reading Natural History at the British Museum and the *Pictorial Museum*'. In Fyfe and Lightman, eds., *Science in the Marketplace*, pp. 196–230.

Steam-Powered Knowledge: William Chambers and the Business of Publishing. Chicago: University of Chicago Press, 2012.

Fyfe, Aileen, and Bernard Lightman, eds. *Science in the Marketplace: Nineteenth-Century Sites and Experiences.* Chicago: University of Chicago Press, 2007.

Galton, Francis. *Hereditary Genius: An Inquiry into Its Laws and Consequences*. London: Macmillan, 1869.
Natural Inheritance. London: Macmillan, 1889.
Gayon, Jean. *Darwinism's Struggle for Survival: Heredity and the Hypothesis of Natural Selection*. Cambridge: Cambridge University Press, 1998.
Geddes, Patrick, and J. Arthur Thomson. *Evolution*. London: Williams and Norgate, 1911.
Gee, Henry. *In Search of Deep Time: Beyond the Fossil Record to a New History of Life*. New York: Free Press, 1999.
Gish, Duane T. *Evolution: The Fossils Say No!* San Diego: Creation Life Publishers, 1972.
Gissis, Snait B., and Eva Jablonka, eds. *Transformations of Lamarckism: From Subtle Fluids to Molecular Biology*. Cambridge, MA: MIT Press, 2011.
Glass, Bentley, Owsei Temkin and William Straus, Jr., eds. *Forerunners of Darwin: 1745–1859*. Baltimore: Johns Hopkins University Press, 1959.
Gliboff, Sander. 'The Case of Paul Kammerer: Evolution and Experimentation in the Early Twentieth Century'. *Journal of the History of Biology*, 39 (2006): 525–63.
Goldsmith, Oliver. *Oliver Goldsmith's History of the Natural World*. Reprinted London: Studio Editions, 1990.
Goodrich, Edwin S. *The Evolution of Living Organisms*. London: T. C. and E. C. Jack, 1912.
Living Organisms: An Account of Their Origin and Evolution. Oxford: Clarendon Press, 1914.
Gould, Stephen Jay, ed. *The Book of Life*. London: Ebury Hutchinson, 1993.
Bully for Brontosaurus. London: Vintage, 2001.
The Mismeasure of Man. New York: Norton, 1981.
Ontogeny and Phylogeny. Cambridge, MA: Harvard University Press, 1977.
Grant, Madison. *The Passing of the Great Race, or The Racial Basis of European History*. 4th ed., introduced by H. F. Osborn. London: George Bell, 1924 [1916].
Gray, Asa. *Darwiniana: Essays and Reviews pertaining to Darwinism*. New York: Appleton, 1871.
Grayson, Donald K. *The Establishment of Human Antiquity*. New York: Academic Press, 1983.
Greene, John C. *The Death of Adam: Evolution and Its Impact on Western Thought*. Ames: Iowa State University Press, 1959.
Greenslade, William, and Terrence Rogers, eds., *Grant Allen: Literature and Cultural Politics in the Fin de Siècle*. Aldershot: Ashgate, 2005.
Gregory, Jane, and Steve Miller. *Science in Public: Communication, Culture and Credibility*. Cambridge, MA: Perseus Publishing, 2000.
Gregory, William King. *Our Face from Fish to Man: A Portrait Gallery of Our Ancient Ancestors and Kinfolk, Together with a Concise History of Our Best Features*. New York: G. P. Putnam's Sons, 1929.
Gresswell, Albert, and George Gresswell. *The Wonderland of Evolution*. London: Field and Tuer; New York: Scribner and Welford, 1884.
Haar, Charles. 'E. L. Youmans: A Chapter in the Diffusion of Science in America'. *Journal of the History of Ideas*, 9 (1948): 193–213.

Haeckel, Ernst. *The Evolution of Man: A Popular Exposition of the Principal Points of Human Ontogeny and Phylogeny*. London: Kegan Paul, 1879. 2 vols.

The Evolution of Man: A Popular Scientific Study. Trans. Joseph McCabe. London: Watts, 1907. 2 vols. in one.

The History of Creation, or The Development of the Earth and Its Inhabitants by the Action of Natural Causes: A Popular Exposition of the Doctrine of Evolution in General and of That of Darwin, Goethe and Lamarck in Particular. Translation revised by E. Ray Lankester. New York: Appleton, 1883 [1876].

The Last Link: Our Present Knowledge of the Descent of Man. London: A. and C. Black, 1898.

The Pedigree of Man and Other Essays. Trans. Edward Aveling. London: Freethought Publishing, 1883.

The Riddle of the Universe at the Close of the Nineteenth Century. London: Watts, 1900.

The Wonders of Life: A Popular Study of Biological Philosophy. New York: Harper, 1905.

Haldane, J. B. S. *The Causes of Evolution*. London: Longmans, Green, 1932.

The Inequality of Man and Other Essays. London: Chatto and Windus, 1932.

Possible Worlds and Other Essays. Reprinted London: Chatto and Windus, 1930.

Hale, Piers J. 'Monkeys into Men and Men into Monkeys: Chance and Contingency in the Evolution of Man, Mind and Morals in Charles Kingsley's *Water Babies*'. *Journal of the History of Biology*, 46 (2013): 551–95.

'Of Mice and Men: Evolution and the Socialist Utopia: William Morris, H. G. Wells and George Bernard Shaw'. *Journal of the History of Biology*, 43 (2010): 17–66.

Political Descent: Malthus, Mutualism, and the Politics of Evolution in Victorian England. Chicago: University of Chicago Press, 2014.

'The Search for Purpose in a Post-Darwinian Universe: George Bernard Shaw, "Creative Evolution", and Shavian Eugenics: "The Dark Side of the Force"'. *History and Philosophy of the Life Sciences*, 28 (2006): 191–214.

Hall, Alex. *Evolution on British Television and Radio: Transmissions and Transmutations*. London: Palgrave Macmillan, 2021.

Hall, G. Stanley. *Adolescence: Its Psychology and Its Relation to Physiology, Anthropology, Sociology, Sex, Crime, Religion and Education*. New York: Appleton, 1904. 2 vols.

Haller, John S. *Outcasts from Evolution: Scientific Attitudes of Racial Inferiority, 1859–1900*. Urbana: University of Illinois Press, 1975.

Hammond, Michael. 'The Expulsion of the Neanderthals from Human Ancestry: Marcellin Boule and the Social Context of Scientific Research'. *Social Studies of Science*, 12 (1982): 1–36.

Haraway, Donna. *Primate Visions: Gender, Race, and Nature in the World of Modern Science*. New York: Routledge, 1989.

Harmer, S. F., and A. E. Shipley, eds. *The Cambridge Natural History*. Cambridge: Cambridge University Press, 1895–1909. 10 vols.

Harrison, Harry. *West of Eden*. New York: Bantam Books, 1984.

Hawkins, Mike. *Social Darwinism in European and American Thought, 1860–1945: Nature as Model, Nature as Threat*. Cambridge: Cambridge University Press, 1997.

Hawks, Ellison. *The Marvels and Mysteries of Science*. London: Odhams Press, 1938.

Heard, Gerald. *The Emergence of Man*. London: Jonathan Cape, 1931.

Hedeen, Stanley. *Big Bone Lick: The Cradle of American Paleontology*. Lexington: University Press of Kentucky, 2008.

Henslow, George. *The Origin of Floral Structures through Insect and Other Agencies*. London: Kegan Paul, 1888.

Herschel, Sir J. F. W. *Physical Geography: From the Encyclopaedia Britannica*. Edinburgh: A. and C. Black, 1862.

Hesketh, Ian, ed. *Imagining the Darwinian Revolution: Historical Narratives of Evolution from the Nineteenth Century to the Present*. Pittsburgh: Pittsburgh University Press, 2022.

Hird, Dennis. *A Picture Book of Evolution*. London: Watts, 1906. Revised ed., ed. C. M. Beadnell. London: Watts, 1934.

Hitchcock, Edward. *The Religion of Geology and Its Connected Sciences*. Boston: Phillips, Sampson, 1851.

Hodge, M. J. S. 'Lamarck's Science of Living Bodies'. *British Journal for the History of Science*, 5 (1971): 323–52.

'The Universal Gestation of Nature: Chambers' Vestiges and Explanations'. *Journal of the History of Biology*, 5 (1972): 127–52.

Hofstadter, Richard. *Social Darwinism in American Thought*. Revised ed. New York: George Braziller, 1959.

Hogben, Lancelot. 'The Present Status of the Evolutionary Hypothesis'. *Discovery*, 5 (1924): 11–15, 61–5 and 102–5.

Holmes, John. 'The Challenge of Evolution in Victorian Poetry'. In Lightman and Zon, eds., *Evolution and Victorian Culture*, pp. 39–63.

Hooker, Joseph Dalton. 'Address of the President'. In *Report of the Thirty-Eighth Meeting of the British Association for the Advancement of Science Held at Norwich [1868]*. London: John Murray, 1869, pp. lviii–lxxv.

Hooton, Earnest A. *Up from the Ape*. London: Allen and Unwin, 1931.

Hopwood, Nick. *Haeckel's Embryos: Images, Evolution and Fraud*. Chicago: University of Chicago Press, 2015.

Howson, Leslie. 'An Experiment with Science for the Nineteenth-Century Book Trade: The International Science Series'. *British Journal for the History of Science*, 33 (2000): 63–93.

Hull, David L. ed. *Darwin and His Critics: The Reception of Darwin's Theory of Evolution by the Scientific Community*. Cambridge, MA: Harvard University Press, 1973; reprinted Chicago: University of Chicago Press, 1985.

Science as a Process: An Evolutionary Account of the Social and Conceptual Development of Science. Chicago: University of Chicago Press, 1988.

Huskinson, Benjamin. *American Creationism, Creation Science, and Intelligent Design in the Evangelical Market*. London: Palgrave Macmillan, 2020.

Hutchinson, H. N. *Creatures of Other Days: Popular Studies in Palaeontology*. New ed. London: Chapman and Hall, 1895.

Extinct Monsters and Creatures of Other Days: A Popular Account of Some of the Larger Forms of Ancient Animal Life. New ed. London: Chapman and Hall, 1910 [1892].

ed. *Living Races of Mankind*. London: Hutchinson, 1905.
Huxley, Julian S. *Essays of a Biologist*. London: Chatto and Windus, 1923.
Evolution: The Modern Synthesis. London: Allen and Unwin, 1942.
ed. *The Humanist Frame*. London: Allen and Unwin, 1961.
Memories. New York: Harper and Row, 1970.
Memories II. London: Allen and Unwin, 1973.
Religion without Revelation. London: Ernest Benn, 1927.
The Stream of Life. London: Watts, 1926.
Huxley, Julian S., and A. C. Haddon, with A. M. Carr-Saunders. *We Europeans: A Survey of 'Racial' Problems*. Harmondsworth: Penguin, 1939.
Huxley, Leonard, ed. *The Life and Letters of Thomas Henry Huxley*. London: Macmillan, 1900. 2 vols.
Huxley, Thomas Henry. *Collected Essays*. London: Macmillan, 1895–6. 9 vols.
Evidence as to Man's Place in Nature. London: Williams and Norgate, 1863.
Hyman, Stanley E. *The Tangled Bank: Darwin, Marx, Fraser and Freud as Imaginative Writers*. New York: Athenaeum, 1962.
James, Frank A. L. 'An "Open Clash between Science and the Church"?: Wilberforce, Huxley and Hooker on Darwin at the British Association, Oxford, 1860'. In David M. Knight and Matthew D. Eddy, eds., *Science and Beliefs: From Natural Philosophy to Natural Science*. Aldershot: Ashgate, 2005, pp. 171–93.
James, Matthew J. *Collecting Evolution: The Galapagos Expedition That Vindicated Darwin*. Oxford: Oxford University Press, 2017.
Jenkins, Bill. *Evolution before Darwin: Theories of the Transmutation of Species in Edinburgh, 1804–1834*. Edinburgh: Edinburgh University Press, 2019.
Jevons, W. Stanley. 'Evolution and the Doctrine of Design'. *Popular Science Monthly*, 5 (1874): 98–100.
Johnson, Lawrence. 'The Chain of Species'. *Popular Science Monthly*, 5 (1874): 313–22 and 460–70.
Johnson, Philip E. *Darwin on Trial*. New York: Intervarsity Press and Regnery Gateway, 1991.
Jones, Elizabeth D. *Ancient DNA: The Making of a Celebrity Science*. New Haven, CT: Yale University Press, 2022.
Jones, Greta. *Social Darwinism and English Thought: The Interaction between Biological and Social Theory*. London: Harvester Press, 1980.
Jones, H. Festing. *Samuel Butler (1835–1902): A Memoir*. London: Macmillan, 1919. 2 vols.
Jordan, David Starr. *Foot-Notes to Evolution*. New York: Appleton, 1905 [1898].
Jordanova, Ludmilla. *Lamarck*. Oxford: Oxford University Press, 1984.
Kammerer, Paul. *The Inheritance of Acquired Characteristics*. New York: Boni and Liveright, 1924.
Rejuvenation and the Prolongation of Human Efficiency: Experiences with the Steinach Operation on Man and Animals. London: Methuen, 1924.
Karlinsky, Harry. *The Origin of Species by Charles Darwin*. Toronto: Davis Press, 1991.
Katz, Wendy R. *Rider Haggard and the Fiction of Empire: A Critical Study of British Imperial Fiction*. Cambridge: Cambridge University Press, 1987.

Keane, Augustus Henry. *The World's People: A Popular Account of Their Bodily Form and Mental Character, Beliefs, Traditions, Political and Social Institutions.* London: Hutchinson; New York: Putnam's, 1908.
Keith, Arthur. *Ancient Types of Man.* London: Harper, 1911.
The Antiquity of Man. London: Williams and Norgate, 1915.
Concerning Man's Origin: Being the President's Address at the Meeting of the British Association Held in Leeds on August 21 1927, and Recent Essays on Darwinian Subjects. London: Watts, 1927.
Darwinism and Its Critics. London: Watts, 1935.
Darwinism and What It Implies. London: Watts, 1928.
A New Theory of Human Evolution. London: Watts, 1948.
Kellogg, Vernon L. *Darwinism Today: A Discussion of Present-Day Scientific Criticism of the Darwinian Selection Theories, Together with a Brief Account of the Principal Other Proposed Auxiliary and Alternative Theories of Species Formation.* New York: Henry Holt; London: George Bell, 1908.
Headquarters Nights: A Record of Conversations and Experiences at the Headquarters of the German Army in France and Belgium. Boston: Atlantic Monthly Press, 1917.
Kempton, Miles. 'Commercial Television and Primate Ethology: Facial Expression between Granada and London Zoo'. *British Journal for the History of Science*, 56 (2023): 83–102.
Kevles, Daniel. *In the Name of Eugenics: Genetics and the Uses of Human Heredity.* Harmondsworth: Pelican, 1986.
Kidd, Benjamin. *Social Evolution.* New York: Macmillan, 1894.
King-Hele, Desmond. *Erasmus Darwin.* New York: Scribners, 1963.
Kingsland, Sharon E. 'The Battling Botanist: Daniel Trembly MacDougall, Mutation Theory, and the Rise of Experimental Evolutionary Biology in America, 1900–1912'. *Isis*, 82 (1991): 479–509.
Kingsley, Charles. *Scientific Lectures and Essays.* London: Macmillan, 1890.
The Water Babies. New ed. London: Macmillan 1889 [1862].
Kipling, Rudyard. *Collected Poems of Rudyard Kipling.* London: Wordsworth Poetry Library, 1994.
Just So Stories. London: Macmillan, 1902.
Kim. London: Macmillan, 1901.
Knox, Robert. *The Races of Men: A Philosophical Enquiry into the Influence of Race on the Destiny of Nations.* 2nd ed. London: Henry Renshaw, 1862 [1850].
Koestler, Arthur. *The Case of the Midwife Toad.* London: Hutchinson, 1971.
Kohler, Michèle, and Chris Kohler. 'The *Origin of Species* as a Book'. In Michael Ruse and Robert J. Richards, eds., *The Cambridge Companion to the 'Origin of Species'.* Cambridge: Cambridge University Press, 2009, pp. 333–51.
Kottler, Malcolm. 'Alfred Russel Wallace, the Origin of Man, and Spiritualism'. *Isis*, 65 (1974): 145–92.
Kropotkin, Peter. *Mutual Aid: A Factor of Evolution.* London: Heinemann, 1908 [1902].
Kuper, Adam. *Anthropologists and Anthropology: The British School 1922–1972.* London: Allan Lane, 1972.
'The Development of Lewis Henry Morgan's Evolutionism'. *Journal of the History of the Behavioral Sciences*, 21 (1985): 3–21.

Laats, Adam, and Harvey Siegal. *Teaching Evolution in a Creation Nation.* Chicago: University of Chicago Press, 2016.

Lack, David. *Darwin's Finches: An Essay on the General Biological Theory of Evolution.* Cambridge: Cambridge University Press, 1947.

LaFollette, Marcel Chotkowski. *Making Science Our Own: Public Images of Science, 1910–1955.* Chicago: University of Chicago Press, 1990.

Science on the Air: Popularizers and Personalities on Radio and Early Television. Chicago: University of Chicago Press, 2008.

Laing, Roisin. 'Victorian Autobiography, Child Study, and the Origins of Child Psychology'. In Bernard Lightman and Bennett Zon, eds., *Victorian Cultures and the Origin of Disciplines.* London: Routledge, 2021, pp. 244–72.

Lamarck, J. B. *Zoological Philosophy.* Trans. Hugh Elliot. London: Macmillan, 1914.

Landau, Misia. *Narratives of Human Evolution.* New Haven, CT: Yale University Press, 1990.

Lankester, E. Ray. *Degeneration: A Chapter in Darwinism.* London: Macmillan, 1880.

Extinct Animals. London: Constable, 1905.

'Inheritance of Acquired Characters'. *Nature*, 39 (1888–9): 485.

Science from an Easy Chair. London: Methuen, 1910.

ed., *Zoological Articles Contributed to the 'Encyclopaedia Britannica'.* London: A. and C. Black, 1891.

'Zoology'. In *Encyclopaedia Britannica*, 9th ed., vol 24. Edinburgh: A. and C. Black, 1888, pp. 799–820.

Largent, Mark A. 'The So-Called Eclipse of Darwinism'. In Joe Cain and Michael Ruse, eds., *Descended from Darwin: Insights into the History of Evolutionary Studies, 1900–1970.* Philadelphia: American Philosophical Society, 2009, pp. 3–21.

Larson, Edward J. *Evolution's Workshop: God and Scripture on the Galapagos Islands.* London: Penguin Books, 2001.

Summer for the Gods: The Scopes Trial and America's Continuing Debate over Science and Religion. New York: Basic Books, 1998.

Laublichler, Manfred D., and Jane Maienschein, eds. *From Embryology to Evo-Devo: A History of Developmental Evolutionism.* Cambridge, MA: MIT Press, 2007.

Leakey, Louis. *Adam's Ancestors: The Evolution of Man and His Culture.* 4th ed., reprinted New York: Harper, 1960.

LeConte, Joseph. *Evolution: Its Nature, Its Evidences, and Its Relation to Religious Thought.* 2nd ed. New York: Appleton, 1898.

Lester, Joe. *E. Ray Lankester and the Making of Modern British Biology.* Ed. Peter J. Bowler. Faringdon: British Society for the History of Science, 1995.

Levine, George. *Darwin and the Novelists: Patterns of Science in Victorian Fiction.* New Haven, CT: Yale University Press, 1998.

Lewes, George Henry. 'Studies in Animal Life'. *Cornhill Review*, 1 (1860): 61–74, 198–27, 283–95, 438–47, 598–607 and 682–90.

Lewin, Roger. *Bones of Contention: Controversies in the Search for Human Origins.* New York: Simon and Schuster, 1987.

Lightman, Bernard. 'Darwin and the Popularization of Evolution'. *Notes and Records of the Royal Society*, 64 (2009): 5–24.
Global Spencerism: The Communication and Appropriation of a British Evolutionist. Leiden: Brill, 2015.
'The "Greatest Living Philosopher" and the Useful Biologist: How Spencer and Darwin Viewed Each Other's Contributions to Evolutionary Theory'. In Ian Hesketh, ed., *Imagining the Darwinian Revolution.* Pittsburgh: University of Pittsburgh Press, 2022, pp. 37–57.
'The Many Lives of Charles Darwin: Biographies and the Definitive Evolutionist'. *Notes and Records of the Royal Society*, 64 (2010): 339–58.
'The Popularization of Evolution and Victorian Culture'. In Lightman and Zon, eds., Evolution and Victorian Culture, pp. 286–311.
'Spencer's American Disciples: Fiske, Youmans and the Appropriation of the System'. in Lightman, ed., *Global Spencerism: The Communication and Appropriation of a British Evolutionist.* Leiden: Brill, 2015, pp. 123–48.
Victorian Popularizers of Science: Designing Nature for New Audiences. Chicago: University of Chicago Press, 2007.
Victorian Science in Context. Chicago: University of Chicago Press, 1997.
Lightman, Bernard, and Bennett Zon, eds. *Evolution and Victorian Culture.* Cambridge: Cambridge University Press, 2014.
Livingstone, David N. *Adam's Ancestors: Race, Religion and the Politics of Human Origins.* Baltimore: Johns Hopkins University Press, 2008.
Darwin's Forgotten Defenders: The Encounter between Evangelical Theology and Evolutionary Thought. Edinburgh: Scottish Universities Press; Grand Rapids, MI: Eerdmans, 1987.
Livingstone, David N., D. G. Hart and Mark A. Noll, eds. *Evangelicalism and Science in Historical Perspective.* New York: Oxford University Press, 1999.
Lodge, Sir Oliver. *Evolution and Creation.* London: Hodder and Stoughton, 1926.
Making of Man: A Study in Evolution. London: Hodder and Stoughton, 1929 [1925].
Lorenz, Konrad. *On Aggression.* New York: Harcourt, Brace and World, 1966.
Lorimer, Douglas A. 'Science and the Secularization of Victorian Images of Race'. In Lightman, ed., *Victorian Science in Context*, pp. 212–35.
Lovejoy, Arthur O. *The Great Chain of Being: A Study in the History of an Idea.* Reprinted New York: Harper, 1960 [1936].
Lubbock, Sir John. *Pre-historic Times: As Illustrated by Ancient Remains and the Manners and Customs of Modern Savages.* London: Williams and Norgate, 1865.
Lull, Richard Swan. *Organic Evolution.* New York: Macmillan, 1917.
Lurie, Edward. *Louis Agassiz: A Life in Science.* Chicago: University of Chicago Press, 1960.
Lydekker, Richard. *Life and Rock: A Collection of Zoological and Geological Essays.* London: Universal Press, 1894.
Phases of Animal Life: Past and Present. London: Longman, Green, 1892.
Lyell, Charles. *The Geological Evidences of the Antiquity of Man: With Remarks on Theories of the Origin of Species by Variation.* London: John Murray, 1863.
Principles of Geology. First ed. [1830–3], reprinted Chicago: University of Chicago Press, 1991. 3 vols.

Lynskey, William. 'Goldsmith and the Chain of Being'. *Journal of the History of Ideas*, 6 (1945): 363–74.

Lyons, Sherrie L. 'Convincing Men They Are Monkeys'. In Alan P. Barr, ed., *Thomas Henry Huxley's Place in Science and Letters: Centenary Essays*. Athens: University of Georgia Press, 1997, pp. 94–118.

MacBride, Ernest William. *Evolution*. London: Ernest Benn, 1927.

An Introduction to the Study of Heredity. New York: Henry Holt, 1924.

MacGregor, Arthur. 'Exhibiting Evolutionism: Darwinism and Pseudo-Darwinism in Museum Practice after 1859'. *Journal of the History of Collections*, 21 (2009): 77–94.

MacLeod, Roy. 'Evolutionism, Internationalism, and Commercial Enterprise: The International Science Series, 1871–1910'. In A. J. Meadows, ed., *The Development of Science Publishing in Europe*. Amsterdam: Elsevier, 1980, pp. 63–93.

Macmillan, Robert. *The Origin of the World*. London: Watts, 1930.

Macnicol, John. 'Eugenics and the Campaign for Voluntary Sterilization in Britain between the Wars'. *Social History of Medicine*, 2 (1989): 147–69.

Maienschein, Jane. *Transforming Traditions in American Biology, 1880–1915*. Baltimore: Johns Hopkins University Press, 1991.

Mantell, Gideon. 'The Geological Age of Reptiles'. *Edinburgh New Philosophical Journal*, 11 (1831): 181–5.

The Medals of Creation, or First Lessons in Geology and in the Study of Organic Remains. London: Henry C. Bohn, 1844. 2 vols.

Petrifactions and Their Teachings, or A Handbook to the Gallery of Organic Remains of the British Museum. London: Henry C. Bohn, 1851.

Marsh, Othniel C. 'Introduction and Succession of Vertebrate Life in America'. *American Journal of Science*, 3rd series, 9 (1877): 337–78; also *Popular Science Monthly*, 12 (1877–78): 513–27 and 674–97; *Nature*, 16 (1877): 448–50, 470–2, 489–91.

Mason, Frances, ed. *Creation by Evolution: A Consensus of Present-Day Knowledge as Set Forth by Leading Authorities in Non-technical Language That All May Understand*. New York: Macmillan, 1928.

ed. *The Great Design: Order and Progress in Nature*. London: Duckworth, 1934.

Matthew, William Diller. *Climate and Evolution*. 2nd ed. New York: New York Academy of Sciences, 1939 [1914].

Dinosaurs: With Especial Reference to the American Museum Collections. New York: American Museum of Natural History, 1915.

The Evolution of the Horse: Guide Leaflet No. 9. New York: American Museum of Natural History, 1903.

'Life on Other Worlds'. *Science*, 53 (1921): 239–41.

Matthews, Shaler. *The Church and the Changing Order*. New York: Macmillan, 1922.

Mayo, Eileen. *The Story of Living Things*. London: Waverly Books, 1944.

Mayr, Ernst. *Systematics and the Origin of Species*. New York: Columbia University Press, 1942.

Mayr, Ernst, and William B. Provine, eds. *The Evolutionary Synthesis: Perspectives on the Unification of Biology*. Cambridge, MA: Harvard University Press, 1980.

McCabe, Joseph. *The ABC of Evolution*. London: Watts, 1920.
Evolution: A General Sketch from Nebula to Man. London: Milner, 1910.
The Riddle of the Universe Today. London: Watts, 1934.
The Story of a Religious Controversy. Boston: Stratford, 1929.
McLaughlin-Jenkins, Erin. 'Common Knowledge: Science and the Late Victorian Working-Class Press'. *History of Science*, 39 (2001): 445–65.
Medawar, Peter. *The Art of the Soluble: Creativity and Originality in Science*. Reprinted Harmondsworth: Penguin Books, 1969.
Mee, Arthur, ed. *The Children's Encyclopedia*. London: Educational Book Co., n. d. 10 vols.
Harmsworth Popular Science. London: Educational Book Co., 1914. 7 vols. [originally 43 parts, 1911–13].
Millar, Ronald. *The Piltdown Men: A Case of Archaeological Fraud*. Reprinted Frogmore: Paladin, 1974.
Millhauser, Milton. *Just before Darwin: Robert Chambers and 'Vestiges'*. Middletown, CT: Wesleyan University Press, 1959.
Mitchell, W. J. T. *The Last Dinosaur Book: The Life and Times of a Cultural Icon*. Chicago: University of Chicago Press, 1998.
Mivart, St. George Jackson. *On the Genesis of Species*. New York: Appleton, 1874.
Moore, James R. 'Herbert Spencer's Henchmen: The Evolution of Protestant Liberals in Late Nineteenth-Century America'. In John Durant, ed., *Darwinism and Divinity: Essays on Evolution and Religious Thought*. Oxford: Basil Blackwell, 1985, pp. 76–100.
The Post-Darwinian Controversies: A Study of the Protestant Struggle to Come to Terms with Darwin in Britain and America, 1870–1900. New York: Cambridge University Press, 1979.
Moorehead, Alan. *Darwin and the Beagle*. London: Hamish Hamilton, 1969.
Moran, Jeffrey P. *American Genesis: The Evolution Controversies from Scopes to Creation Science*. Oxford: Oxford University Press, 2012.
Morgan, Conwy Lloyd. *Emergent Evolution: The Gifford Lectures Delivered in the University of St. Andrews in the Year 1922*. London: Williams and Norgate, 1923.
Morgan, Lewis Henry. *Ancient Society, or Researches in the Lines of Human Progress from Savagery through Barbarism to Civilization*. Chicago: Charles H. Kerr, 1909 [1877].
Morgan, Thomas Hunt. *A Critique of the Theory of Evolution*. Princeton, NJ: Princeton University Press, 1916.
Evolution and Genetics. Princeton, NJ: Princeton University Press; London: Oxford University Press/Humphrey Milford, 1925.
The Scientific Basis of Evolution. London: Faber and Faber, 1932.
Morris, Desmond. *The Naked Ape*. London: Jonathan Cape, 1967.
Morss, J. R. *The Biologizing of Childhood: Developmental Psychology and the Darwinian Myth*. Hove: Laurence Erlbaum, 1990.
Müller-Wille, Staffan, and Hans-Jörg Rheinberger. *A Cultural History of Heredity*. Chicago: University of Chicago Press, 2012.
Muschinske, David. 'The Nonwhite as Child: G. Stanley Hall on the Education of Nonwhite Peoples'. *Journal of the History of the Behavioral Sciences*, 13 (1977): 328–36.

Nelkin, Dorothy. *The Creation Controversy: Science or Scripture in the Public Schools*. New York: Norton, 1983.

Science Textbook Controversies and the Politics of Equal Time. Cambridge, MA: MIT Press, 1977.

Selling Science: How the Press Covers Science and Technology. Revised ed., New York: W. H. Freeman, 1995.

Nieuwland, Ilja. *American Dinosaur Abroad: A Cultural History of Carnegie's Plaster Diplodocus*. Pittsburgh: University of Pittsburgh Press, 2019.

Noakes, Richard. '*Punch* and Comic Journalism in Mid-Victorian Britain'. In Geoffrey Cantor et al., eds., *Science in the Nineteenth-Century Periodical*. Cambridge: Cambridge University Press, 2004, pp. 91–122.

Nott, J. C., and G. R. Gliddon. *Indigenous Races of the Earth, or New Chapters of Ethnological Inquiry*. Philadelphia: Lippincott, 1857.

Types of Mankind, or Ethnological Researches. Philadelphia: Lippincott, Gambo, 1854.

Numbers, Ronald L. *The Creationists*. New ed. Cambridge, MA: Harvard University Press, 2006 [1992].

Darwinism Comes to America. Cambridge, MA: Harvard University Press, 1998.

Numbers, Ronald L., and John Stenhouse, eds. *Disseminating Darwinism: The Role of Place, Race, Religion and Gender*. Cambridge: Cambridge University Press, 1999.

Numbers, Ronald L., and Lester D. Stephens. 'Darwinism in the American South'. In Numbers and Stenhouse, eds., *Disseminating Darwinism*, pp. 123–44.

et al., eds. *Creationism in Twentieth-Century America*. New York: Goddard, 1994–5. 10 vols.

O'Connor, Ralph. *The Earth on Show: Fossils and the Politics of Popular Science, 1802–1856*. Chicago: University of Chicago Press, 2007.

Olby, Robert. *Origins of Mendelism*. 2nd ed. Chicago: University of Chicago Press, 1985.

Oldroyd, David R. *Darwinian Impacts: An Introduction to the Darwinian Revolution*. Milton Keynes: Open University Press, 1980.

Oldroyd, David R., and Ian Langham, eds. *The Wider Domain of Evolutionary Thought*. Dordrecht: D. Reidel, 1983.

O'Leary, Don. *Roman Catholicism and Modern Science: A History*. New York: Continuum, 2006.

Oleson, Alexandra, and Sanborn C. Brown, eds. *The Pursuit of Knowledge in the Early American Republic*. Baltimore: Johns Hopkins University Press, 1976.

Oppenheim, Janet. *The Other World: Spiritualism and Psychic Research in England, 1850–1914*. Cambridge: Cambridge University Press, 1995.

Osborn, Henry Fairfield. *The Age of Mammals in Europe, Asia and North America*. New York: Macmillan, 1910, reissued 1921.

Man Rises to Parnassus: Critical Epochs in the Prehistory of Man. 2nd ed. Princeton, NJ: Princeton University Press, 1928 [1927].

Men of the Old Stone Age: Their Environment, Life and Art. London: George Bell, 1916.

The Origin and Evolution of Life on the Theory of the Action, Reaction and Interaction of Energy. New York: Charles Scribner's Sons, 1917.

The Titanotheres of Ancient Wyoming, Dakota and Nebraska. Washington, DC: United States Geological Survey Monograph 55, 1924, 2 vols.

Ospovat, Dov. 'The Influence of Karl Ernst von Baer's Embryology, 1828–1859: A Reappraisal in Light of Richard Owen and William B. Carpenter's "Paleontological Application of von Baer's Law"'. *Journal of the History of Biology*, 9 (1976): 1–28.

Owen, Richard. *On the Nature of Limbs: A Discourse*. Preface by Brian K. Hall, introduction by Ron Amundson. Reprinted Chicago: University of Chicago Press, 2007 [1849].

Palaeontology, or A Systematic Summary of Extinct Animals and Their Geological Relations. Edinburgh: A. and C. Black, 1860.

'Report on British Fossil Reptiles, Part 2'. *Report ... of the British Association for the Advancement of Science ... for 1841*. London: John Murray, 1842, pp. 60–204.

Pandora, Kathleen. 'The Children's Republic of Science in the Antebellum Literature of Samuel Griswold Goodrich and Jacob Abbott'. *Osiris*, 24 (2009): 75–98.

Paradis, James G. 'Science and Satire in Victorian Culture'. In Bernard Lightman, ed., *Victorian Science in Context*. Chicago: University of Chicago Press, 1997, pp. 143–75.

Pavuk, Alexander. 'Biologist Edwin Grant Conklin and the Idea of a Religious Direction in Human Evolution in the Early 1920s'. *Annals of Science*, 74 (2017): 64–82.

Paylor, Susan. 'Edward B. Aveling: The People's Darwin'. *Endeavour*, 29 (2005): 66–71.

Pearson, Karl. *The Grammar of Science*. 2nd ed. London: A. and C. Black, 1900.

National Life from the Standpoint of Science. London: A. and C. Black, 1901.

Phillips, John. *Life on Earth: Its Origin and Succession*. London: Macmillan, 1860.

Philmus, Robert M., and David Y. Hughes, eds. *H. G. Wells: Early Writings in Science and Science Fiction*. Berkeley: University of California Press, 1975.

Pick, Daniel. *Faces of Degeneration: Aspects of European Cultural Disorder, 1848–1918*. Cambridge: Cambridge University Press, 1989.

Piel, Helen. 'Scientific Broadcasting as a Social Responsibility: John Maynard Smith on Radio, and Television in the 1960s and 1970s'. *British Journal for the History of Science*, 53 (2020): 89–108.

Pietsch, Theodore W. *Trees of Life: A Visual History of Evolution*. Baltimore: Johns Hopkins University Press, 2012.

Pitman, James Hall. *Goldsmith's Animated Nature: A Study of Goldsmith*. New Haven, CT: Yale University Press, 1924.

Poliquin, Rachel. *The Breathless Zoo: Taxidermy and the Cultures of Longing*. University Park: Penn State University Press, 2012.

Porter, Roy. 'Erasmus Darwin: Doctor of Evolution?' In James R. Moore, ed., *History, Humanity and Evolution*. Cambridge: Cambridge University Press, 1989, pp. 39–70.

Poulton, Edward B. *The Colours of Animals: Their Meaning and Use, Especially Considered in the Case of Insects*. New York: Appleton, 1890.

Praeger, Robert Lloyd. *Weeds: Simple Lessons for Children*. Cambridge: Cambridge University Press, 1913.

Price, Geoge McCready, and Joseph McCabe. *Is Evolution True?: A Verbatim Report of the Debate between George McCready Price and Joseph McCabe Held at the Queen's Hall, Langdon Place, London W on September 6 1925*. London: Watts, 1925.

Provine, William B. *The Origins of Theoretical Population Genetics*. Chicago: University of Chicago Press, 1971.

Punnett, Reginald. *Mendelism*. Cambridge: Bowes and Bowes, 1905; 5th ed. London: Macmillan, 1919.

Qureshi, Sadiah. 'Dramas of Development: Exhibitions and Evolution in Victorian Britain'. In Lightman and Zon, eds., *Evolution and Victorian Culture*, pp. 261–85.

Peoples on Parade: Exhibitions, Empire and Anthropology in Nineteenth-Century Britain. Chicago: University of Chicago Press, 2011.

Rader, Karen A., and Victoria E. M. Cain. *Life on Display: Revolutionizing U.S. Museums of Science and Natural History in the Twentieth Century*. Chicago: University of Chicago Press, 2014.

Radick, Gregory. *Disputed Inheritance: The Battle over Mendelism and the Future of Biology*. Chicago: University of Chicago Press, 2023.

The Simian Tongue: The Long Debate about Animal Language. Chicago: University of Chicago Press, 2007.

Rainger, Ronald. *An Agenda for Antiquity: Henry Fairfield Osborn and Vertebrate Paleontology at the American Museum of Natural History, 1890–1935*. Tuscaloosa: University of Alabama Press, 1991.

Raven, Charles. *The Creator Spirit: A Survey of Christian Doctrine in the Light of Biology, Psychology and Mysticism*. London: Martin Hopkinson, 1927.

Rea, Tom. *Bone Wars: The Excavation and Celebrity of Andrew Carnegie's Dinosaur*. Pittsburgh: University of Pittsburgh Press, 2021.

Reader, John. *Missing Links: The Hunt for Earliest Man*. London: Collins, 1981.

Reed, R. T. *Mr. Punch's Prehistoric Peeps*. London: Bradbury, Agnew, 1902.

Regal, Brian. *Henry Fairfield Osborn: Race and the Search for the Origins of Man*. Aldershot: Ashgate, 2002.

Regensburg, Jochen Petzold. 'How like Us Is That Ugly Brute, the Ape: Darwin's "Ape Theory" and Its Traces in Victorian Children's Magazines'. In Eckart Voigts, Barbara Schaff and Monika Pietrzak-Franger, eds., *Reflecting on Darwin*. Farnham: Ashgate, 2014, pp. 57–71.

Reingold, Nathan. 'Definitions and Speculations: The Professionalization of Science in Nineteenth-Century America'. In Alexandra Oleson and Sanborn C. Brown, eds., *The Pursuit of Knowledge in the Early American Republic*. Baltimore: Johns Hopkins University Press, 1976, pp. 33–69.

Renwick, Chris. 'The Practice of Spencerian Science: Patrick Geddes's Biosocial Program, 1876–1889'. *Isis*, 100 (2009): 36–57.

Richards, Morgan. 'Wild Visions'. In H. A. Curry, N. Jardine, J. A. Secord and E. C. Spary, eds., *Worlds of Natural History*. Cambridge: Cambridge University Press, 2018, pp. 518–32.

Richards, Robert J. *Darwin and the Emergence of Evolutionary Theories of Mind and Behavior*. Chicago: University of Chicago Press, 1987.

The Tragic Sense of Life: Ernst Haeckel and the Struggle for Evolutionary Thought. Chicago: University of Chicago Press, 2008.
Richards, Robert J., and Michael Ruse. *Debating Darwin.* Chicago: University of Chicago Press, 2016.
Richmond, Marsha L. 'The 1909 Darwin Celebration: Reexamining Evolution in the Light of Mendel, Mutation and Meiosis'. *Isis*, 97 (2006): 447–84.
Ridley, Matt. *Nature via Nurture: Genes, Experience and What Makes Us Human.* Reprinted London: Fourth Estate, 2011 [2003].
Rieppel, Lukas. *Assembling the Dinosaur: Fossil Hunters, Tycoons, and the Making of a Spectacle.* Cambridge, MA: Harvard University Press, 2019.
Ripley, William Z. *The Races of Europe.* London: Kegan Paul, 1900.
Ritvo, Lucile B. *Darwin's Influence on Freud.* New Haven, CT: Yale University Press, 1990.
Roberts, John H. 'Darwinism, American Protestant Thinkers, and the Puzzle of Motivation'. In Numbers and Stenhouse, eds. *Disseminating Darwinism*, pp. 145–72.
Roger, Jacques. *Buffon: A Life in Natural History.* Trans. Sarah L. Bonnefoi. Ithaca, NY: Cornell University Press, 1997.
The Life Sciences in Eighteenth-Century French Thought. Trans. Robert Ellrich. Stanford, CA: Stanford University Press, 1998.
Les sciences de la vie dans la pensée française du XVIIIe siècle. Paris: Armand Colin, 1963.
Romanes, George John. *The Scientific Evidences of Organic Evolution.* London: Macmillan, 1882.
Rudwick, Martin J. S. *Bursting the Limits of Time: The Reconstruction of Geohistory in the Age of Revolution.* Chicago: University of Chicago Press, 2005.
Earth's Deep History: How It Was Discovered and Why It Matters. Chicago: University of Chicago Press, 2014.
The Meaning of Fossils: Episodes in the History of Paleontology. 2nd ed. New York: Science History Publications, 1976 [1972].
Scenes from Deep Time: Early Pictorial Representations of the Prehistoric World. Chicago: University of Chicago Press, 1992.
Worlds before Adam: The Reconstruction of Geohistory in the Age of Reform. Chicago: University of Chicago Press, 2008.
Rupke, Nicolaas A. *Richard Owen: Victorian Naturalist.* New Haven, CT: Yale University Press, 1994.
Ruse, Michael. *Darwinism as Religion: What Literature Tells Us about Evolution.* Oxford: Oxford University Press, 2017.
Monad to Man: The Concept of Progress in Evolutionary Biology. Cambridge, MA: Harvard University Press, 1996.
Russell, A. Kingsley. *Science Fiction by the Rivals of H. G. Wells.* Secaucus, NJ: Castle Books, 1979.
Russett, Cynthia Eagle. *Darwin in America: The Intellectual Response, 1865–1912.* San Francisco: W. H. Freeman, 1978.
Saleeby, Caleb. *Evolution: The Master Key.* London: Harper Bros., 1906.
Heredity. London: T. C. and E. C. Jack, 1906.
Organic Evolution. London: T. C. and E. C. Jack, 1906.

Salisbury, Robert Cecil, 3rd Marquis. 'Presidential Address'. In *Report of the Meeting of the British Association for the Advancement of Science Held at Oxford, 1894*. London: John Murray, 1895, pp. 2–15.

Schwalbe, Gustav. 'The Descent of Man'. In A. C. Seward, ed., *Darwin and Modern Science*. Cambridge: Cambridge University Press, 1909, pp. 112–36.

Schwartz, Angela, ed. *Streitfall Evolution: Eine Kulturgeschichte*. Cologne: Böhlan, 2017.

Secord, James A. 'Edinburgh Lamarckians: Robert Jameson and Robert E. Grant'. *Journal of the History of Biology*, 24 (1991): 1–18.

Victorian Sensation: The Extraordinary Publication, Reception and Secret Authorship of Vestiges of the Natural History of Creation. Chicago: University of Chicago Press, 2001.

Sedgwick, Adam. 'Vestiges of the Natural History of Creation'. *Edinburgh Review*, 82 (1845): 1–85.

Segerstråle, Ullica. *Defenders of the Truth: The Battle for Science in the Sociobiology Debate and Beyond*. Oxford: Oxford University Press, 2000.

Semonin, Paul. *American Monster: How the Nation's First Prehistoric Creature Became a Symbol of National Identity*. New York: New York University Press, 2000.

Sera-Shriar, Efram. *The Making of British Anthropology, 1813–1871*. London: Pickering and Chatto, 2013.

Seward, A. C., ed. *Darwin and Modern Science: Essays in Commemoration of the Birth of Charles Darwin and the Fiftieth Anniversary of the Publication of the Origin of Species*. Cambridge: Cambridge University Press, 1909.

Shapiro, Adam R. *Trying Biology: The Scopes Trial, Textbooks, and the Antievolution Movement in American Schools*. Chicago: University of Chicago Press, 2013.

Shaw, George Bernard. *Back to Methuselah: A Metabiological Pentateuch*. London: Constable, 1921.

Sheets-Pyenson, Susan. 'Popular Science Periodicals in London and Paris: The Emergence of a Low Scientific Culture, 1820–1875'. *Annals of Science*, 42 (1985): 549–72.

Shuttleworth, Sally. *The Mind of the Child: Child Development in Literature, Science, and Medicine*. Oxford: Oxford University Press, 2010.

Simpson, George Gaylord. *The Meaning of Evolution*. New Haven, CT: Yale University Press, 1949.

Tempo and Mode in Evolution. New York: Columbia University Press, 1944.

This View of Life: The World of an Evolutionist. New York: Harcourt, Brace, and World, 1963.

Skelton, Matthew. 'The Paratext of Everything: The Constructing and Marketing of H. G. Wells' *The Outline of History*'. *Book History*, 4 (2000): 237–75.

Sloan, Phillip R. 'The Buffon-Linnaeus Controversy'. *Isis*, 67 (1976): 356–75.

'Darwin's Invertebrate Program, 1826–1836'. In David Kohn, ed., *The Darwinian Heritage: A Centennial Retrospect*. Princeton, NJ: Princeton University Press, 1985, pp. 71–120.

Smith, Grafton Elliot. *The Evolution of Man: Essays*. 2nd ed. London: Humphrey Milton and Oxford University Press, 1924.

Smith, John Maynard, ed. *Evolution Now: A Century after Darwin*. London: Nature/Macmillan, 1982.
The Theory of Evolution. 2nd ed. Harmondsworth: Penguin, 1966 [1958].
Smocovitis, Vassiliki Betty. 'The 1959 Darwin Centennial Celebration in America'. *Osiris*, 14 (1999): 274–323.
Unifying Biology: The Evolutionary Synthesis and Evolutionary Biology. Princeton, NJ: Princeton University Press, 1996.
Snell, Susan, and Polly Parry, eds. *Museum through the Lens: Photographs from 1880 to 1950*. London: Natural History Museum, 2009.
Sollas, William Johnson. *Ancient Hunters and Their Modern Representatives*. London: Macmillan, 1911.
Sommer, Marianne. 'The Neanderthals'. In Brian Regal, ed., *Icons of Evolution: An Encyclopedia of People, Evidence, and Controversies*. Westport, CT: Greenwood Press, 2008, 2 vols., I: 139–66.
Southwell, Charles, and William Chiltern. 'Theory of Regular Gradation'. *The Oracle of Reason*, 1 and 2 (1841–43), 48 parts. Full text at www.victorianweb.org/victorian/science.darwin.
Spencer, Herbert. *An Autobiography*. London: Williams and Norgate, 1904. 2 vols.
'The Development Hypothesis'. *The Leader*, 20 March 1852: 280.
Education: Intellectual, Moral and Physical. Reprinted New York: A. L. Burt, n.d.
Essays Scientific, Political and Speculative. London: Williams and Norgate, 1883. 3 vols.
'The Factors of Organic Evolution'. *Popular Science Monthly*, 29 (1886): 54–63 and 192–203.
The Factors of Organic Evolution. London: Williams and Norgate, 1886.
First Principles. 5th ed. London: Williams and Norgate, 1898.
Illustrations of Universal Progress: A Series of Discussions. New York: Appleton, 1864.
'The Inadequacy of Natural Selection'. *Contemporary Review*, 63 (1893): 153–66 and 439–56.
The Man versus the State. Harmondsworth: Penguin Books, 1969 [1884].
The Principles of Biology. Vol. 1, London: Williams and Norgate, 1864; vol. 2, 1867.
The Principles of Psychology. London: Longmans, Brown, 1855.
'Professor Weismann's Theories'. *Contemporary Review*, 63 (1893): 743–60.
'A Rejoinder to Professor Weismann'. *Contemporary Review*, 64 (1893): 893–912.
The Study of Sociology. 13th ed. London: Kegan Paul, 1887 [1873].
Stanton, William. *The Leopard's Spots: Scientific Attitudes toward Race in America, 1815–59*. Chicago: University of Chicago Press, 1960.
Stepan, Nancy. *The Idea of Race in Science: Great Britain, 1800–1960*. London: Macmillan Press, 1982.
Stephens, Lester D. *Joseph LeConte: Gentle Prophet of Evolution*. Baton Rouge: Louisiana State University Press, 1982.
Stocking, George W., Jr. *Race, Culture and Evolution: Essays in the History of Anthropology*. New York: Free Press, 1968.

Victorian Anthropology. New York: Free Press, 1987.
Sulloway, Frank J. *Freud, Biologist of the Mind: Beyond the Psychoanalytic Legend.* London: Burnett Books, 1979.
Sully, James. *Studies of Childhood.* New ed., London: Longmans, Green, 1903 [1895].
Swinton, W. E. *The Dinosaurs: A Short History of a Great Group of Extinct Reptiles.* London: T. Murby, 1934.
Teilhard de Chardin, Pierre. *The Phenomenon of Man.* Introduced by Julian Huxley. Reprinted London: Fontana, 1965.
Tennyson, Alfred. *In Memoriam.* Ed. R. H. Ross. New York: Norton, 1973.
Theunissen, Bert. *Eugene Dubois and the Ape-Man from Java: The History of the First 'Missing Link' and Its Discoverer.* Dordrecht: Kluwer, 1989.
Thomson, J. Arthur. *The Gospel of Evolution.* London: Newnes, n.d.
'The Influence of Darwinism on Thought and Life'. In F. S. Marvin, ed., *Science and Civilization.* Oxford: Oxford University Press; London: Humphrey Milford, 1923, pp. 203–20.
ed. *The Outline of Science: A Plain Story Simply Told.* London: Waverly Books, 1922. 2 vols.
Thomson, Keith. *The Legacy of the Mastodon: The Golden Age of Fossils in America.* New Haven, CT: Yale University Press, 2008.
Tobey, Ronald C. *The American Ideology of National Science, 1919–1930.* Pittsburgh: University of Pittsburgh Press, 1981.
Topham, Jonathan. 'Beyond the "Common Context": The Production and Reading of the Bridgewater Treatises'. *Isis*, 89, no. 1 (1998): 233–62.
'Science and Popular Education in the 1830s: The Role of the Bridgewater Treatises'. *British Journal for the History of Science*, 25 (1992): 397–430.
Trautmann, Thomas R. *Lewis Henry Morgan and the Invention of Kinship.* Berkeley: University of California Press, 1987.
Turner, Frank Miller. *Between Science and Religion: The Reaction to Scientific Naturalism in Late Victorian England.* New Haven, CT: Yale University Press, 1974.
Tylor, Edward B. *Anthropology: An Introduction to the Study of Man and Civilization.* Reprinted London: Macmillan, 1913 [1881].
Researches into the Early History of Mankind. London: John Murray, 1865.
Tyndall, John. 'Address of the President'. In *Report of the 44th Meeting of the British Association for the Advancement of Science Held in Belfast in August 1874.* London: John Murray, 1875, pp. lxvi–xcvii.
Fragments of Science. New York: Collier, 1902. 2 vols.
Uglow, Jenny. *The Lunar Men: The Friends Who Made the Future.* London: Faber and Faber, 2002.
Ulett, Mark A. 'Making the Case for Orthogenesis: The Popularization of Definitely-Directed Evolution (1890–1926)'. *Studies in the History and Philosophy of Biological and Biomedical Sciences*, 45 (2014): 124–32.
Van Riper, A. Bowdoin. *Men among the Mammoths: Victorian Science and the Discovery of Human Prehistory.* Chicago: University of Chicago Press, 1993.
Van Wyhe, John. *Phrenology and the Origins of Victorian Scientific Naturalism.* Aldershot: Ashgate, 2004.

'Why There Was No "Darwin's Bulldog"'. *The Linnaean*, 35 (2019): 26–30.
Velikovsky, Immanuel. *Earth in Upheaval*. London: Abacus, 1973 [1956].
Worlds in Collision. London: Abacus, 1972 [1950].
Vogt, Carl. *Lectures on Man: His Place in Creation and in the History of the Earth*. London: For the Anthropological Society, Longmans, Green, 1864.
Wallace, Alfred Russel. *Darwinism: An Exposition of the Theory of Natural Selection*. London: Macmillan, 1890.
The Malay Archipelago: The Land of the Orang-Utan and the Bird of Paradise. London: Macmillan, 1869, 2 vols.
My Life: A Record of Events and Opinions. New York: Dodd, Mead, 1905. 2 vols.
On Natural Selection. 2nd ed., London: Macmillan, 1871.
The World of Life: A Manifestation of Creative Power, Directive Mind and Ultimate Purpose. London: G. Bell, 1911.
Wallace, David Rains. *Beasts of Eden: Walking Whales, Dawn Horses and Other Enigmas of Mammalian Evolution*. Berkeley: University of California Press, 2004.
The Bonehunters' Revenge: Dinosaurs, Greed, and the Greatest Scientific Feud of the Gilded Age. Boston: Houghton Mifflin, 1999.
Waters, C. Kenneth, and Albert Van Helden, eds. *Julian Huxley: Biologist and Statesman of Science*. Houston: Rice University Press, 1992.
Wedgwood, Julia. 'Sir Charles Lyell on the Antiquity of Man'. *Macmillan's Magazine*, 7 (1863): 476–87.
Weiner, J. S. *The Piltdown Forgery*. London: Oxford University Press, 1955.
Weiner, Jonathan. *The Beak of the Finch: A Story of Evolution in Our Time*. New York: Alfred A. Knopf, 1994.
Weismann, August. 'The All-Sufficiency of Natural Selection'. *Contemporary Review*, 64 (1893): 309–38 and 596–610.
The Germ Plasm: A Theory of Heredity. London: Scott; New York: Scribners, 1893.
Wells, Herbert George. *Experiment in Autobiography: Discoveries and Conclusions of a Very Ordinary Brain (since 1866)*. Reprinted London: Faber, 1984. 2 vols.
Mr. Belloc Objects to the 'Outline of History'. London: Watts, 1926.
The Outline of History: Being a Plain History of Life and Mankind. London: Newnes, 1920. 2 vols. Definitive ed., London: Cassell, 1924.
'The Time Machine'. In *The Short Stories of H. G. Wells*. London: Benn, 1927, pp. 9–103 [1895].
Wells, H. G., Julian Huxley and G. P. Wells. *The Science of Life*. London: Cassell, 1931.
Werth, Barry. *Banquet at Delmonico's: The Gilded Age and the Triumph of Evolution in America*. Chicago: University of Chicago Press, 2009.
Whitcomb, J. C., and H. M. Morris. *The Genesis Flood*. Nutley, NJ: Presbyterian and Reformed Publishing, 1961.
Whyte, Adam Gowans. *The Wonder World We Live In*. New York: Knopf, 1921; London: Watts, 1927.
Wilberforce, Samuel. "On the Origin of Species." *Quarterly Review*, 108 (1860): 225–64.

Wilson, Edward O. *Sociobiology: The New Synthesis*. Cambridge, MA: Harvard University Press, 1975.

Woodward, Sir Arthur Smith. *A Guide to the Fossil Remains of Man in the Department of Geology and Palaeontology in the British Museum (Natural History)*. London: British Museum (Natural History), 1915.

'President's Address: Geological Section'. In *Report of the British Association for the Advancement of Science Meeting, 1909*. London: John Murray 1910, pp. 462–71.

Wright, Chauncey. *Darwinism: Being an Examination of Mr. St. George Mivart's 'Genesis of Species' (1871), with an Appendix on Final Causes*. London: John Murray, 1871.

'Spencer's Biology'. *The Nation*, 2 (1866): 729–5.

Wright, Sewall. *Evolution: Selected Papers*. Ed. William B. Provine. Chicago: University of Chicago Press, 1986.

Young, Robert M. *Darwin's Metaphor: Nature's Place in Victorian Culture*. Cambridge: Cambridge University Press, 1985.

Mind, Brain and Adaptation in the Nineteenth Century: Cerebral Localization and Its Biological Context from Gall to Ferrier. Oxford: Clarendon Press, 1970.

Index

acquired characteristics, *See* Lamarckism
adaptation
  and Darwinism, 16, 19, 75, 79, 251
  and Lamarckism, 16, 25, 54
  rejection of, 156, 244, 249
adaptive radiation, 21, 34, 54, 75, 224, 227, 236
Agassiz, Louis, 40–1, 53, 68, 79, 80, 88
aggression, 228, 242–4, 247
All the Year Round, 64
Allen, Garland, 138
Allen, Grant, 12, 14, 22, 104, 107, 125, 135
Altruism, 78, 245–7
Alvarez, Louis, 238
Alvarez, Walter, 238
amateur scientists, 11, 26, 46, 57, 138
America, *See* United States
American Association for the Advancement of Science, 67, 108, 121, 248
American Journal of Science, 69, 121
American Naturalist, 134
American school of neo-Lamarckism, 81, 133–4, 156
ancestral inheritance, law of, 162, 213
Andrews, Roy Chapman, 196
Anglican Church, 151
anniversaries, celebrations of, 144, 158, 168, 192, 225, 236, 255
Ansted, David, 64
Anthropological Society, 89
anthropology, 85, 99, 199–202, 213, 216, *See also* Human race
apes, 52, 62, 71, 85, 90, 188–91, 240–4, *See also* Chimpanzees; Gorillas; Orangutans
archaeology, 85, 91, 191
Archaeopteryx, 73, 120, 127
Ardrey, Robert, 228, 242
Argyll, George Douglas Campbell, 5th Duke of, 79, 92, 132
artificial selection, 17, 58, 157, 212, 214, 244
Ascent of Man, The, TV series, 236
atheism, 27, 31, 107, 136, 234, 247, *See also* Materialism
Athenaeum, 66
Atlantic Monthly, 69, 107
Atlas, newspaper, 52
Attenborough, David, 220, 233, 236
Auel, Jean, 241
Australopithecines, 196, 227, 240
authors, 26, 218, 242, *See also* Science correspondents; Science writers
  amateur experts as, 57
  books, 36, 105–10, 129
  magazines and newspapers, 174, 202
  professional scientists as, 105, 140, 152, 234
Aveling, Edward, 107, 136

Babbage, Charles, 50
Baer, Karl Ernst von, 41
Bagehot, Walter, 103, 206, 207
Bakker, Robert, 237
Balantyne, R. M., 94
Baldwin effect, *See* Organic selection
Bannister, Robert, 204
Barnes, E. W., 149, 150
Barnum, P. T., 87, 90
Barr, James Smith, 28
Bates, Henry Walter, 52, 70
Bateson, William, 118, 145, 155, 157, 159, 160, 213
BBC, 11, 166, 219, 225, 233
Beadnell, C. M., 164
Beagle, voyage of, 54
Beddoes, Thomas, 199
Beecher, Henry Ward, 131, 151, 205
Beer, Gillian, 12, 109
Behe, Michael, 255
Belloc, Hilaire, 149, 165, 221
Bennett, James Gordon Jr., 170
Bergson, Henri, 152, 166, 167, 182
Besant, Annie, 136

bestsellers, 44, 52, 61, 222, 242, 250
biblical flood, 20, 147, 236, 252
biogeography, 59, 67, 206, 221, 236
biometry, 161, 215, 217
bipedalism, 100, 169, 189–90, 192, 196, 226
birds, evolution of, 73–4, 168, 235, 237, 239
Black, Davidson, 196
blending heredity, 70
Boas, Franz, 202, 216
books, 7, 105–6, 109–10, 225, *See also* Bestsellers; names of individual authors
  children's, 35, 129, 164, 237
  literature for education, 140–1
  paperbacks, 111, 141, 163, 224, 233, 246
  pricing of, 8–9, 26, 51, 52, 56, 111, 126, 135, 164
  series, 141, 154, 165
botany, 10, 67, 134, 158
Boule, Marcellin, 193
Bowen, Francis, 53, 69
Bradlaugh, Charles, 136
brains, 43–4, 91, 96, 123, 189–91, 226, 227, 228, *See also* Intelligence; Mind; Phrenology
Brains Trust, radio programme, 224
breeders, influence of, 141
Brewster, David, 53
Bridgewater Treatise, 34, 39
Brinkman, Paul D., 171
Brinton, Daniel, 200
Britain, reaction to *Origin of Species*, 61–7
British Association for the Advancement of Science, 6, 35, 66, 67, 108, 124, 144
British Broadcasting Corporation, 11, 166, 219, 225, 233
broadcasters, 218, 224, 233
Broderip, William, 35
Bronowski, Jacob, 233, 236
Brooks, William Keith, 118
Broom, Robert, 182, 196, 227, 230
Brown, Andrew, 246
Brown, John, 68
Bryan, William Jennings, 148, 216
Buckland, William, 34, 38, 39
Buckley, Arabella, 105, 129–30, 171
Buffon, Georges Louis Leclerc, comte de, 27, 32
Bulwer-Lytton, Sir Henry, 109
Burnham, John C., 141
Burroughs, Edgar Rice, 174, 209
Butler, Samuel, 12, 76, 106, 130, 135, 142, 208
Cain, A. J., 225
Campbell, Reginald, 150
capitalism, 3, 21, 76, 102–3, 131, 203–5, 207
Carnegie, Andrew, 168, 172, 176, 205
Carpenter, William Benjamin, 41, 46, 54, 63, 132
cartoons, 71, 90, 97, 101, 172, 189, 200, 202, 226
catastrophism, 20, 39, 46, 72, 176, 179, 234, 238
Catholicism, 148, 149
Celts, 199
chain of being, 14, 16, 29, 34, 86, 114, 220, 236
Chamber's Encyclopaedia, 132
Chamberlain, Alexander, 209, 211
Chambers, Robert, 6, 26, 43, 46, 48–55, 64, 78, 85–7
chance, in evolution, 17, 18
Chaplin, Charlie, 142
Chardin, Pierre Teilhard de, 149
Chesterton, G. K., 12, 140, 141, 149
children, 10, 102, 117, 207, 209
  books for, 35, 129, 164, 237
Chiltern, William, 46, 53
Chimpanzees, 87, 242, 243
Chomsky, Noam, 242
Christianity, *See also* Anglican Church; Creationism, modern; Design, argument from; Evangelicalism; Roman Catholicism; Theistic evolution
  Darwinism and, 4, 59, 68–9, 93, 149–50
  evolution and, 6, 10, 18, 52, 77–83, 117, 123, 129–31, 230–1
  geology and, 147, 252
chromosomes, 158
cinema, 5, 10, 142, 174, 243
circulation figures, magazines and newspapers, 9, 64
cladism, 239
Clark, Samuel, 35
Clark, W. E. Le Gros, 227
Cleever, George B., 53
climate change, 179, 240
Clodd, Edward, 105, 107, 108, 124, 126–7
Cockburn, William, 52
Combe, George, 43
comparative anatomy, 117, 136, 139, *See also* Morphology
competition, 76, 102–3, 186, 204–7
Compsognathus, 73, 168
Conklin, Edward Grant, 151, 182
Conquest, 149, 221

Conrad, Joseph, 23, 110
Contemporary Review, 132, 144, 204
continuity, 41, 105, 128
cooperation, 204, 245
Cope, Edward Drinker, 81, 129, 133–4, 156, 168, 170, 192, 200
Cornhill Magazine, 64, 104, 107
creation, 18, 25, 32, 35, 50, 72, 251, 253, 255, *See also* Genesis, book of
creationism, modern, 20, 139–40, 143, 146–50, 236, 251–5
creative evolution, 80, 141, 152, 167, 182, 222
culture, 3, 92, 102, 198, *See also* Society, evolution of
Cuvier, Georges, 32, 45

Daily Express, 155
Daily Graphic, 120
Daily Mail, 140
Daily Telegraph, 140, 164
Däniken, Erich von, 252
Darrow, Clarence, 148
Dart, Raymond, 196, 227, 242
Darwin and the General Reader (Ellegård), 5, 7
Darwin, Charles Robert, 10, 39, 45, 54, 74, 104, *See also Descent of Man*; *Origin of Species*
  anniversary celebrations, 144, 158, 192, 225, 236, 255
Darwin, Erasmus, 29, 42
Darwin, Leonard, 214
Darwinian Revolution, 3–5, 7, 56
Darwinism, 2, 7, 12–15, 18, 22, *See also* Social Darwinism
  arguments against, 12, 62, 68, 78–83, 139, 142–6, 152
  arguments for, 62, 68, 69–74, 160–6
  modern, 21, 217–32, 244–51
  progress, 110–20, 160, 228, 231
  Spencer's philosophy, 74–8, 124–8
Davenport, Charles, 215
Davies, J. Barnard, 99
Davison, Thomas, 31
Dawkins, Richard, 220, 234, 235, 245, 247, 250, 257
Dawkins, W. Boyd, 199
De Beer, Gavin, 217, 225
degeneration, 21, 75, 119, 133, 209–11, 215
Descent of Man (Charles Darwin), 7, 56, 73, 74, 84, 99–102
design, argument from, 18, 32, 78–83, 132, 186, 252, *See also* Laws of nature
Desmond, Adrian, 5, 44, 237
determinism, *See* Evolution:predetermined
development, as model for evolution, 20, 40–1, 45, 59, 81, 110–20, *See also* Evolution:predetermined; Ontogeny; Recapitulation theory
D'Holbach, Baron, 31, 42, 44
Dickens, Charles, 37, 64
Dinosaurs, 7, 34–9, 168–79, 235, 237–9
discontinuity, in evolution, 20, 39–42, 53, 175–85, 239, 249, *See also* Catastrophism; Saltations
Discovery, 160, 221, 224
Disraeli, Benjamin, 52
Divergence, 64, 73, 75, 83, 99, 224
DNA, 158, 225, 230, 234, 237, 241, 244, 250
Dobzhansky, Theodosius, 217, 222, 225, 229, 231
Dorlodot, Henri de, 148
Doyle, Sir Arthur Conan, 174, 177
Draper, John William, 117
Drummond, Henry, 146, 150
Du Chaillu, Paul, 23, 71, 87, 94
Dublin Review, 149
Dubois, Eugene, 192–3
Dunn, Leslie, 229

Earth, age of, 25, 32, 91, 144, 147, 175
ecological relationships, 13, 59
Edinburgh, 44, 45, 48, 53
Edinburgh New Philosophical Journal, 46
Edinburgh Review, 62
education
  literature for, 140–1, *See also* Encyclopaedias
  opposition to teaching of evolution, 6, 147, 148, 232, 253–5
Eimer, Theodor, 133, 156–7
Eiseley, Loren, 3, 225
Eldredge, Niles, 248
Eliot, George, 63
Ellegård, Alvar, 5, 7–8, 56
Elsdon-Baker, Fern, 256
embryos, 20, 40–1, 46, 59, 81, 110–20, *See also* Evolution:predetermined; Ontogeny; Recapitulation theory
emergent evolution, 182
Empire, British, 204, 207
empiricism, 68
Encyclopaedia Britannica, 108, 118, 132, 202, 221, 224
English Churchman, 64
enlightenment, philosophy of, 25, 27–31

environment, 179, 180, 250–1, *See also* Adaptation; Ecological relationships
epigenetic evolution, 250–1
equal-time movement, 253
ethics, 100, 102, 204, 207
eugenics, 141, 157, 187, 206, 212–16, 229, 244, *See also* Racism
evangelicalism, 6, 46, 147, 150, 234, 251–5
Evans, Mary Ann, *See* Eliot, George
evening primrose, 158
'Evo-devo', 235, 250–1
evolution, 1, 256, *See also* Creative evolution; Darwinism; Design, argument from; Development, as model for evolution; Emergent evolution; 'Evo-devo'; Lamarckism; Parallel evolution; Progress; Theistic evolution
  discontinuous, 39–42, 53, 175–85, 239, 249
  divergent, 64, 73, 75, 83, 99, 224
  opposition to, 6, 140, 143, 146–50
  pre-Darwinian, 5, 16, 22, 25–55
  predetermined, 48–55, 78–83, 90, 93, 177
  Spencer and, 4, 74–8
Evolution of Man (Haeckel), 111, 117, 164
Evolution Protest Movement, 150
Examiner, 73
exhibitions, 7, 8, 10, 90, 193, 202
experimental biology, 144, 155
expertise, 12, 26, 57
extinction, 32, 45, 51, 169, 177
  mass extinctions, 20, 39, 72, 175, 179, 235, 238–9
Eyre, Edward, 89

Falconer, Hugh, 92
Fenton, Carroll Lane, 165
Fiction, 12, 23, 36, 174
  science, 109, 141, 232, 238
Field, The, 172, 177
Figuier, Louis, 71, 98
Film, 5, 10, 142, 174, 243
Finches, on Galápagos Islands, 54, 224, 236, 251
Fish, 41, 184
Fisher, R. A., 217, 221, 223, 230
Fiske, John, 124, 128, 151, 186, 200, 205
Fleming, Sir Ambrose, 150
Fletcher, John, 46
Flood, biblical, 20, 147, 236, 252
Fortnightly Review, 64, 103, 135
Fosdick, Harry Emerson, 151
Fossey, Dianne, 233, 243
fossil record
  Darwinism and, 23, 70–4, 166, 248
  discontinuity of, 20, 39–42
  early studies of, 25, 49
  evidence for evolution, 31–9, 120–3, 167–85, 235, 237–40
  human origins, 97, 177, 188, 191–8, 226–8, 235, 240–2
  imperfection of, 59, 110
  orthogenesis, 19, 175–9
freak shows, 14, 90, 188, 202
free enterprise, 3, 21, 76, 102–3, 131, 203–5, 207
Free, E. E., 221
Freud, Sigmund, 211
function, change of, *See* Adaptation
fundamentalism, 6, 46, 147, 150, 234, 251–5

Galápagos Islands, 54, 224, 236
Galton, Francis, 157, 162, 212–14
Gardener's Chronicle, 10, 64
Geddes, Patrick, 132, 145, 163
Gee, Henry, 239
gender, 228, 243
Genesis of Species (Mivart), 79
genesis, book of, 28, 32, 35, 251, *See also* Creation
genetic determinism, 213, *See also* Eugenics
genetics, 21, 118, 134, 256, *See also* Heredity; Population genetics
  and Darwinism, 138, 142, 155–60, 166, 217–32, 245
  and eugenics, 160, 161, 212–16, 244
Geoffroy Saint-Hilaire, E., 42, 44
geographical distribution, 69, 112, 136
geology, 25, 32, 147, 176
germ plasm, theory of, 137, 144, 157, 208
Germany, 206
Gish, Duane T., 253
Gliddon, George, 88
Goldsmith, Oliver, 29
Goodall, Jane, 228, 233, 243
Goodrich, Edwin S., 166
Goodrich, Samuel Griswold, 35
Gorillas, 23, 57, 71, 73, 87, 94–9
Gould, Augustus A., 40
Gould, Stephen Jay, 217, 234, 237, 248–9, 251
Gradualism, 41, 105, 128
Grant, Madison, 191, 200, 215
Grant, Peter, 236
Grant, Robert Edmund, 45
Grant, Rosemary, 236

Gray, Asa, 53, 62, 68, 79
Green, John, 3
Gregory, W. K., 189
Gresswell, Albert and George, 129
Gribben, John, 243
group selection, 21, 187, 205, 245, 247
growth, 81, 133, 169, 209, *See also* Development, as model for evolution; Ontogeny; Recapitulation theory

Haddon, A. C., 229
Haeckel, Ernst, 74, 99, 100, 105, 107, 110–20, 145, 164, 192
Haggard, Henry Rider, 188, 202, 206
Haldane, J. B. S., 142, 183, 217, 219, 221, 225, 232
Hall, G. Stanley, 206, 209
Halsted, Beverly, 239
Hamilton, Bill, 246
Haraway, Donna, 244
Hardy, Thomas, 23
Harmsworth Popular Science, 158, 160, 161, 177, 183, 208, 210, 214
Harper's Monthly, 107
Harper's Weekly, 97, 191
Harris, George, 67
Harrison, Harry, 238
Hawkins, Benjamin Waterhouse, 36, 97, 170
Hawks, Ellison, 220
Heard, Gerald, 219, 227
Henslow, George, 134
heredity, *See also* Genetics
  blending, 70
  Darwinism and, 58, 138, 142, 155–60
  determinant of human behaviour, 213, 248
  early views of, 134, 209
Herschel, Sir John, 79
Hill, James J., 205
Hird, Dennis, 164
History of Creation (Haeckel), 74, 105, 107, 111–15
Hitchcock, Edward, 40, 41, 54, 73
Hofstadter, Richard, 203
Hogben, Lancelot, 160, 221
Home University Library, 140, 145, 154, 158, 160, 163, 214
Homo erectus, 198, 227, 228, 241
Hooker, Joseph Dalton, 58, 62, 64, 66, 69, 72
Hooton, E. A., 189
hopeful monster, 218, *See also* Saltations
Hopwood, Nick, 112
horse, evolution of, 111, 120–3, 168, 180
human race, *See also* Anthropology; Races, human; Society, evolution of
  antiquity of, 57, 91–3, 199
  fossils, 97, 177, 188, 191–8, 226–8, 235, 240–2
  origin of, 46, 62, 66, 67, 71, 84–103, 188–91
humanism, 230–1
Hunt, James, 89
Hutchinson, H. N., 171, 172, 202
Huxley, Julian, 143, 184
  Evolution, The Modern Synthesis, 217, 223–4
  humanism, 226, 230–1
  radio broadcaster, 166, 219, 221
  views on race, 202, 215, 229
Huxley, Thomas Henry, 6, 52, 63, 96–9
  and fossils, 72, 73, 120, 168
  support for Darwin, 4, 12, 57, 62, 64, 66, 144, 146
  writing, lecturing and teaching, 8, 65, 108, 118, 120, 132
  X club, 65
Hyatt, Alpheus, 81, 133–4
hybridism, 137, 139, 241

idealism, 79, 111
Illustrated London News, 104, 149, 172, 193, 195, 227
illustrations, 10, 36, 168, 193, 202, 233, 256
imperialism, 204, 207
individualism, 3, 76, 102–3, 131, 203–5, 207
inheritance of acquired characteristics, *See* Lamarckism
intelligence, 18, 88, 92, 100, 183, 190, 229, 237, *See also* Brains
intelligent design, 252, 255
International Scientific Series, 107, 163
Ireland, 191, 200
isolation, geographical, 130, 223, 249, *See also* Geographical distribution

Jameson, Robert, 34, 45
Java man, 192–3, *See also Homo erectus*; *Pithecanthropus*
Jefferson, Thomas, 32
Jenkin, Fleeming, 70
Jensen, Arthur, 248
Johannsen, Donald, 240
Johnson, Philip, 255
Jones, Elizabeth, 234
Jordan, David Starr, 183
Journal of Researches (Charles Darwin), 54

journalism, 11, 22, 218, *See also* Authors:in magazines and newspapers; Newspapers
Juveniles, *See* Children

Kammerer, Paul, 154, 208, 250
Keane, Augustus Henry, 200
Keith, Arthur, 149, 165, 192, 193, 206, 229
Kellogg, Vernon, 145, 163, 206
Kelvin, Lord, *See* Thomson, William, Lord Kelvin
Kettlewell, H. P. D., 225
Kidd, Benjamin, 163, 187, 206
kin selection, 246
Kingsley, Charles, 35, 77, 82, 97, 119, 123
Kipling, Rudyard, 109, 127, 202, 209
Kleinberg, Otto, 229
Knight, Charles R., 172, 177, 194
Knowledge, 107, 126
Knox, Robert, 45, 89
Koestler, Arthur, 154, 250
Kovalevskii, Alexander, 112, 119
Kropotkin, Peter, 146, 207
Kubrick, Stanley, 243

Lack, David, 224
laissez-faire economics, 3, 21, 76, 102–3, 131, 203–5, 207
Lamarck, J. B., 16, 42
Lamarckism, 13, 16, 19, 25, 44–5, 47, 54, 225, *See also* American school of Neo-Lamarckism
  Spencer and, 74–8, 186, 207–9
  support for from Christians, 151–4
Landau, Misia, 189
language, evolution of, 242
Lankester, E. Ray, 117, 118–20, 140, 151, 164, 172, 209–11
laws of nature, 18, 49, 128, *See also* Design, argument from
Leakey, Louis, 227, 240, 243
LeConte, Joseph, 179, 182, 200, 208
lecturing, to public, 8, 46, 57, 65, 108
Leidy, Joseph, 169
Levine, George, 12, 109
Lewes, George Henry, 64, 72
Lewontin, Richard, 248
life
  development of, 8, 25, 40, 110–20, 167–85
  extraterrestrial, 183, 231, 232
  origin of, 49
Life on Earth, TV series, 220, 236
Lightman, Bernard, 5, 12, 81, 129
Lincoln, Abraham, 54
Linnaeus, Carl, 28, 29
Linton, Sally, 243
Literary and Philosophical Societies, 47
literature, and Darwinism, 23, 145, *See also* Fiction
living fossils, 73
Lockyer, J. Norman, 65
Lodge, Sir Oliver, 151
Lombroso, Cesare, 211
London Mercury, 149
Lorenz, Konrad, 242
Lowell, John Amory, 69
Lubbock, John, 67, 92, 99
Lull, Richard Swan, 179
Lydekker, Richard, 171
Lyell, Charles, 39, 58, 72, 85, 92, 105
Lysenko, T.D., 223, 250

MacBride, Ernest William, 154, 160, 200, 214, 229
MacDougal, Daniel Trembly, 158
MacDougal, William, 206
MacLeay, W.S., 50, 86
Macmillan, Robert, 164
Macmillan's Magazine, 63, 64, 72, 82, 92
Magazines, 7, 9, 26, 56, 107, 140, *See also* titles of individual magazines
Malay Archipelago, The, 70
Malthus, Thomas, 3, 13, 59
mammals, evolution of, 71, 175, 178, 180
mammoths, 32–4, 38, 234
man, *See* Human race
Man's Place in Nature (Huxley), 189
Mantell, Gideon, 34, 35, 38, 41
Marsh, Othniel Charles, 74, 108, 168, 169, 171, 193, 205
Martin, John, 36
Marxism, 235, 239
Mason, Frances, 151, 154
mass extinctions, 20, 39, 72, 175, 179, 235, 238–9, *See also* Catastrophism
materialism, 25, 27, 31, 46, 131, 136, 139, 143, *See also* naturalism
Mather, Kirtley, 151
Matthew, William Diller, 179, 183, 232
Matthews, Shaler, 151
Mayr, Ernst, 217, 222, 249
McCabe, Joseph, 150, 164, 165
McGee, W. J., 200
Mechanics Institutes, 26, 46, 47, 52
Medawar, Peter, 231
media, 1, 5, 6, 10–11, 140–2, 218, 233, *See also* Books; Cinema; Radio; Television
Mencken, H. L., 148
Mendel, Gregor, 157

Mendelism, *See* Genetics
Messenger, Ernest, 148
Miller, Hugh, 40, 53
mimicry, 70, 156, 160, 163
mind, 43–4, 211, *See also* Materialism; Psychology
miracles, *See* Creation
missing links, 14, 24, 58, 90, 94, 97, 191–8
Mivart, St. George, 79, 83, 135
models, 36, 170, 172, 174, 193
Modern Review, 132
modern synthesis, in evolution theory, 217–32, 244, 251
molecular biology, 241, 244
Moore, James, 4
Moorhead, Alan, 236
morality, evolution of, 100, 102, 204, 207
Morgan, Conwy Lloyd, 182
Morgan, Elaine, 243
Morgan, Lewis Henry, 198
Morgan, Thomas Hunt, 118, 158, 159, 166, 220
morphology, 110, 117, 138, *See also* Comparative anatomy
Morris, Desmond, 242
Morris, Henry M., 252
Morton, Samuel George, 88
movies, 10, 142, 174, 243
Muller, Hermann, 229, 232
Museums, 10, 35, 108, 168, 171, 172, 176–8, 256, *See also* Natural History Museum, London
mutation theory, of H. de Vries, 158
mutations, genetic, 137, 142, 159–60, 220, 221
mutual aid, 204, 245

narrative structure, of theories, 49, 189
Nation, The, 75
National Geographic, 233, 240, 243
National Review, 63
nationalism, 195
natural history, 9, 10, 11, 47
Natural History (Buffon), 27
Natural History Museum, London, 108, 166, 168, 172, 177, 195, 226, 239
Natural History Review, 65
Natural History, magazine, 248
Natural Science, 163
natural selection, 3, 14, 16, 18, 127, *See also* Darwinism; Group selection; Kin selection; Organic selection; Social Darwinism
  arguments against, 78–83, 131–7, 142–6
  defence of, 69–74, 136–7
  modern theory of, 180–5
  origins of, 54, 58, 62
Natural Theology, 31, 69, 186, *See also* Design, argument from
naturalism, 18, 74–8, 111, 123, *See also* Materialism
Nature, 9, 65, 107, 121, 149, 155, 193, 239, 246
nature *vs.* nurture, 22, 244, 250
Neanderthals, 85, 97, 98, 188, 191–4, 228, 241
nebular hypothesis, 126
neo-Darwinism, 131, 137, 143, 161–3, 212, *See also* Darwinism, modern
neo-Lamarckism, 131–6, 143, 145, *See also* American school of Neo-Lamarckism; Lamarckism
New Biology, 224
New York Herald, 170
New York Review, 27
New York Times, 67, 107, 117, 148, 171, 248
New York Tribune, 120, 171
New York World, 155
newspapers, science in, 7, 9, 26, 64, 107, 226, *See also* Science correspondents
Nineteenth Century, 135, 146
Noble, G. K., 155
nonadaptive characters, 17, 19, 81, 131, 132, 156, 223, *See also* Orthogenesis
Nordau, Max, 210
North American Review, 27, 79
North British Review, 70
Nott, Josiah C., 88
novels, *See* Fiction

Odontornithes, 74
Oenothera lamarckiana, 158
ontogeny, 115, 200
Oracle of Reason, 46, 53
orangutans, 87, 90
organic selection, 152
Origin of Species (Charles Darwin), 4, 6, 7, 12, 15, 22, 23
  centenary of, 225
  publication, 56–61
  reaction to, 61–9
  Vestiges and, 48, 54
orthogenesis, 17, 19, 133, 156–7, 175–9, 180
Osborn, Henry Fairfield, 148, 172, 175–9, 191, 194, 195–6, 215
Outlines of Cosmic Philosophy (Fiske), 128, 198

Owen, Richard, 34, 39, 41, 42, 54, 62, 71, 73, 87, 96–9
Owen, Robert, 47

Page, John, 71
palaeoanthropology, 240–2, *See also* Human race:origin of
palaeontology, *See* Fossil record
Paley, William, 31, 79, 252
pangenesis, 69, 134, 213
paperbacks, 111, 141, 163, 224, 233, 246
parallel evolution, 17, 51, 133, 156–7, 176, 190, *See also* Evolution:predetermined; Orthogenesis
  and humans, 71, 86, 88, 195–6
parallelism, law of, in embryology, 40
parasites, 21, 119
Peabody Museum, Yale, 108, 170, 171
Peale, Charles Wilson, 33
Pearson, Karl, 161, 206, 210, 212, 213, 215
Pearson's, 174
Pekin man, 196, *See also Homo erectus*
Pengelly, William, 92
Penny Cyclopaedia, 35, 38
peppered moth, 221, 225
Phillips, John, 71
philosophy, 27–31, 47, *See also* Idealism; Naturalism
phrenology, 14, 43–4, 49, 52, 53, 87
phylogenies, 112, 114, 115, 117, 121
Piltdown man, 177, 194, 226
Pithecanthropus, 114, 192–3, 196, 227, *See also Homo erectus*
polygenism, 88, 198, 199, *See also* Human race:origin of
popular science, 2, 5–10, 11, 12, 106–10, 218, 233–6, *See also* Books; Magazines; Newspapers, science in
Popular Science Monthly, 107, 108, 121, 127, 131, 135, 140, 221
population genetics, 162, 213, 217, 229, 230
Poulton, E. B., 163
Powell, J. W., 200
Praeger, Robert Lloyd, 166
Price, George, 246
Price, George McCready, 147, 148, 165, 252
pricing, of books, 8–9, 51, 52, 56, 111, 126, 135, 164
primatology, 242
Principles of Biology (Spencer), 76
Proctor, Richard, 107
professionalization of science, 12, 26, 57
progress
  biological evolution, 15, 19, 75–6, 110–20, 123–9, 160, 228, 231
  chain of being, 16, 114
  rejection of, 17
  social evolution, 4, 42, 76, 123
protestants, evangelical, 46, 147, 150, 234, 251–5
psychology, 85, 211
publishing, innovations in, 26–7, 140–2
Punch, 90, 97, 152, 171, 172, 226
punctuated equilibrium theory, 248
Punnett, Reginald, 160
purpose, in evolution, 150–5, *See also* Design, argument from; Progress, in biological evolution

Quarterly Review, 63
Quatrefages, Armand de, 191

Races of Men, The (Knox), 89
races, human, *See also* Human race
  classification of, 86–7, 187, 198–203
  origin of, 51, 87–91, 100, 188, 195–6
racial senility, 133, 177, 178
racism, 87–91, 198–203, 206, 216, 229, 244, 248, *See also* Eugenics
radical evolutionism, 42–8
radio, 5, 10, 140, 166, 219, 221, 224
random variation, 17, 69, 143
Rationalist Press Association, 163
Reader, 65
readership, 5, 7
  of books, 35, 51, 115, 135
  of magazines, 26, 65, 107
recapitulation theory, 20, 105, 110–20, 133, 154, 200, 209–12, 217
Reed, Edward Tennyson, 172
Reingold, Nathan, 27
Religion, *See* Christianity
Richards, Robert, 111
Ridley, Matt, 234, 250
Ripley, W. Z., 200
Rivers, W. H. R., 202
Rockefeller, John D., 205
Rogers, William Barton, 68
Roman Catholicism, 148, 149
Romanes, George John, 135, 136, 209
Romanticism, 30
Rose, Stephen, 247
Royal Institution, London, 8, 65
Ruse, Michael, 12, 111, 237
Russell, Dale, 238

Sagan, Carl, 233
Saleeby, Caleb, 161, 214
sales figures, for books, 51, 52, 53, 61, 101
Salisbury, Marquis of, 144
saltations, 17, 45, 63, 70, 137, 142, 156, 157, 213
Schwalbe, Gustav, 192
Science, 193
science correspondents, 11
science fiction, 10, 109, 141, 232, 238
Science Gossip, 107
Science News, 224
Science Progress, 193
Science Service, US, 219, 226
Science Siftings, 107
science writers, 5, 12, 106–10, 152, 234
science, popular, *See* Popular science
Scientific American, 27, 107, 141, 160, 172, 233, 242
scientific community, 26, *See also* Professionalization of science
  and Darwinism, 12, 57, 138, 224, 225, 234, 253
  and genetics, 219, 229
scientists
  amateur, 11, 26, 46, 57, 138
  attitudes towards popular science, 11
  book authors, 105, 152, 234
  writing for magazines and newspapers, 107, 140
Scopes' 'Monkey trial', 6, 140, 143, 147, 165, 216
Scott, Walter, 107
Secord, James, 5, 48, 51, 87
Sedgwick, Adam, 52, 87
Seeley, Harry Govier, 120
selection, *See also* Natural selection
  artificial, 17, 58, 157, 212, 214, 244
  group, 21, 187, 205, 245, 247
  kin, 246
  sexual, 14, 99–102, 245
self-education literature, 140–1
Selous, Frederick Courtney, 206
Senility, racial, 133, 177, 178
Seward, A. C., 145
sexual selection, 14, 99–102, 245
Shaw, George Bernard, 139, 140, 141, 152, 167, 183, 208, 222
Simpson, George Gaylord, 7, 217, 223, 231
slavery, 62, 88, 89, 205
Smellie, William, 28
Smith, Grafton Elliot, 190, 195
Smith, John Maynard, 225, 246
social Darwinism, 3, 21, 76, 78, 102–3, 186–216, 244
  individualistic, 203–5
  nationalist, 206–7
  racial, 212–16
socialism, 125, 199, 214
society, evolution of, 198
sociobiology, 22, 243, 247
sociology, 107, 198, 208
Soleki, Ralph, 241
Sollas, W. J., 191
Southwell, Charles, 46, 85
specialization, in evolution, 54, 59, 71, 127
speciation, 223
species, 28, 29, 235, 239
Spencer, Herbert, 4, 6, 10, 13, 15, 31, 44, 47, 52, 54, 57, 58, 62, 102–3, 186
  and free-enterprise, 203–5
  and human races, 198, 200, 210
  and Lamarckism, 16, 74–8, 124–8, 135, 145, 207
spontaneous generation, 27, 31, *See also* Life, origin of
Spurzheim, J. C., 43
state of nature, 205, 210
Steele, Ted, 250
Steinach, Eugen, 155
stone age, 91–3
Story of Creation (Clodd), 126
Strand Magazine, 174
struggle for existence, 4, 14, 21, 83, 146, 187, 220, *See also* Social Darwinism
  and Darwinism, 13, 17, 59, 64, 203–7
Sully, James, 209
Sumner, William Graham, 205, 208
survival of the fittest, 76, 77, 83, 186, 203, 205, 207, *See also* Natural selection
swamping, of variations, 161
Swinton, W. E., 178

taxonomy, 239
Taylor, J. L., 163
teaching of evolution, 6, 147, 148, 232, 253–5
Teilhard de Chardin, Pierre, 231
television, 5, 10, 225, 233, 236
Temple of Nature (Erasmus Darwin), 30
Tennyson, Alfred, 13, 37
theistic evolution, 16, 39–42, 80, 129–31, 132, 143, 150–5, 230, *See also* Christianity
Thiry, Paul Henri, Baron d'Holbach, *See* D'Holbach, Baron
Thomson, J. Arthur, 12, 140, 145, 152, 154, 161, 163, 182

Thomson, William, Lord Kelvin, 144
Time, 238
'Time Machine, The' (Wells), 10, 109, 119, 211
Times, 63, 64, 92, 149
toolmaking, 85, 100, 190, 191, 193
transmutationism, 28, 45, 47, 96
tree of life, 19, 41, 230, 231
  branching, 20, 67, 72, 166, 180–5
  with central trunk, 20, 152, 167, 220
Trimmer, Joshua, 36
Tweed, W. M., 170
Tylor, Edward B., 92, 198
Tyndall, John, 108, 124

uniformitarianism, 39, 105
United States
  American Association for the Advancement of Science, 67, 108, 121, 248
  American school of neo-Lamarckism, 81, 133–4, 156
  fossils, 32, 73
  media, 6, 11
  parallelism, 40–1
  publications, 27, 35, 107
  reaction to *Origin of Species*, 67–9
  reaction to *Vestiges*, 53–4
  religion, 81, 252, 253
  Science Service, 219, 226
  Spencer's work, 77, 205
Universal Review, 142
use-inheritance, *See* Lamarckism
utilitarianism, 16, 19, 21

variation, in species, 19, 58
  directed, 17, 69, 114, 163, 176, *See also* Orthogenesis
  genetic, 159, 160, 218, 221
  random, 17, 69, 143
varieties, within species, 157, 159, 160
Velikovsky, Immanuel, 238
Verne, Jules, 174
vertebrates, origin of, 20, 32, 49, 99, 112, 121, 127
Vestiges of the Natural History of Creation (Chambers), 5, 26, 43, 48–55, 70, 78, 85–7
Vogt, Carl, 90
Voronoff, Sergei, 208
Vries, Hugo de, 145, 157, 158
Vyvyan, Sir Richard, 53

Wallace, Alfred Russel, 44, 58, 80, 94, 132, 136, 161
Wallace, David Rains, 171
War of the Worlds, The (Wells), 10, 183, 232
Ward, Lester Frank, 208
warfare, 206–7
Washburn, Sherwood, 228
Washington Post, 171
Water Babies, The (Kingsley), 35, 82, 97, 119
Watts, Charles Albert, 141, 163–5
Wedgwood, Julia, 92
Weiner, J. S., 226
Weiner, Jonathan, 237
Weismann, August, 137, 144, 145, 157, 161, 210, 214
Weldon, W. F .R., 161
Wells, H. G., 12, 118, 161
  fiction works, 10, 109, 119, 183, 211
  Outline of History, 149, 165, 184, 193, 222
Westminster Review, 63, 70
Whitcomb, John C., 252
Whyte, Adam Gowans, 164
Wilberforce, Samuel, 6, 62, 66, 94
Wilson, E. O., 235, 245, 247, 250
Winchell, Alexander, 200
Winners in Life's Race (Buckley), 129–30
Witness, The, 53
Woodward, Sir Arthur Smith, 177, 195
Wright, Chauncey, 75, 79
Wright, Sewall, 217, 224
Wynne-Edwards, V. C., 245

X Club, 65

Youmans, E. L., 107, 127
Young, Robert M., 3–4

Zahm, John, 148
Zallinger, Rudolph, 256
Zoology, 44, 108, 118
Zoonomia (Erasmus Darwin), 29

Printed in the United States
by Baker & Taylor Publisher Services